Abstract Algebra

Theory and Applications

Abstract Algebra
Theory and Applications

Thomas W. Judson
Stephen F. Austin State University

Sage Exercises for Abstract Algebra
Robert A. Beezer
University of Puget Sound

Traducción al español
Antonio Behn
Universidad de Chile

August 9, 2021

Abstract Algebra: Theory and Applications

2021 edition

Orthogonal Publishing L3C
Ann Arbor, Michigan
www.orthogonalpublishing.com

Typeset in 10pt Adobe Minion Pro using PDFLaTeX.

About the cover: *Birds Of Play*, by Robert Chaffer, is a demonstration that the Euclidean plane can be completely tiled by means of an iterated function system (IFS) acting on a single image element. In this case, the element figure lies mostly within a square with opposite vertices at the wingtips of the bird figure. Nine transformations were selected from the group of symmetries of the square to be the constituents employed in the IFS.

The background detail of the cover-image displays several iterations in the orbital path created by this IFS. The orbital images converge to completely fill the interior of the original square since the birds are contracted to one-third size at each iteration. The square is the fixed set of this IFS. If the contractive feature is omitted then the image will grow in scale and, after infinitely many iterations, tile the entire plane.

One additional feature in *Birds of Play* is a continuous curve that can be traced from the lower-left corner of the image to the upper-right corner. This curve is a first approximation of a Peano space-filling curve that can fill the entire square with a single continuous curve.

Robert Chaffer is Professor Emeritus at Central Michigan University. His academic interests are in abstract algebra, combinatorics, geometry, and computer applications. Since retirement from teaching he has devoted much of his time to applying those interests to creation of art images.
http://people.cst.cmich.edu/chaff1ra/Art_From_Mathematics/

Acknowledgements

I would like to acknowledge the following reviewers for their helpful comments and suggestions.

- David Anderson, University of Tennessee, Knoxville
- Robert Beezer, University of Puget Sound
- Myron Hood, California Polytechnic State University
- Herbert Kasube, Bradley University
- John Kurtzke, University of Portland
- Inessa Levi, University of Louisville
- Geoffrey Mason, University of California, Santa Cruz
- Bruce Mericle, Mankato State University
- Kimmo Rosenthal, Union College
- Mark Teply, University of Wisconsin

I would also like to thank Steve Quigley, Marnie Pommett, Cathie Griffin, Kelle Karshick, and the rest of the staff at PWS Publishing for their guidance throughout this project. It has been a pleasure to work with them.

Robert Beezer encouraged me to make *Abstract Algebra: Theory and Applications* available as an open source textbook, a decision that I have never regretted. With his assistance, the book has been rewritten in PreTeXt (pretextbook.org), making it possible to quickly output print, web, PDF versions and more from the same source. The open source version of this book has received support from the National Science Foundation (Awards #DUE-1020957, #DUE-1625223, and #DUE-1821329).

Preface

This text is intended for a one or two-semester undergraduate course in abstract algebra. Traditionally, these courses have covered the theoretical aspects of groups, rings, and fields. However, with the development of computing in the last several decades, applications that involve abstract algebra and discrete mathematics have become increasingly important, and many science, engineering, and computer science students are now electing to minor in mathematics. Though theory still occupies a central role in the subject of abstract algebra and no student should go through such a course without a good notion of what a proof is, the importance of applications such as coding theory and cryptography has grown significantly.

Until recently most abstract algebra texts included few if any applications. However, one of the major problems in teaching an abstract algebra course is that for many students it is their first encounter with an environment that requires them to do rigorous proofs. Such students often find it hard to see the use of learning to prove theorems and propositions; applied examples help the instructor provide motivation.

This text contains more material than can possibly be covered in a single semester. Certainly there is adequate material for a two-semester course, and perhaps more; however, for a one-semester course it would be quite easy to omit selected chapters and still have a useful text. The order of presentation of topics is standard: groups, then rings, and finally fields. Emphasis can be placed either on theory or on applications. A typical one-semester course might cover groups and rings while briefly touching on field theory, using Chapters 1 through 6, 9, 10, 11, 13 (the first part), 16, 17, 18 (the first part), 20, and 21. Parts of these chapters could be deleted and applications substituted according to the interests of the students and the instructor. A two-semester course emphasizing theory might cover Chapters 1 through 6, 9, 10, 11, 13 through 18, 20, 21, 22 (the first part), and 23. On the other hand, if applications are to be emphasized, the course might cover Chapters 1 through 14, and 16 through 22. In an applied course, some of the more theoretical results could be assumed or omitted. A chapter dependency chart appears below. (A broken line indicates a partial dependency.)

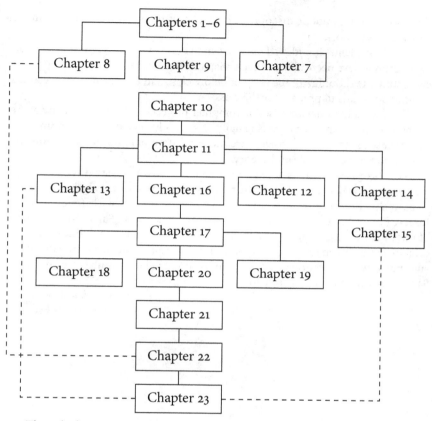

Though there are no specific prerequisites for a course in abstract algebra, students who have had other higher-level courses in mathematics will generally be more prepared than those who have not, because they will possess a bit more mathematical sophistication. Occasionally, we shall assume some basic linear algebra; that is, we shall take for granted an elementary knowledge of matrices and determinants. This should present no great problem, since most students taking a course in abstract algebra have been introduced to matrices and determinants elsewhere in their career, if they have not already taken a sophomore or junior-level course in linear algebra.

Exercise sections are the heart of any mathematics text. An exercise set appears at the end of each chapter. The nature of the exercises ranges over several categories; computational, conceptual, and theoretical problems are included. A section presenting hints and solutions to many of the exercises appears at the end of the text. Often in the solutions a proof is only sketched, and it is up to the student to provide the details. The exercises range in difficulty from very easy to very challenging. Many of the more substantial problems require careful thought, so the student should not be discouraged if the solution is not forthcoming after a few minutes of work.

Ideally, students should read the relavent material before attending class. Reading questions have been added to each chapter before the exercises. To prepare for class, students should read the chapter before class and then answer the section's reading questions to prepare for the class.

There are additional exercises or computer projects at the ends of many of the chapters. The computer projects usually require a knowledge of programming. All of these exercises and projects are more substantial in nature and allow the exploration of new results and theory.

Sage (sagemath.org) is a free, open source, software system for advanced mathematics, which is ideal for assisting with a study of abstract algebra. Sage can be used either on your own computer, a local server, or on CoCalc (cocalc.com). Robert Beezer has written a comprehensive introduction to Sage and a selection of relevant exercises that appear at the end of each chapter, including live Sage cells in the web version of the book. All of the Sage code has been subject to automated tests of accuracy, using the most recent version available at this time: SageMath Version 9.3 (released 2021-05-09).

Thomas W. Judson
Nacogdoches, Texas 2021

Contents

Preliminaries

$\mathscr{A}$ certain amount of mathematical maturity is necessary to find and study applications of abstract algebra. A basic knowledge of set theory, mathematical induction, equivalence relations, and matrices is a must. Even more important is the ability to read and understand mathematical proofs. In this chapter we will outline the background needed for a course in abstract algebra.

1.1 A Short Note on Proofs

Abstract mathematics is different from other sciences. In laboratory sciences such as chemistry and physics, scientists perform experiments to discover new principles and verify theories. Although mathematics is often motivated by physical experimentation or by computer simulations, it is made rigorous through the use of logical arguments. In studying abstract mathematics, we take what is called an axiomatic approach; that is, we take a collection of objects S and assume some rules about their structure. These rules are called **axioms**. Using the axioms for S, we wish to derive other information about S by using logical arguments. We require that our axioms be consistent; that is, they should not contradict one another. We also demand that there not be too many axioms. If a system of axioms is too restrictive, there will be few examples of the mathematical structure.

A **statement** in logic or mathematics is an assertion that is either true or false. Consider the following examples:

- $3 + 56 - 13 + 8/2$.

- All cats are black.

- $2 + 3 = 5$.

- $2x = 6$ exactly when $x = 4$.

- If $ax^2 + bx + c = 0$ and $a \neq 0$, then

$$x = \frac{-b \pm \sqrt{b^2 - 4ac}}{2a}.$$

- $x^3 - 4x^2 + 5x - 6$.

All but the first and last examples are statements, and must be either true or false.

A *mathematical proof* is nothing more than a convincing argument about the accuracy of a statement. Such an argument should contain enough detail to convince the audience; for instance, we can see that the statement "$2x = 6$ exactly when $x = 4$" is false by evaluating $2 \cdot 4$ and noting that $6 \neq 8$, an argument that would satisfy anyone. Of course, audiences may vary widely: proofs can be addressed to another student, to a professor, or to the reader of a text. If more detail than needed is presented in the proof, then the explanation will be either long-winded or poorly written. If too much detail is omitted, then the proof may not be convincing. Again it is important to keep the audience in mind. High school students require much more detail than do graduate students. A good rule of thumb for an argument in an introductory abstract algebra course is that it should be written to convince one's peers, whether those peers be other students or other readers of the text.

Let us examine different types of statements. A statement could be as simple as "$10/5 = 2$;" however, mathematicians are usually interested in more complex statements such as "If p, then q," where p and q are both statements. If certain statements are known or assumed to be true, we wish to know what we can say about other statements. Here p is called the **hypothesis** and q is known as the **conclusion**. Consider the following statement: If $ax^2 + bx + c = 0$ and $a \neq 0$, then

$$x = \frac{-b \pm \sqrt{b^2 - 4ac}}{2a}.$$

The hypothesis is $ax^2 + bx + c = 0$ and $a \neq 0$; the conclusion is

$$x = \frac{-b \pm \sqrt{b^2 - 4ac}}{2a}.$$

Notice that the statement says nothing about whether or not the hypothesis is true. However, if this entire statement is true and we can show that $ax^2 + bx + c = 0$ with $a \neq 0$ is true, then the conclusion *must* be true. A proof of this statement might simply be a series of equations:

$$ax^2 + bx + c = 0$$

$$x^2 + \frac{b}{a}x = -\frac{c}{a}$$

$$x^2 + \frac{b}{a}x + \left(\frac{b}{2a}\right)^2 = \left(\frac{b}{2a}\right)^2 - \frac{c}{a}$$

$$\left(x + \frac{b}{2a}\right)^2 = \frac{b^2 - 4ac}{4a^2}$$

$$x + \frac{b}{2a} = \frac{\pm\sqrt{b^2 - 4ac}}{2a}$$

$$x = \frac{-b \pm \sqrt{b^2 - 4ac}}{2a}.$$

If we can prove a statement true, then that statement is called a **proposition**. A proposition of major importance is called a **theorem**. Sometimes instead of proving a theorem or proposition all at once, we break the proof down into modules; that is, we prove several supporting propositions, which are called *lemmas*, and use the results of these propositions to prove the main result. If we can prove a proposition or a theorem, we will often, with very little effort, be able to derive other related propositions called *corollaries*.

Some Cautions and Suggestions

There are several different strategies for proving propositions. In addition to using different methods of proof, students often make some common mistakes when they are first learning how to prove theorems. To aid students who are studying abstract mathematics for the first time, we list here some of the difficulties that they may encounter and some of the strategies of proof available to them. It is a good idea to keep referring back to this list as a reminder. (Other techniques of proof will become apparent throughout this chapter and the remainder of the text.)

- A theorem cannot be proved by example; however, the standard way to show that a statement is not a theorem is to provide a counterexample.

- Quantifiers are important. Words and phrases such as *only, for all, for every,* and *for some* possess different meanings.

- Never assume any hypothesis that is not explicitly stated in the theorem. *You cannot take things for granted.*

- Suppose you wish to show that an object *exists* and is *unique*. First show that there actually is such an object. To show that it is unique, assume that there are two such objects, say r and s, and then show that $r = s$.

- Sometimes it is easier to prove the contrapositive of a statement. Proving the statement "If p, then q" is exactly the same as proving the statement "If not q, then not p."

- Although it is usually better to find a direct proof of a theorem, this task can sometimes be difficult. It may be easier to assume that the theorem that you are trying to prove is false, and to hope that in the course of your argument you are forced to make some statement that cannot possibly be true.

Remember that one of the main objectives of higher mathematics is proving theorems. Theorems are tools that make new and productive applications of mathematics possible. We use examples to give insight into existing theorems and to foster intuitions as to what new theorems might be true. Applications, examples, and proofs are tightly interconnected—much more so than they may seem at first appearance.

1.2 Sets and Equivalence Relations

Set Theory

A *set* is a well-defined collection of objects; that is, it is defined in such a manner that we can determine for any given object x whether or not x belongs to the set. The objects that belong to a set are called its *elements* or *members*. We will denote sets by capital letters, such as A or X; if a is an element of the set A, we write $a \in A$.

A set is usually specified either by listing all of its elements inside a pair of braces or by stating the property that determines whether or not an object x belongs to the set. We might write

$$X = \{x_1, x_2, \ldots, x_n\}$$

for a set containing elements $x_1, x_2, \ldots, x_n$ or

$$X = \{x : x \text{ satisfies } \mathcal{P}\}$$

if each x in X satisfies a certain property $\mathcal{P}$. For example, if E is the set of even positive integers, we can describe E by writing either

$$E = \{2, 4, 6, \ldots\} \quad \text{or} \quad E = \{x : x \text{ is an even integer and } x > 0\}.$$

We write $2 \in E$ when we want to say that 2 is in the set E, and $-3 \notin E$ to say that -3 is not in the set E.

Some of the more important sets that we will consider are the following:

$$\mathbb{N} = \{n : n \text{ is a natural number}\} = \{1, 2, 3, \ldots\};$$
$$\mathbb{Z} = \{n : n \text{ is an integer}\} = \{\ldots, -1, 0, 1, 2, \ldots\};$$
$$\mathbb{Q} = \{r : r \text{ is a rational number}\} = \{p/q : p, q \in \mathbb{Z} \text{ where } q \neq 0\};$$
$$\mathbb{R} = \{x : x \text{ is a real number}\};$$
$$\mathbb{C} = \{z : z \text{ is a complex number}\}.$$

We can find various relations between sets as well as perform operations on sets. A set A is a *subset* of B, written $A \subset B$ or $B \supset A$, if every element of A is also an element of B. For example,

$$\{4, 5, 8\} \subset \{2, 3, 4, 5, 6, 7, 8, 9\}$$

and

$$\mathbb{N} \subset \mathbb{Z} \subset \mathbb{Q} \subset \mathbb{R} \subset \mathbb{C}.$$

Trivially, every set is a subset of itself. A set B is a *proper subset* of a set A if $B \subset A$ but $B \neq A$. If A is not a subset of B, we write $A \not\subset B$; for example, $\{4, 7, 9\} \not\subset \{2, 4, 5, 8, 9\}$. Two sets are *equal*, written $A = B$, if we can show that $A \subset B$ and $B \subset A$.

It is convenient to have a set with no elements in it. This set is called the *empty set* and is denoted by $\varnothing$. Note that the empty set is a subset of every set.

To construct new sets out of old sets, we can perform certain operations: the *union* $A \cup B$ of two sets A and B is defined as

$$A \cup B = \{x : x \in A \text{ or } x \in B\};$$

the *intersection* of A and B is defined by

$$A \cap B = \{x : x \in A \text{ and } x \in B\}.$$

If $A = \{1, 3, 5\}$ and $B = \{1, 2, 3, 9\}$, then

$$A \cup B = \{1, 2, 3, 5, 9\} \quad \text{and} \quad A \cap B = \{1, 3\}.$$

We can consider the union and the intersection of more than two sets. In this case we write

$$\bigcup_{i=1}^{n} A_i = A_1 \cup \ldots \cup A_n$$

and

$$\bigcap_{i=1}^{n} A_i = A_1 \cap \ldots \cap A_n$$

for the union and intersection, respectively, of the sets $A_1, \ldots, A_n$.

When two sets have no elements in common, they are said to be *disjoint*; for example, if E is the set of even integers and O is the set of odd integers, then E and O are disjoint. Two sets A and B are disjoint exactly when $A \cap B = \varnothing$.

Sometimes we will work within one fixed set U, called the *universal set*. For any set $A \subset U$, we define the *complement* of A, denoted by A', to be the set

$$A' = \{x : x \in U \text{ and } x \notin A\}.$$

We define the *difference* of two sets A and B to be

$$A \smallsetminus B = A \cap B' = \{x : x \in A \text{ and } x \notin B\}.$$

Example 1.1. Let $\mathbb{R}$ be the universal set and suppose that

$$A = \{x \in \mathbb{R} : 0 < x \le 3\} \quad \text{and} \quad B = \{x \in \mathbb{R} : 2 \le x < 4\}.$$

Then

$$A \cap B = \{x \in \mathbb{R} : 2 \le x \le 3\}$$
$$A \cup B = \{x \in \mathbb{R} : 0 < x < 4\}$$
$$A \smallsetminus B = \{x \in \mathbb{R} : 0 < x < 2\}$$
$$A' = \{x \in \mathbb{R} : x \le 0 \text{ or } x > 3\}.$$

$\square$

Proposition 1.2. Let A, B, and C be sets. Then

1. $A \cup A = A$, $A \cap A = A$, and $A \smallsetminus A = \varnothing$;

2. $A \cup \varnothing = A$ and $A \cap \varnothing = \varnothing$;

3. $A \cup (B \cup C) = (A \cup B) \cup C$ and $A \cap (B \cap C) = (A \cap B) \cap C$;

4. $A \cup B = B \cup A$ and $A \cap B = B \cap A$;

5. $A \cup (B \cap C) = (A \cup B) \cap (A \cup C)$;

6. $A \cap (B \cup C) = (A \cap B) \cup (A \cap C)$.

Proof. We will prove (1) and (3) and leave the remaining results to be proven in the exercises.

(1) Observe that

$$A \cup A = \{x : x \in A \text{ or } x \in A\}$$
$$= \{x : x \in A\}$$
$$= A$$

and

$$A \cap A = \{x : x \in A \text{ and } x \in A\}$$

$$= \{x : x \in A\}$$
$$= A.$$

Also, $A \smallsetminus A = A \cap A' = \varnothing$.

(3) For sets A, B, and C,

$$
\begin{aligned}
A \cup (B \cup C) &= A \cup \{x : x \in B \text{ or } x \in C\} \\
&= \{x : x \in A \text{ or } x \in B, \text{ or } x \in C\} \\
&= \{x : x \in A \text{ or } x \in B\} \cup C \\
&= (A \cup B) \cup C.
\end{aligned}
$$

A similar argument proves that $A \cap (B \cap C) = (A \cap B) \cap C$. ∎

Theorem 1.3. De Morgan's Laws. Let A and B be sets. Then

1. $(A \cup B)' = A' \cap B'$;

2. $(A \cap B)' = A' \cup B'$.

Proof. (1) If $A \cup B = \varnothing$, then the theorem follows immediately since both A and B are the empty set. Otherwise, we must show that $(A \cup B)' \subset A' \cap B'$ and $(A \cup B)' \supset A' \cap B'$. Let $x \in (A \cup B)'$. Then $x \notin A \cup B$. So x is neither in A nor in B, by the definition of the union of sets. By the definition of the complement, $x \in A'$ and $x \in B'$. Therefore, $x \in A' \cap B'$ and we have $(A \cup B)' \subset A' \cap B'$.

To show the reverse inclusion, suppose that $x \in A' \cap B'$. Then $x \in A'$ and $x \in B'$, and so $x \notin A$ and $x \notin B$. Thus $x \notin A \cup B$ and so $x \in (A \cup B)'$. Hence, $(A \cup B)' \supset A' \cap B'$ and so $(A \cup B)' = A' \cap B'$.

The proof of (2) is left as an exercise. ∎

Example 1.4. Other relations between sets often hold true. For example,

$$(A \smallsetminus B) \cap (B \smallsetminus A) = \varnothing.$$

To see that this is true, observe that

$$
\begin{aligned}
(A \smallsetminus B) \cap (B \smallsetminus A) &= (A \cap B') \cap (B \cap A') \\
&= A \cap A' \cap B \cap B' \\
&= \varnothing.
\end{aligned}
$$

□

Cartesian Products and Mappings

Given sets A and B, we can define a new set $A \times B$, called the ***Cartesian product***
of A and B, as a set of ordered pairs. That is,

$$A \times B = \{(a, b) : a \in A \text{ and } b \in B\}.$$

Example 1.5. If $A = \{x, y\}$, $B = \{1, 2, 3\}$, and $C = \varnothing$, then $A \times B$ is the set

$$\{(x, 1), (x, 2), (x, 3), (y, 1), (y, 2), (y, 3)\}$$

and

$$A \times C = \varnothing.$$

$\square$

We define the ***Cartesian product of n sets*** to be

$$A_1 \times \cdots \times A_n = \{(a_1, \ldots, a_n) : a_i \in A_i \text{ for } i = 1, \ldots, n\}.$$

If $A = A_1 = A_2 = \cdots = A_n$, we often write A^n for $A \times \cdots \times A$ (where A would be
written n times). For example, the set $\mathbb{R}^3$ consists of all of 3-tuples of real numbers.

Subsets of $A \times B$ are called ***relations***. We will define a ***mapping*** or ***function***
$f \subset A \times B$ from a set A to a set B to be the special type of relation where each
element $a \in A$ has a unique element $b \in B$ such that $(a, b) \in f$. Another way
of saying this is that for every element in A, f assigns a unique element in B.
We usually write $f : A \to B$ or $A \xrightarrow{f} B$. Instead of writing down ordered pairs
$(a, b) \in A \times B$, we write $f(a) = b$ or $f : a \mapsto b$. The set A is called the ***domain*** of
f and

$$f(A) = \{f(a) : a \in A\} \subset B$$

is called the ***range*** or ***image*** of f. We can think of the elements in the function's
domain as input values and the elements in the function's range as output values.

Example 1.6. Suppose $A = \{1, 2, 3\}$ and $B = \{a, b, c\}$. In Figure 1.7 we define
relations f and g from A to B. The relation f is a mapping, but g is not because
$1 \in A$ is not assigned to a unique element in B; that is, $g(1) = a$ and $g(1) = b$. $\square$

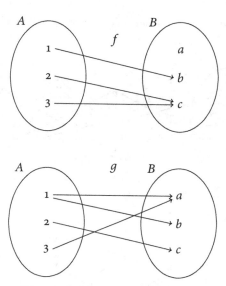

Figure 1.7. Mappings and relations

Given a function $f : A \rightarrow B$, it is often possible to write a list describing what the function does to each specific element in the domain. However, not all functions can be described in this manner. For example, the function $f : \mathbb{R} \rightarrow \mathbb{R}$ that sends each real number to its cube is a mapping that must be described by writing $f(x) = x^3$ or $f : x \mapsto x^3$.

Consider the relation $f : \mathbb{Q} \rightarrow \mathbb{Z}$ given by $f(p/q) = p$. We know that $1/2 = 2/4$, but is $f(1/2) = 1$ or 2? This relation cannot be a mapping because it is not well-defined. A relation is **well-defined** if each element in the domain is assigned to a *unique* element in the range.

If $f : A \rightarrow B$ is a map and the image of f is B, i.e., $f(A) = B$, then f is said to be **onto** or **surjective**. In other words, if there exists an $a \in A$ for each $b \in B$ such that $f(a) = b$, then f is onto. A map is **one-to-one** or **injective** if $a_1 \neq a_2$ implies $f(a_1) \neq f(a_2)$. Equivalently, a function is one-to-one if $f(a_1) = f(a_2)$ implies $a_1 = a_2$. A map that is both one-to-one and onto is called **bijective**.

Example 1.8. Let $f : \mathbb{Z} \rightarrow \mathbb{Q}$ be defined by $f(n) = n/1$. Then f is one-to-one but not onto. Define $g : \mathbb{Q} \rightarrow \mathbb{Z}$ by $g(p/q) = p$ where p/q is a rational number expressed in its lowest terms with a positive denominator. The function g is onto but not one-to-one. $\square$

Given two functions, we can construct a new function by using the range of the first function as the domain of the second function. Let $f : A \rightarrow B$ and

$g : B \to C$ be mappings. Define a new map, the *composition* of f and g from A to C, by $(g \circ f)(x) = g(f(x))$.

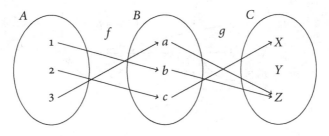

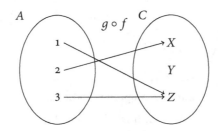

Figure 1.9. Composition of maps

Example 1.10. Consider the functions $f : A \to B$ and $g : B \to C$ that are defined in Figure 1.9 (top). The composition of these functions, $g \circ f : A \to C$, is defined in Figure 1.9 (bottom). □

Example 1.11. Let $f(x) = x^2$ and $g(x) = 2x + 5$. Then

$$(f \circ g)(x) = f(g(x)) = (2x + 5)^2 = 4x^2 + 20x + 25$$

and

$$(g \circ f)(x) = g(f(x)) = 2x^2 + 5.$$

In general, order makes a difference; that is, in most cases $f \circ g \neq g \circ f$. □

Example 1.12. Sometimes it is the case that $f \circ g = g \circ f$. Let $f(x) = x^3$ and $g(x) = \sqrt[3]{x}$. Then

$$(f \circ g)(x) = f(g(x)) = f(\sqrt[3]{x}) = (\sqrt[3]{x})^3 = x$$

and

$$(g \circ f)(x) = g(f(x)) = g(x^3) = \sqrt[3]{x^3} = x.$$

□

Example 1.13. Given a 2×2 matrix

$$A = \begin{pmatrix} a & b \\ c & d \end{pmatrix},$$

we can define a map $T_A : \mathbb{R}^2 \to \mathbb{R}^2$ by

$$T_A(x, y) = (ax + by, cx + dy)$$

for (x, y) in $\mathbb{R}^2$. This is actually matrix multiplication; that is,

$$\begin{pmatrix} a & b \\ c & d \end{pmatrix} \begin{pmatrix} x \\ y \end{pmatrix} = \begin{pmatrix} ax + by \\ cx + dy \end{pmatrix}.$$

Maps from $\mathbb{R}^n$ to $\mathbb{R}^m$ given by matrices are called *linear maps* or *linear transformations*. □

Example 1.14. Suppose that $S = \{1, 2, 3\}$. Define a map $\pi : S \to S$ by

$$\pi(1) = 2, \qquad \pi(2) = 1, \qquad \pi(3) = 3.$$

This is a bijective map. An alternative way to write π is

$$\begin{pmatrix} 1 & 2 & 3 \\ \pi(1) & \pi(2) & \pi(3) \end{pmatrix} = \begin{pmatrix} 1 & 2 & 3 \\ 2 & 1 & 3 \end{pmatrix}.$$

For any set S, a one-to-one and onto mapping $\pi : S \to S$ is called a *permutation* of S. □

Theorem 1.15. Let $f : A \to B$, $g : B \to C$, and $h : C \to D$. Then

1. The composition of mappings is associative; that is, $(h \circ g) \circ f = h \circ (g \circ f)$;

2. If f and g are both one-to-one, then the mapping $g \circ f$ is one-to-one;

3. If f and g are both onto, then the mapping $g \circ f$ is onto;

4. If f and g are bijective, then so is $g \circ f$.

Proof. We will prove (1) and (3). Part (2) is left as an exercise. Part (4) follows directly from (2) and (3).

(1) We must show that

$$h \circ (g \circ f) = (h \circ g) \circ f.$$

For $a \in A$ we have

$$\begin{aligned}
(h \circ (g \circ f))(a) &= h((g \circ f)(a)) \\
&= h(g(f(a))) \\
&= (h \circ g)(f(a)) \\
&= ((h \circ g) \circ f)(a).
\end{aligned}$$

(3) Assume that f and g are both onto functions. Given $c \in C$, we must show that there exists an $a \in A$ such that $(g \circ f)(a) = g(f(a)) = c$. However, since g is onto, there is an element $b \in B$ such that $g(b) = c$. Similarly, there is an $a \in A$ such that $f(a) = b$. Accordingly,

$$(g \circ f)(a) = g(f(a)) = g(b) = c.$$

$\blacksquare$

If S is any set, we will use id_S or id to denote the **identity mapping** from S to itself. Define this map by $id(s) = s$ for all $s \in S$. A map $g : B \to A$ is an **inverse mapping** of $f : A \to B$ if $g \circ f = id_A$ and $f \circ g = id_B$; in other words, the inverse function of a function simply "undoes" the function. A map is said to be **invertible** if it has an inverse. We usually write f^{-1} for the inverse of f.

Example 1.16. The function $f(x) = x^3$ has inverse $f^{-1}(x) = \sqrt[3]{x}$ by Example 1.12 on page 10. $\qquad\square$

Example 1.17. The natural logarithm and the exponential functions, $f(x) = \ln x$ and $f^{-1}(x) = e^x$, are inverses of each other provided that we are careful about choosing domains. Observe that

$$f(f^{-1}(x)) = f(e^x) = \ln e^x = x$$

and

$$f^{-1}(f(x)) = f^{-1}(\ln x) = e^{\ln x} = x$$

whenever composition makes sense. $\qquad\square$

Example 1.18. Suppose that

$$A = \begin{pmatrix} 3 & 1 \\ 5 & 2 \end{pmatrix}.$$

Then A defines a map from $\mathbb{R}^2$ to $\mathbb{R}^2$ by

$$T_A(x, y) = (3x + y, 5x + 2y).$$

We can find an inverse map of T_A by simply inverting the matrix A; that is, $T_A^{-1} = T_{A^{-1}}$. In this example,

$$A^{-1} = \begin{pmatrix} 2 & -1 \\ -5 & 3 \end{pmatrix};$$

hence, the inverse map is given by

$$T_A^{-1}(x, y) = (2x - y, -5x + 3y).$$

It is easy to check that

$$T_A^{-1} \circ T_A(x, y) = T_A \circ T_A^{-1}(x, y) = (x, y).$$

Not every map has an inverse. If we consider the map

$$T_B(x, y) = (3x, 0)$$

given by the matrix

$$B = \begin{pmatrix} 3 & 0 \\ 0 & 0 \end{pmatrix},$$

then an inverse map would have to be of the form

$$T_B^{-1}(x, y) = (ax + by, cx + dy)$$

and

$$(x, y) = T_B \circ T_B^{-1}(x, y) = (3ax + 3by, 0)$$

for all x and y. Clearly this is impossible because y might not be 0. □

Example 1.19. Given the permutation

$$\pi = \begin{pmatrix} 1 & 2 & 3 \\ 2 & 3 & 1 \end{pmatrix}$$

on $S = \{1, 2, 3\}$, it is easy to see that the permutation defined by

$$\pi^{-1} = \begin{pmatrix} 1 & 2 & 3 \\ 3 & 1 & 2 \end{pmatrix}$$

is the inverse of π. In fact, any bijective mapping possesses an inverse, as we will see in the next theorem. □

Theorem 1.20. A mapping is invertible if and only if it is both one-to-one and onto.

Proof. Suppose first that $f : A \to B$ is invertible with inverse $g : B \to A$. Then $g \circ f = id_A$ is the identity map; that is, $g(f(a)) = a$. If $a_1, a_2 \in A$ with $f(a_1) = f(a_2)$, then $a_1 = g(f(a_1)) = g(f(a_2)) = a_2$. Consequently, f is one-to-one. Now suppose that $b \in B$. To show that f is onto, it is necessary to find an $a \in A$ such that $f(a) = b$, but $f(g(b)) = b$ with $g(b) \in A$. Let $a = g(b)$.

Conversely, let f be bijective and let $b \in B$. Since f is onto, there exists an $a \in A$ such that $f(a) = b$. Because f is one-to-one, a must be unique. Define g by letting $g(b) = a$. We have now constructed the inverse of f.   ∎

Equivalence Relations and Partitions

A fundamental notion in mathematics is that of equality. We can generalize equality with equivalence relations and equivalence classes. An *equivalence relation* on a set X is a relation $R \subset X \times X$ such that

- $(x, x) \in R$ for all $x \in X$ (*reflexive property*);

- $(x, y) \in R$ implies $(y, x) \in R$ (*symmetric property*);

- (x, y) and $(y, z) \in R$ imply $(x, z) \in R$ (*transitive property*).

Given an equivalence relation R on a set X, we usually write $x \sim y$ instead of $(x, y) \in R$. If the equivalence relation already has an associated notation such as $=, \equiv$, or $\cong$, we will use that notation.

Example 1.21. Let p, q, r, and s be integers, where q and s are nonzero. Define $p/q \sim r/s$ if $ps = qr$. Clearly $\sim$ is reflexive and symmetric. To show that it is also transitive, suppose that $p/q \sim r/s$ and $r/s \sim t/u$, with q, s, and u all nonzero. Then $ps = qr$ and $ru = st$. Therefore,

$$psu = qru = qst.$$

Since $s \neq 0$, $pu = qt$. Consequently, $p/q \sim t/u$.   □

Example 1.22. Suppose that f and g are differentiable functions on $\mathbb{R}$. We can define an equivalence relation on such functions by letting $f(x) \sim g(x)$ if $f'(x) = g'(x)$. It is clear that $\sim$ is both reflexive and symmetric. To demonstrate transitivity, suppose that $f(x) \sim g(x)$ and $g(x) \sim h(x)$. From calculus we know that $f(x) - g(x) = c_1$ and $g(x) - h(x) = c_2$, where c_1 and c_2 are both constants. Hence,

$$f(x) - h(x) = (f(x) - g(x)) + (g(x) - h(x)) = c_1 + c_2$$

and $f'(x) - h'(x) = 0$. Therefore, $f(x) \sim h(x)$.   □

Example 1.23. For (x_1, y_1) and (x_2, y_2) in $\mathbb{R}^2$, define $(x_1, y_1) \sim (x_2, y_2)$ if $x_1^2 + y_1^2 = x_2^2 + y_2^2$. Then $\sim$ is an equivalence relation on $\mathbb{R}^2$. $\quad\square$

Example 1.24. Let A and B be 2×2 matrices with entries in the real numbers. We can define an equivalence relation on the set of 2×2 matrices, by saying $A \sim B$ if there exists an invertible matrix P such that $PAP^{-1} = B$. For example, if

$$A = \begin{pmatrix} 1 & 2 \\ -1 & 1 \end{pmatrix} \quad \text{and} \quad B = \begin{pmatrix} -18 & 33 \\ -11 & 20 \end{pmatrix},$$

then $A \sim B$ since $PAP^{-1} = B$ for

$$P = \begin{pmatrix} 2 & 5 \\ 1 & 3 \end{pmatrix}.$$

Let I be the 2×2 identity matrix; that is,

$$I = \begin{pmatrix} 1 & 0 \\ 0 & 1 \end{pmatrix}.$$

Then $IAI^{-1} = IAI = A$; therefore, the relation is reflexive. To show symmetry, suppose that $A \sim B$. Then there exists an invertible matrix P such that $PAP^{-1} = B$. So

$$A = P^{-1}BP = P^{-1}B(P^{-1})^{-1}.$$

Finally, suppose that $A \sim B$ and $B \sim C$. Then there exist invertible matrices P and Q such that $PAP^{-1} = B$ and $QBQ^{-1} = C$. Since

$$C = QBQ^{-1} = QPAP^{-1}Q^{-1} = (QP)A(QP)^{-1},$$

the relation is transitive. Two matrices that are equivalent in this manner are said to be *similar*. $\quad\square$

A *partition* $\mathcal{P}$ of a set X is a collection of nonempty sets $X_1, X_2, \ldots$ such that $X_i \cap X_j = \varnothing$ for $i \neq j$ and $\bigcup_k X_k = X$. Let $\sim$ be an equivalence relation on a set X and let $x \in X$. Then $[x] = \{y \in X : y \sim x\}$ is called the *equivalence class* of x. We will see that an equivalence relation gives rise to a partition via equivalence classes. Also, whenever a partition of a set exists, there is some natural underlying equivalence relation, as the following theorem demonstrates.

Theorem 1.25. Given an equivalence relation $\sim$ on a set X, the equivalence classes of X form a partition of X. Conversely, if $\mathcal{P} = \{X_i\}$ is a partition of a set X, then there is an equivalence relation on X with equivalence classes X_i.

Proof. Suppose there exists an equivalence relation ~ on the set X. For any $x \in X$, the reflexive property shows that $x \in [x]$ and so $[x]$ is nonempty. Clearly $X = \bigcup_{x \in X} [x]$. Now let $x, y \in X$. We need to show that either $[x] = [y]$ or $[x] \cap [y] = \varnothing$. Suppose that the intersection of $[x]$ and $[y]$ is not empty and that $z \in [x] \cap [y]$. Then $z \sim x$ and $z \sim y$. By symmetry and transitivity $x \sim y$; hence, $[x] \subset [y]$. Similarly, $[y] \subset [x]$ and so $[x] = [y]$. Therefore, any two equivalence classes are either disjoint or exactly the same.

Conversely, suppose that $\mathcal{P} = \{X_i\}$ is a partition of a set X. Let two elements be equivalent if they are in the same partition. Clearly, the relation is reflexive. If x is in the same partition as y, then y is in the same partition as x, so $x \sim y$ implies $y \sim x$. Finally, if x is in the same partition as y and y is in the same partition as z, then x must be in the same partition as z, and transitivity holds. ∎

Corollary 1.26. Two equivalence classes of an equivalence relation are either disjoint or equal.

Let us examine some of the partitions given by the equivalence classes in the last set of examples.

Example 1.27. In the equivalence relation in Example 1.21 on page 14, two pairs of integers, (p, q) and (r, s), are in the same equivalence class when they reduce to the same fraction in its lowest terms. □

Example 1.28. In the equivalence relation in Example 1.22 on page 14, two functions $f(x)$ and $g(x)$ are in the same partition when they differ by a constant. □

Example 1.29. We defined an equivalence class on $\mathbb{R}^2$ by $(x_1, y_1) \sim (x_2, y_2)$ if $x_1^2 + y_1^2 = x_2^2 + y_2^2$. Two pairs of real numbers are in the same partition when they lie on the same circle about the origin. □

Example 1.30. Let r and s be two integers and suppose that $n \in \mathbb{N}$. We say that r is *congruent* to s *modulo* n, or r is congruent to s mod n, if $r - s$ is evenly divisible by n; that is, $r - s = nk$ for some $k \in \mathbb{Z}$. In this case we write $r \equiv s \pmod{n}$. For example, $41 \equiv 17 \pmod{8}$ since $41 - 17 = 24$ is divisible by 8. We claim that congruence modulo n forms an equivalence relation of $\mathbb{Z}$. Certainly any integer r is equivalent to itself since $r - r = 0$ is divisible by n. We will now show that the relation is symmetric. If $r \equiv s \pmod{n}$, then $r - s = -(s - r)$ is divisible by n. So $s - r$ is divisible by n and $s \equiv r \pmod{n}$. Now suppose that $r \equiv s \pmod{n}$ and $s \equiv t \pmod{n}$. Then there exist integers k and l such that $r - s = kn$ and $s - t = ln$. To show transitivity, it is necessary to prove that $r - t$ is divisible by n. However,

$$r - t = r - s + s - t = kn + ln = (k + l)n,$$

and so $r - t$ is divisible by n.

If we consider the equivalence relation established by the integers modulo 3, then

$$[0] = \{\ldots, -3, 0, 3, 6, \ldots\},$$
$$[1] = \{\ldots, -2, 1, 4, 7, \ldots\},$$
$$[2] = \{\ldots, -1, 2, 5, 8, \ldots\}.$$

Notice that $[0] \cup [1] \cup [2] = \mathbb{Z}$ and also that the sets are disjoint. The sets $[0]$, $[1]$, and $[2]$ form a partition of the integers.

The integers modulo n are a very important example in the study of abstract algebra and will become quite useful in our investigation of various algebraic structures such as groups and rings. In our discussion of the integers modulo n we have actually assumed a result known as the division algorithm, which will be stated and proved in Chapter 2.   □

Sage.   Sage is a powerful, open source, system for exact, numerical, and symbolic mathematical computations. Electronic versions of this text contain comprehensive introductions to the use of Sage to study abstract algebra, and include a set of exercises. These can be found at the book's website. Due to the format of this version of the text, at the end of each chapter we have just included brief suggestions of how Sage might be employed.

1.3 Reading Questions

1. What do relations and mappings have in common?

2. What makes relations and mappings different?

3. State carefully the three defining properties of an equivalence relation. In other words, do not just *name* the properties, give their definitions.

4. What is the big deal about equivalence relations? (Hint: Partitions.)

5. Describe a general technique for proving that two sets are equal.

1.4 Exercises

1. Suppose that

$$A = \{x : x \in \mathbb{N} \text{ and } x \text{ is even}\},$$
$$B = \{x : x \in \mathbb{N} \text{ and } x \text{ is prime}\},$$
$$C = \{x : x \in \mathbb{N} \text{ and } x \text{ is a multiple of } 5\}.$$

Describe each of the following sets.

   (a) $A \cap B$ (c) $A \cup B$

   (b) $B \cap C$ (d) $A \cap (B \cup C)$

2. If $A = \{a, b, c\}$, $B = \{1, 2, 3\}$, $C = \{x\}$, and $D = \varnothing$, list all of the elements in each of the following sets.

   (a) $A \times B$ (c) $A \times B \times C$

   (b) $B \times A$ (d) $A \times D$

3. Find an example of two nonempty sets A and B for which $A \times B = B \times A$ is true.

4. Prove $A \cup \varnothing = A$ and $A \cap \varnothing = \varnothing$.

5. Prove $A \cup B = B \cup A$ and $A \cap B = B \cap A$.

6. Prove $A \cup (B \cap C) = (A \cup B) \cap (A \cup C)$.

7. Prove $A \cap (B \cup C) = (A \cap B) \cup (A \cap C)$.

8. Prove $A \subset B$ if and only if $A \cap B = A$.

9. Prove $(A \cap B)' = A' \cup B'$.

10. Prove $A \cup B = (A \cap B) \cup (A \smallsetminus B) \cup (B \smallsetminus A)$.

11. Prove $(A \cup B) \times C = (A \times C) \cup (B \times C)$.

12. Prove $(A \cap B) \smallsetminus B = \varnothing$.

13. Prove $(A \cup B) \smallsetminus B = A \smallsetminus B$.

14. Prove $A \smallsetminus (B \cup C) = (A \smallsetminus B) \cap (A \smallsetminus C)$.

15. Prove $A \cap (B \smallsetminus C) = (A \cap B) \smallsetminus (A \cap C)$.

16. Prove $(A \smallsetminus B) \cup (B \smallsetminus A) = (A \cup B) \smallsetminus (A \cap B)$.

17. Which of the following relations $f : \mathbb{Q} \to \mathbb{Q}$ define a mapping? In each case, supply a reason why f is or is not a mapping.

(a) $f(p/q) = \dfrac{p+1}{p-2}$

(c) $f(p/q) = \dfrac{p+q}{q^2}$

(b) $f(p/q) = \dfrac{3p}{3q}$

(d) $f(p/q) = \dfrac{3p^2}{7q^2} - \dfrac{p}{q}$

18. Determine which of the following functions are one-to-one and which are onto. If the function is not onto, determine its range.

(a) $f : \mathbb{R} \to \mathbb{R}$ defined by $f(x) = e^x$

(b) $f : \mathbb{Z} \to \mathbb{Z}$ defined by $f(n) = n^2 + 3$

(c) $f : \mathbb{R} \to \mathbb{R}$ defined by $f(x) = \sin x$

(d) $f : \mathbb{Z} \to \mathbb{Z}$ defined by $f(x) = x^2$

19. Let $f : A \to B$ and $g : B \to C$ be invertible mappings; that is, mappings such that f^{-1} and g^{-1} exist. Show that $(g \circ f)^{-1} = f^{-1} \circ g^{-1}$.

20.

(a) Define a function $f : \mathbb{N} \to \mathbb{N}$ that is one-to-one but not onto.

(b) Define a function $f : \mathbb{N} \to \mathbb{N}$ that is onto but not one-to-one.

21. Prove the relation defined on $\mathbb{R}^2$ by $(x_1, y_1) \sim (x_2, y_2)$ if $x_1^2 + y_1^2 = x_2^2 + y_2^2$ is an equivalence relation.

22. Let $f : A \to B$ and $g : B \to C$ be maps.

(a) If f and g are both one-to-one functions, show that $g \circ f$ is one-to-one.

(b) If $g \circ f$ is onto, show that g is onto.

(c) If $g \circ f$ is one-to-one, show that f is one-to-one.

(d) If $g \circ f$ is one-to-one and f is onto, show that g is one-to-one.

(e) If $g \circ f$ is onto and g is one-to-one, show that f is onto.

23. Define a function on the real numbers by

$$f(x) = \frac{x+1}{x-1}.$$

What are the domain and range of f? What is the inverse of f? Compute $f \circ f^{-1}$ and $f^{-1} \circ f$.

24. Let $f : X \to Y$ be a map with $A_1, A_2 \subset X$ and $B_1, B_2 \subset Y$.

(a) Prove $f(A_1 \cup A_2) = f(A_1) \cup f(A_2)$.

(b) Prove $f(A_1 \cap A_2) \subset f(A_1) \cap f(A_2)$. Give an example in which equality fails.

(c) Prove $f^{-1}(B_1 \cup B_2) = f^{-1}(B_1) \cup f^{-1}(B_2)$, where

$$f^{-1}(B) = \{x \in X : f(x) \in B\}.$$

(d) Prove $f^{-1}(B_1 \cap B_2) = f^{-1}(B_1) \cap f^{-1}(B_2)$.

(e) Prove $f^{-1}(Y \smallsetminus B_1) = X \smallsetminus f^{-1}(B_1)$.

25. Determine whether or not the following relations are equivalence relations on the given set. If the relation is an equivalence relation, describe the partition given by it. If the relation is not an equivalence relation, state why it fails to be one.

(a) $x \sim y$ in $\mathbb{R}$ if $x \geq y$

(b) $m \sim n$ in $\mathbb{Z}$ if $mn > 0$

(c) $x \sim y$ in $\mathbb{R}$ if $|x - y| \leq 4$

(d) $m \sim n$ in $\mathbb{Z}$ if $m \equiv n \pmod{6}$

26. Define a relation $\sim$ on $\mathbb{R}^2$ by stating that $(a, b) \sim (c, d)$ if and only if $a^2 + b^2 \leq c^2 + d^2$. Show that $\sim$ is reflexive and transitive but not symmetric.

27. Show that an $m \times n$ matrix gives rise to a well-defined map from $\mathbb{R}^n$ to $\mathbb{R}^m$.

28. Find the error in the following argument by providing a counterexample. "The reflexive property is redundant in the axioms for an equivalence relation. If $x \sim y$, then $y \sim x$ by the symmetric property. Using the transitive property, we can deduce that $x \sim x$."

29. **Projective Real Line.** Define a relation on $\mathbb{R}^2 \smallsetminus \{(0, 0)\}$ by letting $(x_1, y_1) \sim (x_2, y_2)$ if there exists a nonzero real number λ such that $(x_1, y_1) = (\lambda x_2, \lambda y_2)$. Prove that $\sim$ defines an equivalence relation on $\mathbb{R}^2 \smallsetminus (0, 0)$. What are the corresponding equivalence classes? This equivalence relation defines the projective line, denoted by $\mathbb{P}(\mathbb{R})$, which is very important in geometry.

1.5 References and Suggested Readings

[1] Artin, M. *Algebra (Classic Version)*. 2nd ed. Pearson, Upper Saddle River, NJ, 2018.

[2] Childs, L. *A Concrete Introduction to Higher Algebra*. 2nd ed. Springer-Verlag, New York, 1995.

[3] Dummit, D. and Foote, R. *Abstract Algebra*. 3rd ed. Wiley, New York, 2003.

[4] Ehrlich, G. *Fundamental Concepts of Algebra*. PWS-KENT, Boston, 1991.

[5] Fraleigh, J. B. *A First Course in Abstract Algebra*. 7th ed. Pearson, Upper Saddle River, NJ, 2003.

[6] Gallian, J. A. *Contemporary Abstract Algebra*. 7th ed. Brooks/Cole, Belmont, CA, 2009.

[7] Halmos, P. *Naive Set Theory*. Springer, New York, 1991. One of the best references for set theory.

[8] Herstein, I. N. *Abstract Algebra*. 3rd ed. Wiley, New York, 1996.

[9] Hungerford, T. W. *Algebra*. Springer, New York, 1974. One of the standard graduate algebra texts.

[10] Lang, S. *Algebra*. 3rd ed. Springer, New York, 2002. Another standard graduate text.

[11] Lidl, R. and Pilz, G. *Applied Abstract Algebra*. 2nd ed. Springer, New York, 1998.

[12] Mackiw, G. *Applications of Abstract Algebra*. Wiley, New York, 1985.

[13] Nickelson, W. K. *Introduction to Abstract Algebra*. 3rd ed. Wiley, New York, 2006.

[14] Solow, D. *How to Read and Do Proofs*. 5th ed. Wiley, New York, 2009.

[15] van der Waerden, B. L. *A History of Algebra*. Springer-Verlag, New York, 1985. An account of the historical development of algebra.

The Integers

The integers are the building blocks of mathematics. In this chapter we will investigate the fundamental properties of the integers, including mathematical induction, the division algorithm, and the Fundamental Theorem of Arithmetic.

2.1 Mathematical Induction

Suppose we wish to show that

$$1 + 2 + \cdots + n = \frac{n(n+1)}{2}$$

for any natural number n. This formula is easily verified for small numbers such as $n = 1, 2, 3,$ or 4, but it is impossible to verify for all natural numbers on a case-by-case basis. To prove the formula true in general, a more generic method is required.

Suppose we have verified the equation for the first n cases. We will attempt to show that we can generate the formula for the $(n+1)$th case from this knowledge. The formula is true for $n = 1$ since

$$1 = \frac{1(1+1)}{2}.$$

If we have verified the first n cases, then

$$1 + 2 + \cdots + n + (n+1) = \frac{n(n+1)}{2} + n + 1$$
$$= \frac{n^2 + 3n + 2}{2}$$
$$= \frac{(n+1)[(n+1)+1]}{2}.$$

This is exactly the formula for the $(n+1)$th case.

This method of proof is known as **mathematical induction**. Instead of attempting to verify a statement about some subset S of the positive integers $\mathbb{N}$ on a case-by-case basis, an impossible task if S is an infinite set, we give a specific proof for the smallest integer being considered, followed by a generic argument showing that if the statement holds for a given case, then it must also hold for the next case in the sequence. We summarize mathematical induction in the following axiom.

Principle 2.1. First Principle of Mathematical Induction. Let $S(n)$ be a statement about integers for $n \in \mathbb{N}$ and suppose $S(n_0)$ is true for some integer n_0. If for all integers k with $k \geq n_0$, $S(k)$ implies that $S(k+1)$ is true, then $S(n)$ is true for all integers n greater than or equal to n_0.

Example 2.2. For all integers $n \geq 3$, $2^n > n + 4$. Since

$$8 = 2^3 > 3 + 4 = 7,$$

the statement is true for $n_0 = 3$. Assume that $2^k > k + 4$ for $k \geq 3$. Then $2^{k+1} = 2 \cdot 2^k > 2(k+4)$. But

$$2(k+4) = 2k + 8 > k + 5 = (k+1) + 4$$

since k is positive. Hence, by induction, the statement holds for all integers $n \geq 3$.

$\square$

Example 2.3. Every integer $10^{n+1} + 3 \cdot 10^n + 5$ is divisible by 9 for $n \in \mathbb{N}$. For $n = 1$,

$$10^{1+1} + 3 \cdot 10 + 5 = 135 = 9 \cdot 15$$

is divisible by 9. Suppose that $10^{k+1} + 3 \cdot 10^k + 5$ is divisible by 9 for $k \geq 1$. Then

$$10^{(k+1)+1} + 3 \cdot 10^{k+1} + 5 = 10^{k+2} + 3 \cdot 10^{k+1} + 50 - 45$$
$$= 10(10^{k+1} + 3 \cdot 10^k + 5) - 45$$

is divisible by 9.

$\square$

Example 2.4. We will prove the binomial theorem using mathematical induction; that is,

$$(a+b)^n = \sum_{k=0}^{n} \binom{n}{k} a^k b^{n-k},$$

where a and b are real numbers, $n \in \mathbb{N}$, and

$$\binom{n}{k} = \frac{n!}{k!(n-k)!}$$

is the binomial coefficient. We first show that

$$\binom{n+1}{k} = \binom{n}{k} + \binom{n}{k-1}.$$

This result follows from

$$\binom{n}{k} + \binom{n}{k-1} = \frac{n!}{k!(n-k)!} + \frac{n!}{(k-1)!(n-k+1)!}$$

$$= \frac{(n+1)!}{k!(n+1-k)!}$$

$$= \binom{n+1}{k}.$$

If $n = 1$, the binomial theorem is easy to verify. Now assume that the result is true for n greater than or equal to 1. Then

$$(a+b)^{n+1} = (a+b)(a+b)^n$$

$$= (a+b)\left(\sum_{k=0}^{n} \binom{n}{k} a^k b^{n-k}\right)$$

$$= \sum_{k=0}^{n} \binom{n}{k} a^{k+1} b^{n-k} + \sum_{k=0}^{n} \binom{n}{k} a^k b^{n+1-k}$$

$$= a^{n+1} + \sum_{k=1}^{n} \binom{n}{k-1} a^k b^{n+1-k} + \sum_{k=1}^{n} \binom{n}{k} a^k b^{n+1-k} + b^{n+1}$$

$$= a^{n+1} + \sum_{k=1}^{n} \left[\binom{n}{k-1} + \binom{n}{k}\right] a^k b^{n+1-k} + b^{n+1}$$

$$= \sum_{k=0}^{n+1} \binom{n+1}{k} a^k b^{n+1-k}.$$

$\square$

We have an equivalent statement of the Principle of Mathematical Induction that is often very useful.

Principle 2.5. Second Principle of Mathematical Induction. Let $S(n)$ be a statement about integers for $n \in \mathbb{N}$ and suppose $S(n_0)$ is true for some integer n_0. If $S(n_0), S(n_0 + 1), \ldots, S(k)$ imply that $S(k+1)$ for $k \geq n_0$, then the statement $S(n)$ is true for all integers $n \geq n_0$.

A nonempty subset S of $\mathbb{Z}$ is **well-ordered** if S contains a least element. Notice that the set $\mathbb{Z}$ is not well-ordered since it does not contain a smallest element. However, the natural numbers are well-ordered.

Principle 2.6. Principle of Well-Ordering. Every nonempty subset of the natural numbers is well-ordered.

The Principle of Well-Ordering is equivalent to the Principle of Mathematical Induction.

Lemma 2.7. The Principle of Mathematical Induction implies that 1 is the least positive natural number.

Proof. Let $S = \{n \in \mathbb{N} : n \geq 1\}$. Then $1 \in S$. Assume that $n \in S$. Since $0 < 1$, it must be the case that $n = n + 0 < n + 1$. Therefore, $1 \leq n < n + 1$. Consequently, if $n \in S$, then $n + 1$ must also be in S, and by the Principle of Mathematical Induction, and we have $S = \mathbb{N}$. ∎

Theorem 2.8. The Principle of Mathematical Induction implies the Principle of Well-Ordering. That is, every nonempty subset of $\mathbb{N}$ contains a least element.

Proof. We must show that if S is a nonempty subset of the natural numbers, then S contains a least element. If S contains 1, then the theorem is true by Lemma 2.7. Assume that if S contains an integer k such that $1 \leq k \leq n$, then S contains a least element. We will show that if a set S contains an integer less than or equal to $n + 1$, then S has a least element. If S does not contain an integer less than $n + 1$, then $n + 1$ is the smallest integer in S. Otherwise, since S is nonempty, S must contain an integer less than or equal to n. In this case, by induction, S contains a least element. ∎

Induction can also be very useful in formulating definitions. For instance, there are two ways to define $n!$, the factorial of a positive integer n.

- The *explicit* definition: $n! = 1 \cdot 2 \cdot 3 \cdots (n - 1) \cdot n$.
- The *inductive* or *recursive* definition: $1! = 1$ and $n! = n(n - 1)!$ for $n > 1$.

Every good mathematician or computer scientist knows that looking at problems recursively, as opposed to explicitly, often results in better understanding of complex issues.

2.2 The Division Algorithm

An application of the Principle of Well-Ordering that we will use often is the division algorithm.

Theorem 2.9. Division Algorithm. Let a and b be integers, with $b > 0$. Then there exist unique integers q and r such that

$$a = bq + r$$

where $0 \leq r < b$.

Proof. This is a perfect example of the existence-and-uniqueness type of proof. We must first prove that the numbers q and r actually exist. Then we must show that if q' and r' are two other such numbers, then $q = q'$ and $r = r'$.

Existence of q and r. Let

$$S = \{a - bk : k \in \mathbb{Z} \text{ and } a - bk \geq 0\}.$$

If $0 \in S$, then b divides a, and we can let $q = a/b$ and $r = 0$. If $0 \notin S$, we can use the Well-Ordering Principle. We must first show that S is nonempty. If $a > 0$, then $a - b \cdot 0 \in S$. If $a < 0$, then $a - b(2a) = a(1 - 2b) \in S$. In either case $S \neq \emptyset$. By the Well-Ordering Principle, S must have a smallest member, say $r = a - bq$. Therefore, $a = bq + r, r \geq 0$. We now show that $r < b$. Suppose that $r > b$. Then

$$a - b(q + 1) = a - bq - b = r - b > 0.$$

In this case we would have $a - b(q+1)$ in the set S. But then $a - b(q+1) < a - bq$, which would contradict the fact that $r = a - bq$ is the smallest member of S. So $r \leq b$. Since $0 \notin S$, $r \neq b$ and so $r < b$.

Uniqueness of q and r. Suppose there exist integers r, r', q, and q' such that

$$a = bq + r, 0 \leq r < b \quad \text{and} \quad a = bq' + r', 0 \leq r' < b.$$

Then $bq + r = bq' + r'$. Assume that $r' \geq r$. From the last equation we have $b(q - q') = r' - r$; therefore, b must divide $r' - r$ and $0 \leq r' - r \leq r' < b$. This is possible only if $r' - r = 0$. Hence, $r = r'$ and $q = q'$. ∎

Let a and b be integers. If $b = ak$ for some integer k, we write $a \mid b$. An integer d is called a ***common divisor*** of a and b if $d \mid a$ and $d \mid b$. The ***greatest common divisor*** of integers a and b is a positive integer d such that d is a common divisor of a and b and if d' is any other common divisor of a and b, then $d' \mid d$. We write $d = \gcd(a, b)$; for example, $\gcd(24, 36) = 12$ and $\gcd(120, 102) = 6$. We say that two integers a and b are ***relatively prime*** if $\gcd(a, b) = 1$.

Theorem 2.10. Let a and b be nonzero integers. Then there exist integers r and s such that

$$\gcd(a, b) = ar + bs.$$

Furthermore, the greatest common divisor of a and b is unique.

Proof. Let

$$S = \{am + bn : m, n \in \mathbb{Z} \text{ and } am + bn > 0\}.$$

Clearly, the set S is nonempty; hence, by the Well-Ordering Principle S must have a smallest member, say $d = ar + bs$. We claim that $d = \gcd(a, b)$. Write $a = dq + r'$ where $0 \leq r' < d$.

If $r' > 0$, then

$$r' = a - dq$$
$$= a - (ar + bs)q$$
$$= a - arq - bsq$$
$$= a(1 - rq) + b(-sq),$$

which is in S. But this would contradict the fact that d is the smallest member of S. Hence, $r' = 0$ and d divides a. A similar argument shows that d divides b. Therefore, d is a common divisor of a and b.

Suppose that d' is another common divisor of a and b, and we want to show that $d' \mid d$. If we let $a = d'h$ and $b = d'k$, then

$$d = ar + bs = d'hr + d'ks = d'(hr + ks).$$

So d' must divide d. Hence, d must be the unique greatest common divisor of a and b. ∎

Corollary 2.11. Let a and b be two integers that are relatively prime. Then there exist integers r and s such that $ar + bs = 1$.

The Euclidean Algorithm

Among other things, Theorem 2.10 allows us to compute the greatest common divisor of two integers.

Example 2.12. Let us compute the greatest common divisor of 945 and 2415. First observe that

$$2415 = 945 \cdot 2 + 525$$
$$945 = 525 \cdot 1 + 420$$
$$525 = 420 \cdot 1 + 105$$
$$420 = 105 \cdot 4 + 0.$$

Reversing our steps, 105 divides 420, 105 divides 525, 105 divides 945, and 105 divides 2415. Hence, 105 divides both 945 and 2415. If d were another common divisor of 945 and 2415, then d would also have to divide 105. Therefore, $\gcd(945, 2415) = 105$.

If we work backward through the above sequence of equations, we can also obtain numbers r and s such that $945r + 2415s = 105$. Observe that

$$
\begin{aligned}
105 &= 525 + (-1) \cdot 420 \\
&= 525 + (-1) \cdot [945 + (-1) \cdot 525] \\
&= 2 \cdot 525 + (-1) \cdot 945 \\
&= 2 \cdot [2415 + (-2) \cdot 945] + (-1) \cdot 945 \\
&= 2 \cdot 2415 + (-5) \cdot 945.
\end{aligned}
$$

So $r = -5$ and $s = 2$. Notice that r and s are not unique, since $r = 41$ and $s = -16$ would also work.   □

To compute $\gcd(a, b) = d$, we are using repeated divisions to obtain a decreasing sequence of positive integers $r_1 > r_2 > \cdots > r_n = d$; that is,

$$
\begin{aligned}
b &= aq_1 + r_1 \\
a &= r_1q_2 + r_2 \\
r_1 &= r_2q_3 + r_3 \\
&\;\;\vdots \\
r_{n-2} &= r_{n-1}q_n + r_n \\
r_{n-1} &= r_nq_{n+1}.
\end{aligned}
$$

To find r and s such that $ar + bs = d$, we begin with this last equation and substitute results obtained from the previous equations:

$$
\begin{aligned}
d &= r_n \\
&= r_{n-2} - r_{n-1}q_n \\
&= r_{n-2} - q_n(r_{n-3} - q_{n-1}r_{n-2}) \\
&= -q_nr_{n-3} + (1 + q_nq_{n-1})r_{n-2} \\
&\;\;\vdots \\
&= ra + sb.
\end{aligned}
$$

The algorithm that we have just used to find the greatest common divisor d of two integers a and b and to write d as the linear combination of a and b is known as the *Euclidean algorithm*.

Prime Numbers

Let p be an integer such that $p > 1$. We say that p is a *prime number*, or simply p is *prime*, if the only positive numbers that divide p are 1 and p itself. An integer $n > 1$ that is not prime is said to be *composite*.

Lemma 2.13. Euclid. Let a and b be integers and p be a prime number. If $p \mid ab$, then either $p \mid a$ or $p \mid b$.

Proof. Suppose that p does not divide a. We must show that $p \mid b$. Since $\gcd(a, p) = 1$, there exist integers r and s such that $ar + ps = 1$. So

$$b = b(ar + ps) = (ab)r + p(bs).$$

Since p divides both ab and itself, p must divide $b = (ab)r + p(bs)$. ∎

Theorem 2.14. Euclid. There exist an infinite number of primes.

Proof. We will prove this theorem by contradiction. Suppose that there are only a finite number of primes, say $p_1, p_2, \ldots, p_n$. Let $P = p_1 p_2 \cdots p_n + 1$. Then P must be divisible by some p_i for $1 \leq i \leq n$. In this case, p_i must divide $P - p_1 p_2 \cdots p_n = 1$, which is a contradiction. Hence, either P is prime or there exists an additional prime number $p \neq p_i$ that divides P. ∎

Theorem 2.15. Fundamental Theorem of Arithmetic. Let n be an integer such that $n > 1$. Then

$$n = p_1 p_2 \cdots p_k,$$

where $p_1, \ldots, p_k$ are primes (not necessarily distinct). Furthermore, this factorization is unique; that is, if

$$n = q_1 q_2 \cdots q_l,$$

then $k = l$ and the q_i's are just the p_i's rearranged.

Proof. Uniqueness. To show uniqueness we will use induction on n. The theorem is certainly true for $n = 2$ since in this case n is prime. Now assume that the result holds for all integers m such that $1 \leq m < n$, and

$$n = p_1 p_2 \cdots p_k = q_1 q_2 \cdots q_l,$$

where $p_1 \leq p_2 \leq \cdots \leq p_k$ and $q_1 \leq q_2 \leq \cdots \leq q_l$. By Lemma 2.13, $p_1 \mid q_i$ for some $i = 1, \ldots, l$ and $q_1 \mid p_j$ for some $j = 1, \ldots, k$. Since all of the p_i's and q_i's are prime, $p_1 = q_i$ and $q_1 = p_j$. Hence, $p_1 = q_1$ since $p_1 \leq p_j = q_1 \leq q_i = p_1$. By the induction hypothesis,

$$n' = p_2 \cdots p_k = q_2 \cdots q_l$$

has a unique factorization. Hence, $k = l$ and $q_i = p_i$ for $i = 1, \ldots, k$.

Existence. To show existence, suppose that there is some integer that cannot be written as the product of primes. Let S be the set of all such numbers. By the Principle of Well-Ordering, S has a smallest number, say a. If the only positive factors of a are a and 1, then a is prime, which is a contradiction. Hence, $a = a_1 a_2$ where $1 < a_1 < a$ and $1 < a_2 < a$. Neither $a_1 \in S$ nor $a_2 \in S$, since a is the smallest element in S. So

$$a_1 = p_1 \cdots p_r$$
$$a_2 = q_1 \cdots q_s.$$

Therefore,

$$a = a_1 a_2 = p_1 \cdots p_r q_1 \cdots q_s.$$

So $a \notin S$, which is a contradiction.   ∎

↪ Historical Note ↩

Prime numbers were first studied by the ancient Greeks. Two important results from antiquity are Euclid's proof that an infinite number of primes exist and the Sieve of Eratosthenes, a method of computing all of the prime numbers less than a fixed positive integer n. One problem in number theory is to find a function f such that $f(n)$ is prime for each integer n. Pierre Fermat (1601?–1665) conjectured that $2^{2^n} + 1$ was prime for all n, but later it was shown by Leonhard Euler (1707–1783) that

$$2^{2^5} + 1 = 4{,}294{,}967{,}297$$

is a composite number. One of the many unproven conjectures about prime numbers is Goldbach's Conjecture. In a letter to Euler in 1742, Christian Goldbach stated the conjecture that every even integer with the exception of 2 seemed to be the sum of two primes: $4 = 2 + 2$, $6 = 3 + 3$, $8 = 3 + 5$, ... Although the conjecture has been verified for the numbers up through 4×10^{18}, it has yet to be proven in general. Since prime numbers play an important role in public key cryptography, there is currently a great deal of interest in determining whether or not a large number is prime.

Sage. Sage's original purpose was to support research in number theory, so it is perfect for the types of computations with the integers that we have in this chapter.

2.3 Reading Questions

1. Use Sage to express 123456792 as a product of prime numbers.
2. Find the greatest common divisor of 84 and 52.
3. Find integers r and s so that $r(84) + s(52) = \gcd(84, 52)$.
4. Explain the use of the term "induction hypothesis."
5. What is Goldbach's Conjecture? And why is it called a "conjecture"?

2.4 Exercises

1. Prove that
    $$1^2 + 2^2 + \cdots + n^2 = \frac{n(n+1)(2n+1)}{6}$$
    for $n \in \mathbb{N}$.

2. Prove that
    $$1^3 + 2^3 + \cdots + n^3 = \frac{n^2(n+1)^2}{4}$$
    for $n \in \mathbb{N}$.

3. Prove that $n! > 2^n$ for $n \geq 4$.

4. Prove that
    $$x + 4x + 7x + \cdots + (3n-2)x = \frac{n(3n-1)x}{2}$$
    for $n \in \mathbb{N}$.

5. Prove that $10^{n+1} + 10^n + 1$ is divisible by 3 for $n \in \mathbb{N}$.

6. Prove that $4 \cdot 10^{2n} + 9 \cdot 10^{2n-1} + 5$ is divisible by 99 for $n \in \mathbb{N}$.

7. Show that
    $$\sqrt[n]{a_1 a_2 \cdots a_n} \leq \frac{1}{n} \sum_{k=1}^{n} a_k.$$

8. Prove the Leibniz rule for $f^{(n)}(x)$, where $f^{(n)}$ is the nth derivative of f; that is, show that
    $$(fg)^{(n)}(x) = \sum_{k=0}^{n} \binom{n}{k} f^{(k)}(x) g^{(n-k)}(x).$$

9. Use induction to prove that $1 + 2 + 2^2 + \cdots + 2^n = 2^{n+1} - 1$ for $n \in \mathbb{N}$.

10. Prove that
    $$\frac{1}{2} + \frac{1}{6} + \cdots + \frac{1}{n(n+1)} = \frac{n}{n+1}$$
    for $n \in \mathbb{N}$.

11. If x is a nonnegative real number, then show that $(1 + x)^n - 1 \geq nx$ for $n = 0, 1, 2, \ldots$.

12. **Power Sets.** Let X be a set. Define the *power set* of X, denoted $\mathcal{P}(X)$, to be the set of all subsets of X. For example,

$$\mathcal{P}(\{a, b\}) = \{\varnothing, \{a\}, \{b\}, \{a, b\}\}.$$

For every positive integer n, show that a set with exactly n elements has a power set with exactly 2^n elements.

13. Prove that the two principles of mathematical induction stated in Section 2.1 are equivalent.

14. Show that the Principle of Well-Ordering for the natural numbers implies that 1 is the smallest natural number. Use this result to show that the Principle of Well-Ordering implies the Principle of Mathematical Induction; that is, show that if $S \subset \mathbb{N}$ such that $1 \in S$ and $n + 1 \in S$ whenever $n \in S$, then $S = \mathbb{N}$.

15. For each of the following pairs of numbers a and b, calculate $\gcd(a, b)$ and find integers r and s such that $\gcd(a, b) = ra + sb$.

    (a) 14 and 39   (d) 471 and 562

    (b) 234 and 165   (e) 23771 and 19945

    (c) 1739 and 9923   (f) −4357 and 3754

16. Let a and b be nonzero integers. If there exist integers r and s such that $ar + bs = 1$, show that a and b are relatively prime.

17. **Fibonacci Numbers.** The Fibonacci numbers are

$$1, 1, 2, 3, 5, 8, 13, 21, \ldots.$$

We can define them inductively by $f_1 = 1$, $f_2 = 1$, and $f_{n+2} = f_{n+1} + f_n$ for $n \in \mathbb{N}$.

    (a) Prove that $f_n < 2^n$.

    (b) Prove that $f_{n+1} f_{n-1} = f_n^2 + (-1)^n$, $n \geq 2$.

    (c) Prove that $f_n = [(1 + \sqrt{5})^n - (1 - \sqrt{5})^n]/2^n \sqrt{5}$.

    (d) Show that $\phi = \lim_{n \to \infty} f_{n+1}/f_n = (\sqrt{5} + 1)/2$. The constant ϕ is known as the *golden ratio*.

    (e) Prove that f_n and f_{n+1} are relatively prime.

18. Let a and b be integers such that $\gcd(a, b) = 1$. Let r and s be integers such that $ar + bs = 1$. Prove that

$$\gcd(a, s) = \gcd(r, b) = \gcd(r, s) = 1.$$

19. Let $x, y \in \mathbb{N}$ be relatively prime. If xy is a perfect square, prove that x and y must both be perfect squares.

20. Using the division algorithm, show that every perfect square is of the form $4k$ or $4k + 1$ for some nonnegative integer k.

21. Suppose that a, b, r, s are pairwise relatively prime and that

$$a^2 + b^2 = r^2$$
$$a^2 - b^2 = s^2.$$

Prove that a, r, and s are odd and b is even.

22. Let $n \in \mathbb{N}$. Use the division algorithm to prove that every integer is congruent mod n to precisely one of the integers $0, 1, \ldots, n - 1$. Conclude that if r is an integer, then there is exactly one s in $\mathbb{Z}$ such that $0 \le s < n$ and $[r] = [s]$. Hence, the integers are indeed partitioned by congruence mod n.

23. Define the **least common multiple** of two nonzero integers a and b, denoted by $\operatorname{lcm}(a, b)$, to be the nonnegative integer m such that both a and b divide m, and if a and b divide any other integer n, then m also divides n. Prove there exists a unique least common multiple for any two integers a and b.

24. If $d = \gcd(a, b)$ and $m = \operatorname{lcm}(a, b)$, prove that $dm = |ab|$.

25. Show that $\operatorname{lcm}(a, b) = ab$ if and only if $\gcd(a, b) = 1$.

26. Prove that $\gcd(a, c) = \gcd(b, c) = 1$ if and only if $\gcd(ab, c) = 1$ for integers a, b, and c.

27. Let $a, b, c \in \mathbb{Z}$. Prove that if $\gcd(a, b) = 1$ and $a \mid bc$, then $a \mid c$.

28. Let $p \ge 2$. Prove that if $2^p - 1$ is prime, then p must also be prime.

29. Prove that there are an infinite number of primes of the form $6n + 5$.

30. Prove that there are an infinite number of primes of the form $4n - 1$.

31. Using the fact that 2 is prime, show that there do not exist integers p and q such that $p^2 = 2q^2$. Demonstrate that therefore $\sqrt{2}$ cannot be a rational number.

2.5 Programming Exercises

1. **The Sieve of Eratosthenes.** One method of computing all of the prime numbers less than a certain fixed positive integer N is to list all of the numbers n such that $1 < n < N$. Begin by eliminating all of the multiples of 2. Next eliminate all of the multiples of 3. Now eliminate all of the multiples of 5. Notice that 4 has already been crossed out. Continue in this manner, noticing that we do not have to go all the way to N; it suffices to stop at $\sqrt{N}$. Using this method, compute all of the prime numbers less than $N = 250$. We can also use this method to find all of the integers that are relatively prime to an integer N. Simply eliminate the prime factors of N and all of their multiples. Using this method, find all of the numbers that are relatively prime to $N = 120$. Using the Sieve of Eratosthenes, write a program that will compute all of the primes less than an integer N.

2. Let $\mathbb{N}^0 = \mathbb{N} \cup \{0\}$. Ackermann's function is the function $A : \mathbb{N}^0 \times \mathbb{N}^0 \to \mathbb{N}^0$ defined by the equations

$$
\begin{aligned}
A(0, y) &= y + 1, \\
A(x + 1, 0) &= A(x, 1), \\
A(x + 1, y + 1) &= A(x, A(x + 1, y)).
\end{aligned}
$$

   Use this definition to compute $A(3, 1)$. Write a program to evaluate Ackermann's function. Modify the program to count the number of statements executed in the program when Ackermann's function is evaluated. How many statements are executed in the evaluation of $A(4, 1)$? What about $A(5, 1)$?

3. Write a computer program that will implement the Euclidean algorithm. The program should accept two positive integers a and b as input and should output $\gcd(a, b)$ as well as integers r and s such that

$$
\gcd(a, b) = ra + sb.
$$

2.6 References and Suggested Readings

[1] Brookshear, J. G. *Theory of Computation: Formal Languages, Automata, and Complexity.* Benjamin/Cummings, Redwood City, CA, 1989. Shows the relationships of the theoretical aspects of computer science to set theory and the integers.

[2] Hardy, G. H. and Wright, E. M. *An Introduction to the Theory of Numbers.* 6th ed. Oxford University Press, New York, 2008.

[3]   Niven, I. and Zuckerman, H. S. *An Introduction to the Theory of Numbers.* 5th ed. Wiley, New York, 1991.

[4]   Vanden Eynden, C. *Elementary Number Theory.* 2nd ed. Waveland Press, Long Grove IL, 2001.

3

Groups

$\mathcal{W}$e begin our study of algebraic structures by investigating sets associated with single operations that satisfy certain reasonable axioms; that is, we want to define an operation on a set in a way that will generalize such familiar structures as the integers $\mathbb{Z}$ together with the single operation of addition, or invertible 2×2 matrices together with the single operation of matrix multiplication. The integers and the 2×2 matrices, together with their respective single operations, are examples of algebraic structures known as groups.

The theory of groups occupies a central position in mathematics. Modern group theory arose from an attempt to find the roots of a polynomial in terms of its coefficients. Groups now play a central role in such areas as coding theory, counting, and the study of symmetries; many areas of biology, chemistry, and physics have benefited from group theory.

3.1 Integer Equivalence Classes and Symmetries

Let us now investigate some mathematical structures that can be viewed as sets with single operations.

The Integers mod n

The integers mod n have become indispensable in the theory and applications of algebra. In mathematics they are used in cryptography, coding theory, and the detection of errors in identification codes.

We have already seen that two integers a and b are equivalent mod n if n divides $a - b$. The integers mod n also partition $\mathbb{Z}$ into n different equivalence classes; we will denote the set of these equivalence classes by $\mathbb{Z}_n$. Consider the integers modulo 12 and the corresponding partition of the integers:

$$[0] = \{\ldots, -12, 0, 12, 24, \ldots\},$$
$$[1] = \{\ldots, -11, 1, 13, 25, \ldots\},$$
$$\vdots$$
$$[11] = \{\ldots, -1, 11, 23, 35, \ldots\}.$$

When no confusion can arise, we will use $0, 1, \ldots, 11$ to indicate the equivalence classes $[0], [1], \ldots, [11]$ respectively. We can do arithmetic on $\mathbb{Z}_n$. For two integers a and b, define addition modulo n to be $(a + b)$ (mod n); that is, the remainder when $a + b$ is divided by n. Similarly, multiplication modulo n is defined as (ab) (mod n), the remainder when ab is divided by n.

Example 3.1. The following examples illustrate integer arithmetic modulo n:

$$7 + 4 \equiv 1 \quad (\text{mod } 5) \qquad 7 \cdot 3 \equiv 1 \quad (\text{mod } 5)$$
$$3 + 5 \equiv 0 \quad (\text{mod } 8) \qquad 3 \cdot 5 \equiv 7 \quad (\text{mod } 8)$$
$$3 + 4 \equiv 7 \quad (\text{mod } 12) \qquad 3 \cdot 4 \equiv 0 \quad (\text{mod } 12).$$

In particular, notice that it is possible that the product of two nonzero numbers modulo n can be equivalent to 0 modulo n. □

Example 3.2. Most, but not all, of the usual laws of arithmetic hold for addition and multiplication in $\mathbb{Z}_n$. For instance, it is not necessarily true that there is a multiplicative inverse. Consider the multiplication table for $\mathbb{Z}_8$ in Figure 3.3. Notice that 2, 4, and 6 do not have multiplicative inverses; that is, for $n = 2, 4,$ or 6, there is no integer k such that $kn \equiv 1$ (mod 8). □

·	0	1	2	3	4	5	6	7
0	0	0	0	0	0	0	0	0
1	0	1	2	3	4	5	6	7
2	0	2	4	6	0	2	4	6
3	0	3	6	1	4	7	2	5
4	0	4	0	4	0	4	0	4
5	0	5	2	7	4	1	6	3
6	0	6	4	2	0	6	4	2
7	0	7	6	5	4	3	2	1

Figure 3.3. Multiplication table for $\mathbb{Z}_8$

Proposition 3.4. Let $\mathbb{Z}_n$ be the set of equivalence classes of the integers mod n and $a, b, c \in \mathbb{Z}_n$.

1. Addition and multiplication are commutative:

$$a + b \equiv b + a \quad (\text{mod } n)$$
$$ab \equiv ba \quad (\text{mod } n).$$

2. Addition and multiplication are associative:

$$(a + b) + c \equiv a + (b + c) \pmod{n}$$
$$(ab)c \equiv a(bc) \pmod{n}.$$

3. There are both additive and multiplicative identities:

$$a + 0 \equiv a \pmod{n}$$
$$a \cdot 1 \equiv a \pmod{n}.$$

4. Multiplication distributes over addition:

$$a(b + c) \equiv ab + ac \pmod{n}.$$

5. For every integer a there is an additive inverse $-a$:

$$a + (-a) \equiv 0 \pmod{n}.$$

6. Let a be a nonzero integer. Then $\gcd(a, n) = 1$ if and only if there exists a multiplicative inverse b for $a \pmod{n}$; that is, a nonzero integer b such that

$$ab \equiv 1 \pmod{n}.$$

Proof. We will prove (1) and (6) and leave the remaining properties to be proven in the exercises.

(1) Addition and multiplication are commutative modulo n since the remainder of $a + b$ divided by n is the same as the remainder of $b + a$ divided by n.

(6) Suppose that $\gcd(a, n) = 1$. Then there exist integers r and s such that $ar + ns = 1$. Since $ns = 1 - ar$, it must be the case that $ar \equiv 1 \pmod{n}$. Letting b be the equivalence class of r, $ab \equiv 1 \pmod{n}$.

Conversely, suppose that there exists an integer b such that $ab \equiv 1 \pmod{n}$. Then n divides $ab - 1$, so there is an integer k such that $ab - nk = 1$. Let $d = \gcd(a, n)$. Since d divides $ab - nk$, d must also divide 1; hence, $d = 1$. ∎

Symmetries

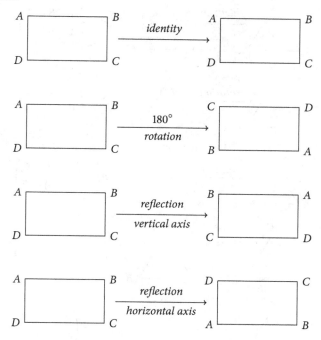

Figure 3.5. Rigid motions of a rectangle

A *symmetry* of a geometric figure is a rearrangement of the figure preserving the arrangement of its sides and vertices as well as its distances and angles. A map from the plane to itself preserving the symmetry of an object is called a *rigid motion*. For example, if we look at the rectangle in Figure 3.5, it is easy to see that a rotation of 180° or 360° returns a rectangle in the plane with the same orientation as the original rectangle and the same relationship among the vertices. A reflection of the rectangle across either the vertical axis or the horizontal axis can also be seen to be a symmetry. However, a 90° rotation in either direction cannot be a symmetry unless the rectangle is a square.

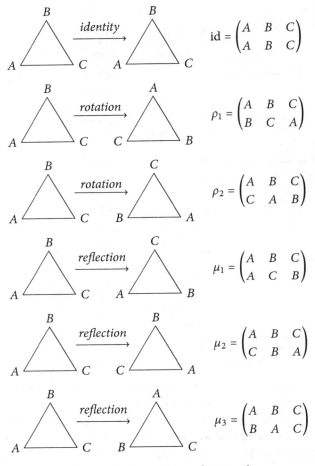

Figure 3.6. Symmetries of a triangle

Let us find the symmetries of the equilateral triangle $\triangle ABC$. To find a symmetry of $\triangle ABC$, we must first examine the permutations of the vertices A, B, and C and then ask if a permutation extends to a symmetry of the triangle. Recall that a **permutation** of a set S is a one-to-one and onto map $\pi : S \to S$. The three vertices have $3! = 6$ permutations, so the triangle has at most six symmetries. To see that there are six permutations, observe there are three different possibilities for the first vertex, and two for the second, and the remaining vertex is determined by the placement of the first two. So we have $3 \cdot 2 \cdot 1 = 3! = 6$ different arrangements. To denote the permutation of the vertices of an equilateral triangle that sends A

to B, B to C, and C to A, we write the array

$$\begin{pmatrix} A & B & C \\ B & C & A \end{pmatrix}.$$

Notice that this particular permutation corresponds to the rigid motion of rotating the triangle by $120°$ in a clockwise direction. In fact, every permutation gives rise to a symmetry of the triangle. All of these symmetries are shown in Figure 3.6.

A natural question to ask is what happens if one motion of the triangle $\triangle ABC$ is followed by another. Which symmetry is $\mu_1 \rho_1$; that is, what happens when we do the permutation ρ_1 and then the permutation μ_1? *Remember that we are composing functions here. Although we usually multiply left to right, we compose functions right to left.* We have

$$(\mu_1 \rho_1)(A) = \mu_1(\rho_1(A)) = \mu_1(B) = C$$
$$(\mu_1 \rho_1)(B) = \mu_1(\rho_1(B)) = \mu_1(C) = B$$
$$(\mu_1 \rho_1)(C) = \mu_1(\rho_1(C)) = \mu_1(A) = A.$$

This is the same symmetry as μ_2. Suppose we do these motions in the opposite order, ρ_1 then μ_1. It is easy to determine that this is the same as the symmetry μ_3; hence, $\rho_1 \mu_1 \neq \mu_1 \rho_1$. A multiplication table for the symmetries of an equilateral triangle $\triangle ABC$ is given in Figure 3.7.

Notice that in the multiplication table for the symmetries of an equilateral triangle, for every motion of the triangle α there is another motion β such that $\alpha\beta = $ id; that is, for every motion there is another motion that takes the triangle back to its original orientation.

$\circ$	id	ρ_1	ρ_2	μ_1	μ_2	μ_3
id	id	ρ_1	ρ_2	μ_1	μ_2	μ_3
ρ_1	ρ_1	ρ_2	id	μ_3	μ_1	μ_2
ρ_2	ρ_2	id	ρ_1	μ_2	μ_3	μ_1
μ_1	μ_1	μ_2	μ_3	id	ρ_1	ρ_2
μ_2	μ_2	μ_3	μ_1	ρ_2	id	ρ_1
μ_3	μ_3	μ_1	μ_2	ρ_1	ρ_2	id

Figure 3.7. Symmetries of an equilateral triangle

3.2 Definitions and Examples

The integers mod n and the symmetries of a triangle or a rectangle are examples of groups. A *binary operation* or *law of composition* on a set G is a function

$G \times G \to G$ that assigns to each pair $(a, b) \in G \times G$ a unique element $a \circ b$, or ab in G, called the composition of a and b. A **group** $(G, \circ)$ is a set G together with a law of composition $(a, b) \mapsto a \circ b$ that satisfies the following axioms.

- The law of composition is **associative**. That is,

$$(a \circ b) \circ c = a \circ (b \circ c)$$

  for $a, b, c \in G$.

- There exists an element $e \in G$, called the **identity element**, such that for any element $a \in G$

$$e \circ a = a \circ e = a.$$

- For each element $a \in G$, there exists an **inverse element** in G, denoted by a^{-1}, such that

$$a \circ a^{-1} = a^{-1} \circ a = e.$$

A group G with the property that $a \circ b = b \circ a$ for all $a, b \in G$ is called **abelian** or **commutative**. Groups not satisfying this property are said to be **nonabelian** or **noncommutative**.

Example 3.8. The integers $\mathbb{Z} = \{\ldots, -1, 0, 1, 2, \ldots\}$ form a group under the operation of addition. The binary operation on two integers $m, n \in \mathbb{Z}$ is just their sum. Since the integers under addition already have a well-established notation, we will use the operator $+$ instead of $\circ$; that is, we shall write $m + n$ instead of $m \circ n$. The identity is 0, and the inverse of $n \in \mathbb{Z}$ is written as $-n$ instead of n^{-1}. Notice that the set of integers under addition have the additional property that $m + n = n + m$ and therefore form an abelian group. $\square$

Most of the time we will write ab instead of $a \circ b$; however, if the group already has a natural operation such as addition in the integers, we will use that operation. That is, if we are adding two integers, we still write $m + n$, $-n$ for the inverse, and 0 for the identity as usual. We also write $m - n$ instead of $m + (-n)$.

It is often convenient to describe a group in terms of an addition or multiplication table. Such a table is called a *Cayley table*.

Example 3.9. The integers mod n form a group under addition modulo n. Consider $\mathbb{Z}_5$, consisting of the equivalence classes of the integers 0, 1, 2, 3, and 4. We define the group operation on $\mathbb{Z}_5$ by modular addition. We write the binary operation on the group additively; that is, we write $m + n$. The element 0 is the identity of the group and each element in $\mathbb{Z}_5$ has an inverse. For instance,

$2 + 3 = 3 + 2 = 0$. Figure 3.10 is a Cayley table for $\mathbb{Z}_5$. By Proposition 3.4 on page 37, $\mathbb{Z}_n = \{0, 1, \ldots, n-1\}$ is a group under the binary operation of addition mod n. $\square$

+	0	1	2	3	4
0	0	1	2	3	4
1	1	2	3	4	0
2	2	3	4	0	1
3	3	4	0	1	2
4	4	0	1	2	3

Figure 3.10. Cayley table for $(\mathbb{Z}_5, +)$

Example 3.11. Not every set with a binary operation is a group. For example, if we let modular multiplication be the binary operation on $\mathbb{Z}_n$, then $\mathbb{Z}_n$ fails to be a group. The element 1 acts as a group identity since $1 \cdot k = k \cdot 1 = k$ for any $k \in \mathbb{Z}_n$; however, a multiplicative inverse for 0 does not exist since $0 \cdot k = k \cdot 0 = 0$ for every k in $\mathbb{Z}_n$. Even if we consider the set $\mathbb{Z}_n \setminus \{0\}$, we still may not have a group. For instance, let $2 \in \mathbb{Z}_6$. Then 2 has no multiplicative inverse since

$$0 \cdot 2 = 0 \qquad 1 \cdot 2 = 2$$
$$2 \cdot 2 = 4 \qquad 3 \cdot 2 = 0$$
$$4 \cdot 2 = 2 \qquad 5 \cdot 2 = 4.$$

By Proposition 3.4 on page 37, every nonzero k does have an inverse in $\mathbb{Z}_n$ if k is relatively prime to n. Denote the set of all such nonzero elements in $\mathbb{Z}_n$ by $U(n)$. Then $U(n)$ is a group called the ***group of units*** of $\mathbb{Z}_n$. Figure 3.12 is a Cayley table for the group $U(8)$. $\square$

·	1	3	5	7
1	1	3	5	7
3	3	1	7	5
5	5	7	1	3
7	7	5	3	1

Figure 3.12. Multiplication table for $U(8)$

Example 3.13. The symmetries of an equilateral triangle described in Section 3.1 form a nonabelian group. As we observed, it is not necessarily true that $\alpha\beta = \beta\alpha$ for two symmetries α and β. Using Figure 3.7 on page 41, which is a Cayley table for this group, we can easily check that the symmetries of an equilateral triangle are indeed a group. We will denote this group by either S_3 or D_3, for reasons that will be explained later.   □

Example 3.14. We use $\mathbb{M}_2(\mathbb{R})$ to denote the set of all 2×2 matrices. Let $GL_2(\mathbb{R})$ be the subset of $\mathbb{M}_2(\mathbb{R})$ consisting of invertible matrices; that is, a matrix

$$A = \begin{pmatrix} a & b \\ c & d \end{pmatrix}$$

is in $GL_2(\mathbb{R})$ if there exists a matrix A^{-1} such that $AA^{-1} = A^{-1}A = I$, where I is the 2×2 identity matrix. For A to have an inverse is equivalent to requiring that the determinant of A be nonzero; that is, $\det A = ad - bc \neq 0$. The set of invertible matrices forms a group called the ***general linear group***. The identity of the group is the identity matrix

$$I = \begin{pmatrix} 1 & 0 \\ 0 & 1 \end{pmatrix}.$$

The inverse of $A \in GL_2(\mathbb{R})$ is

$$A^{-1} = \frac{1}{ad - bc} \begin{pmatrix} d & -b \\ -c & a \end{pmatrix}.$$

The product of two invertible matrices is again invertible. Matrix multiplication is associative, satisfying the other group axiom. For matrices it is not true in general that $AB = BA$; hence, $GL_2(\mathbb{R})$ is another example of a nonabelian group.   □

Example 3.15. Let

$$1 = \begin{pmatrix} 1 & 0 \\ 0 & 1 \end{pmatrix} \qquad I = \begin{pmatrix} 0 & 1 \\ -1 & 0 \end{pmatrix}$$

$$J = \begin{pmatrix} 0 & i \\ i & 0 \end{pmatrix} \qquad K = \begin{pmatrix} i & 0 \\ 0 & -i \end{pmatrix},$$

where $i^2 = -1$. Then the relations $I^2 = J^2 = K^2 = -1$, $IJ = K$, $JK = I$, $KI = J$, $JI = -K$, $KJ = -I$, and $IK = -J$ hold. The set $Q_8 = \{\pm 1, \pm I, \pm J, \pm K\}$ is a group called the ***quaternion group***. Notice that Q_8 is noncommutative.   □

Example 3.16. Let $\mathbb{C}^*$ be the set of nonzero complex numbers. Under the operation of multiplication $\mathbb{C}^*$ forms a group. The identity is 1. If $z = a + bi$ is a nonzero complex number, then

$$z^{-1} = \frac{a - bi}{a^2 + b^2}$$

is the inverse of z. It is easy to see that the remaining group axioms hold. □

A group is *finite*, or has *finite order*, if it contains a finite number of elements; otherwise, the group is said to be *infinite* or to have *infinite order*. The *order* of a finite group is the number of elements that it contains. If G is a group containing n elements, we write $|G| = n$. The group $\mathbb{Z}_5$ is a finite group of order 5; the integers $\mathbb{Z}$ form an infinite group under addition, and we sometimes write $|\mathbb{Z}| = \infty$.

Basic Properties of Groups

Proposition 3.17. The identity element in a group G is unique; that is, there exists only one element $e \in G$ such that $eg = ge = g$ for all $g \in G$.

Proof. Suppose that e and e' are both identities in G. Then $eg = ge = g$ and $e'g = ge' = g$ for all $g \in G$. We need to show that $e = e'$. If we think of e as the identity, then $ee' = e'$; but if e' is the identity, then $ee' = e$. Combining these two equations, we have $e = ee' = e'$. ■

Inverses in a group are also unique. If g' and g'' are both inverses of an element g in a group G, then $gg' = g'g = e$ and $gg'' = g''g = e$. We want to show that $g' = g''$, but $g' = g'e = g'(gg'') = (g'g)g'' = eg'' = g''$. We summarize this fact in the following proposition.

Proposition 3.18. If g is any element in a group G, then the inverse of g, denoted by g^{-1}, is unique.

Proposition 3.19. Let G be a group. If $a, b \in G$, then $(ab)^{-1} = b^{-1}a^{-1}$.

Proof. Let $a, b \in G$. Then $abb^{-1}a^{-1} = aea^{-1} = aa^{-1} = e$. Similarly, $b^{-1}a^{-1}ab = e$. But by the previous proposition, inverses are unique; hence, $(ab)^{-1} = b^{-1}a^{-1}$. ■

Proposition 3.20. Let G be a group. For any $a \in G$, $(a^{-1})^{-1} = a$.

Proof. Observe that $a^{-1}(a^{-1})^{-1} = e$. Consequently, multiplying both sides of this equation by a, we have

$$(a^{-1})^{-1} = e(a^{-1})^{-1} = aa^{-1}(a^{-1})^{-1} = ae = a.$$

■

It makes sense to write equations with group elements and group operations. If a and b are two elements in a group G, does there exist an element $x \in G$ such that $ax = b$? If such an x does exist, is it unique? The following proposition answers both of these questions positively.

Proposition 3.21. Let G be a group and a and b be any two elements in G. Then the equations $ax = b$ and $xa = b$ have unique solutions in G.

Proof. Suppose that $ax = b$. We must show that such an x exists. We can multiply both sides of $ax = b$ by a^{-1} to find $x = ex = a^{-1}ax = a^{-1}b$.

To show uniqueness, suppose that x_1 and x_2 are both solutions of $ax = b$; then $ax_1 = b = ax_2$. So $x_1 = a^{-1}ax_1 = a^{-1}ax_2 = x_2$. The proof for the existence and uniqueness of the solution of $xa = b$ is similar. ∎

Proposition 3.22. *If G is a group and $a, b, c \in G$, then $ba = ca$ implies $b = c$ and $ab = ac$ implies $b = c$.*

This proposition tells us that the *right and left cancellation laws* are true in groups. We leave the proof as an exercise.

We can use exponential notation for groups just as we do in ordinary algebra. If G is a group and $g \in G$, then we define $g^0 = e$. For $n \in \mathbb{N}$, we define

$$g^n = \underbrace{g \cdot g \cdots g}_{n \text{ times}}$$

and

$$g^{-n} = \underbrace{g^{-1} \cdot g^{-1} \cdots g^{-1}}_{n \text{ times}}.$$

Theorem 3.23. In a group, the usual laws of exponents hold; that is, for all $g, h \in G$,

1. $g^m g^n = g^{m+n}$ for all $m, n \in \mathbb{Z}$;

2. $(g^m)^n = g^{mn}$ for all $m, n \in \mathbb{Z}$;

3. $(gh)^n = (h^{-1}g^{-1})^{-n}$ for all $n \in \mathbb{Z}$. Furthermore, if G is abelian, then $(gh)^n = g^n h^n$.

We will leave the proof of this theorem as an exercise. Notice that $(gh)^n \neq g^n h^n$ in general, since the group may not be abelian. If the group is $\mathbb{Z}$ or $\mathbb{Z}_n$, we write the group operation additively and the exponential operation multiplicatively; that is, we write ng instead of g^n. The laws of exponents now become

1. $mg + ng = (m + n)g$ for all $m, n \in \mathbb{Z}$;

2. $m(ng) = (mn)g$ for all $m, n \in \mathbb{Z}$;

3. $m(g + h) = mg + mh$ for all $n \in \mathbb{Z}$.

It is important to realize that the last statement can be made only because $\mathbb{Z}$ and $\mathbb{Z}_n$ are commutative groups.

Although the first clear axiomatic definition of a group was not given until the late 1800s, group-theoretic methods had been employed before this time in the development of many areas of mathematics, including geometry and the theory of algebraic equations.

Joseph-Louis Lagrange used group-theoretic methods in a 1770–1771 memoir to study methods of solving polynomial equations. Later, Évariste Galois (1811–1832) succeeded in developing the mathematics necessary to determine exactly which polynomial equations could be solved in terms of the coefficients of the polynomial. Galois' primary tool was group theory.

The study of geometry was revolutionized in 1872 when Felix Klein proposed that geometric spaces should be studied by examining those properties that are invariant under a transformation of the space. Sophus Lie, a contemporary of Klein, used group theory to study solutions of partial differential equations. One of the first modern treatments of group theory appeared in William Burnside's *The Theory of Groups of Finite Order* [1], first published in 1897.

3.3 Subgroups

Definitions and Examples

Sometimes we wish to investigate smaller groups sitting inside a larger group. The set of even integers $2\mathbb{Z} = \{\ldots, -2, 0, 2, 4, \ldots\}$ is a group under the operation of addition. This smaller group sits naturally inside of the group of integers under addition. We define a *subgroup* H of a group G to be a subset H of G such that when the group operation of G is restricted to H, H is a group in its own right. Observe that every group G with at least two elements will always have at least two subgroups, the subgroup consisting of the identity element alone and the entire group itself. The subgroup $H = \{e\}$ of a group G is called the *trivial subgroup*. A subgroup that is a proper subset of G is called a *proper subgroup*. In many of the examples that we have investigated up to this point, there exist other subgroups besides the trivial and improper subgroups.

Example 3.24. Consider the set of nonzero real numbers, $\mathbb{R}^*$, with the group operation of multiplication. The identity of this group is 1 and the inverse of any element $a \in \mathbb{R}^*$ is just $1/a$. We will show that

$$\mathbb{Q}^* = \{p/q : p \text{ and } q \text{ are nonzero integers}\}$$

is a subgroup of $\mathbb{R}^*$. The identity of $\mathbb{R}^*$ is 1; however, $1 = 1/1$ is the quotient of two nonzero integers. Hence, the identity of $\mathbb{R}^*$ is in $\mathbb{Q}^*$. Given two elements in $\mathbb{Q}^*$, say p/q and r/s, their product pr/qs is also in $\mathbb{Q}^*$. The inverse of any element $p/q \in \mathbb{Q}^*$ is again in $\mathbb{Q}^*$ since $(p/q)^{-1} = q/p$. Since multiplication in $\mathbb{R}^*$ is associative, multiplication in $\mathbb{Q}^*$ is associative.   □

Example 3.25. Recall that $\mathbb{C}^*$ is the multiplicative group of nonzero complex numbers. Let $H = \{1, -1, i, -i\}$. Then H is a subgroup of $\mathbb{C}^*$. It is quite easy to verify that H is a group under multiplication and that $H \subset \mathbb{C}^*$.   □

Example 3.26. Let $SL_2(\mathbb{R})$ be the subset of $GL_2(\mathbb{R})$ consisting of matrices of determinant one; that is, a matrix

$$A = \begin{pmatrix} a & b \\ c & d \end{pmatrix}$$

is in $SL_2(\mathbb{R})$ exactly when $ad - bc = 1$. To show that $SL_2(\mathbb{R})$ is a subgroup of the general linear group, we must show that it is a group under matrix multiplication. The 2×2 identity matrix is in $SL_2(\mathbb{R})$, as is the inverse of the matrix A:

$$A^{-1} = \begin{pmatrix} d & -b \\ -c & a \end{pmatrix}.$$

It remains to show that multiplication is closed; that is, that the product of two matrices of determinant one also has determinant one. We will leave this task as an exercise. The group $SL_2(\mathbb{R})$ is called the **special linear group**.   □

Example 3.27. It is important to realize that a subset H of a group G can be a group without being a subgroup of G. For H to be a subgroup of G, it must inherit the binary operation of G. The set of all 2×2 matrices, $\mathbb{M}_2(\mathbb{R})$, forms a group under the operation of addition. The 2×2 general linear group is a subset of $\mathbb{M}_2(\mathbb{R})$ and is a group under matrix multiplication, but it is not a subgroup of $\mathbb{M}_2(\mathbb{R})$. If we add two invertible matrices, we do not necessarily obtain another invertible matrix. Observe that

$$\begin{pmatrix} 1 & 0 \\ 0 & 1 \end{pmatrix} + \begin{pmatrix} -1 & 0 \\ 0 & -1 \end{pmatrix} = \begin{pmatrix} 0 & 0 \\ 0 & 0 \end{pmatrix},$$

but the zero matrix is not in $GL_2(\mathbb{R})$.   □

Example 3.28. One way of telling whether or not two groups are the same is by examining their subgroups. Other than the trivial subgroup and the group itself, the group $\mathbb{Z}_4$ has a single subgroup consisting of the elements 0 and 2. From the group $\mathbb{Z}_2$, we can form another group of four elements as follows. As a set this group is $\mathbb{Z}_2 \times \mathbb{Z}_2$. We perform the group operation coordinatewise; that is, $(a, b) + (c, d) = (a + c, b + d)$. Figure 3.29 is an addition table for $\mathbb{Z}_2 \times \mathbb{Z}_2$. Since there are three nontrivial proper subgroups of $\mathbb{Z}_2 \times \mathbb{Z}_2$, $H_1 = \{(0, 0), (0, 1)\}$, $H_2 = \{(0, 0), (1, 0)\}$, and $H_3 = \{(0, 0), (1, 1)\}$, $\mathbb{Z}_4$ and $\mathbb{Z}_2 \times \mathbb{Z}_2$ must be different groups.   □

+	$(0, 0)$	$(0, 1)$	$(1, 0)$	$(1, 1)$
$(0, 0)$	$(0, 0)$	$(0, 1)$	$(1, 0)$	$(1, 1)$
$(0, 1)$	$(0, 1)$	$(0, 0)$	$(1, 1)$	$(1, 0)$
$(1, 0)$	$(1, 0)$	$(1, 1)$	$(0, 0)$	$(0, 1)$
$(1, 1)$	$(1, 1)$	$(1, 0)$	$(0, 1)$	$(0, 0)$

Figure 3.29. Addition table for $\mathbb{Z}_2 \times \mathbb{Z}_2$

Some Subgroup Theorems

Let us examine some criteria for determining exactly when a subset of a group is a subgroup.

Proposition 3.30. A subset H of G is a subgroup if and only if it satisfies the following conditions.

   1. The identity e of G is in H.

   2. If $h_1, h_2 \in H$, then $h_1 h_2 \in H$.

   3. If $h \in H$, then $h^{-1} \in H$.

Proof. First suppose that H is a subgroup of G. We must show that the three conditions hold. Since H is a group, it must have an identity e_H. We must show that $e_H = e$, where e is the identity of G. We know that $e_H e_H = e_H$ and that $e e_H = e_H e = e_H$; hence, $e e_H = e_H e_H$. By right-hand cancellation, $e = e_H$. The second condition holds since a subgroup H is a group. To prove the third condition, let $h \in H$. Since H is a group, there is an element $h' \in H$ such that $h h' = h' h = e$. By the uniqueness of the inverse in G, $h' = h^{-1}$.

Conversely, if the three conditions hold, we must show that H is a group under the same operation as G; however, these conditions plus the associativity of the binary operation are exactly the axioms stated in the definition of a group. ■

Proposition 3.31. Let H be a subset of a group G. Then H is a subgroup of G if and only if $H \neq \varnothing$, and whenever $g, h \in H$ then gh^{-1} is in H.

Proof. First assume that H is a subgroup of G. We wish to show that $gh^{-1} \in H$ whenever g and h are in H. Since h is in H, its inverse h^{-1} must also be in H. Because of the closure of the group operation, $gh^{-1} \in H$.

Conversely, suppose that $H \subset G$ such that $H \neq \varnothing$ and $gh^{-1} \in H$ whenever $g, h \in H$. If $g \in H$, then $gg^{-1} = e$ is in H. If $g \in H$, then $eg^{-1} = g^{-1}$ is also in H. Now let $h_1, h_2 \in H$. We must show that their product is also in H. However, $h_1(h_2^{-1})^{-1} = h_1 h_2 \in H$. Hence, H is a subgroup of G.   ■

Sage.   The first half of this text is about group theory. Sage includes Groups, Algorithms and Programming (GAP), a program designed primarly for just group theory, and in continuous development since 1986. Many of Sage's computations for groups ultimately are performed by GAP.

3.4 Reading Questions

1. In the group $\mathbb{Z}_8$ compute, (a) $6 + 7$, and (b) 2^{-1}.

2. In the group $U(16)$ compute, (a) $5 \cdot 7$, and (b) 3^{-1}.

3. State the definition of a group.

4. Explain a single method that will decide if a subset of a group is itself a subgroup.

5. Explain the origin of the term "abelian" for a commutative group.

6. Give an example of a group you have seen in your previous mathematical experience, but that is not an example in this chapter.

3.5 Exercises

1. Find all $x \in \mathbb{Z}$ satisfying each of the following equations.

    (a) $3x \equiv 2 \pmod 7$ (d) $9x \equiv 3 \pmod 5$

    (b) $5x + 1 \equiv 13 \pmod{23}$ (e) $5x \equiv 1 \pmod 6$

    (c) $5x + 1 \equiv 13 \pmod{26}$ (f) $3x \equiv 1 \pmod 6$

2. Which of the following multiplication tables defined on the set $G = \{a, b, c, d\}$ form a group? Support your answer in each case.

(a)

◦	a	b	c	d
a	a	c	d	a
b	b	b	c	d
c	c	d	a	b
d	d	a	b	c

(c)

◦	a	b	c	d
a	a	b	c	d
b	b	c	d	a
c	c	d	a	b
d	d	a	b	c

(b)

◦	a	b	c	d
a	a	b	c	d
b	b	a	d	c
c	c	d	a	b
d	d	c	b	a

(d)

◦	a	b	c	d
a	a	b	c	d
b	b	a	c	d
c	c	b	a	d
d	d	d	b	c

3. Write out Cayley tables for groups formed by the symmetries of a rectangle and for $(\mathbb{Z}_4, +)$. How many elements are in each group? Are the groups the same? Why or why not?

4. Describe the symmetries of a rhombus and prove that the set of symmetries forms a group. Give Cayley tables for both the symmetries of a rectangle and the symmetries of a rhombus. Are the symmetries of a rectangle and those of a rhombus the same?

5. Describe the symmetries of a square and prove that the set of symmetries is a group. Give a Cayley table for the symmetries. How many ways can the vertices of a square be permuted? Is each permutation necessarily a symmetry of the square? The symmetry group of the square is denoted by D_4.

6. Give a multiplication table for the group $U(12)$.

7. Let $S = \mathbb{R} \setminus \{-1\}$ and define a binary operation on S by $a * b = a + b + ab$. Prove that $(S, *)$ is an abelian group.

8. Give an example of two elements A and B in $GL_2(\mathbb{R})$ with $AB \neq BA$.

9. Prove that the product of two matrices in $SL_2(\mathbb{R})$ has determinant one.

10. Prove that the set of matrices of the form

$$\begin{pmatrix} 1 & x & y \\ 0 & 1 & z \\ 0 & 0 & 1 \end{pmatrix}$$

is a group under matrix multiplication. This group, known as the *Heisenberg group*, is important in quantum physics. Matrix multiplication in the Heisenberg group is defined by

$$\begin{pmatrix} 1 & x & y \\ 0 & 1 & z \\ 0 & 0 & 1 \end{pmatrix} \begin{pmatrix} 1 & x' & y' \\ 0 & 1 & z' \\ 0 & 0 & 1 \end{pmatrix} = \begin{pmatrix} 1 & x + x' & y + y' + xz' \\ 0 & 1 & z + z' \\ 0 & 0 & 1 \end{pmatrix}.$$

11. Prove that $\det(AB) = \det(A)\det(B)$ in $GL_2(\mathbb{R})$. Use this result to show that the binary operation in the group $GL_2(\mathbb{R})$ is closed; that is, if A and B are in $GL_2(\mathbb{R})$, then $AB \in GL_2(\mathbb{R})$.

12. Let $\mathbb{Z}_2^n = \{(a_1, a_2, \ldots, a_n) : a_i \in \mathbb{Z}_2\}$. Define a binary operation on $\mathbb{Z}_2^n$ by

$$(a_1, a_2, \ldots, a_n) + (b_1, b_2, \ldots, b_n) = (a_1 + b_1, a_2 + b_2, \ldots, a_n + b_n).$$

Prove that $\mathbb{Z}_2^n$ is a group under this operation. This group is important in algebraic coding theory.

13. Show that $\mathbb{R}^* = \mathbb{R} \setminus \{0\}$ is a group under the operation of multiplication.

14. Given the groups $\mathbb{R}^*$ and $\mathbb{Z}$, let $G = \mathbb{R}^* \times \mathbb{Z}$. Define a binary operation $\circ$ on G by $(a, m) \circ (b, n) = (ab, m + n)$. Show that G is a group under this operation.

15. Prove or disprove that every group containing six elements is abelian.

16. Give a specific example of some group G and elements $g, h \in G$ where $(gh)^n \neq g^n h^n$.

17. Give an example of three different groups with eight elements. Why are the groups different?

18. Show that there are $n!$ permutations of a set containing n items.

19. Show that

$$0 + a \equiv a + 0 \equiv a \pmod{n}$$

for all $a \in \mathbb{Z}_n$.

20. Prove that there is a multiplicative identity for the integers modulo n:

$$a \cdot 1 \equiv a \pmod{n}.$$

21. For each $a \in \mathbb{Z}_n$ find an element $b \in \mathbb{Z}_n$ such that

$$a + b \equiv b + a \equiv 0 \pmod{n}.$$

22. Show that addition and multiplication mod n are well defined operations. That is, show that the operations do not depend on the choice of the representative from the equivalence classes mod n.

23. Show that addition and multiplication mod n are associative operations.

24. Show that multiplication distributes over addition modulo n:

$$a(b + c) \equiv ab + ac \pmod{n}.$$

25. Let a and b be elements in a group G. Prove that $ab^n a^{-1} = (aba^{-1})^n$ for $n \in \mathbb{Z}$.

26. Let $U(n)$ be the group of units in $\mathbb{Z}_n$. If $n > 2$, prove that there is an element $k \in U(n)$ such that $k^2 = 1$ and $k \neq 1$.

27. Prove that the inverse of $g_1 g_2 \cdots g_n$ is $g_n^{-1} g_{n-1}^{-1} \cdots g_1^{-1}$.

28. Prove the remainder of Proposition 3.21 on page 45: if G is a group and $a, b \in G$, then the equation $xa = b$ has a unique solution in G.

29. Prove Theorem 3.23 on page 46.

30. Prove the right and left cancellation laws for a group G; that is, show that in the group G, $ba = ca$ implies $b = c$ and $ab = ac$ implies $b = c$ for elements $a, b, c \in G$.

31. Show that if $a^2 = e$ for all elements a in a group G, then G must be abelian.

32. Show that if G is a finite group of even order, then there is an $a \in G$ such that a is not the identity and $a^2 = e$.

33. Let G be a group and suppose that $(ab)^2 = a^2 b^2$ for all a and b in G. Prove that G is an abelian group.

34. Find all the subgroups of $\mathbb{Z}_3 \times \mathbb{Z}_3$. Use this information to show that $\mathbb{Z}_3 \times \mathbb{Z}_3$ is not the same group as $\mathbb{Z}_9$. (See Example 3.28 on page 48 for a short description of the product of groups.)

35. Find all the subgroups of the symmetry group of an equilateral triangle.

36. Compute the subgroups of the symmetry group of a square.

37. Let $H = \{2^k : k \in \mathbb{Z}\}$. Show that H is a subgroup of $\mathbb{Q}^*$.

38. Let $n = 0, 1, 2, \ldots$ and $n\mathbb{Z} = \{nk : k \in \mathbb{Z}\}$. Prove that $n\mathbb{Z}$ is a subgroup of $\mathbb{Z}$. Show that these subgroups are the only subgroups of $\mathbb{Z}$.

39. Let $\mathbb{T} = \{z \in \mathbb{C}^* : |z| = 1\}$. Prove that $\mathbb{T}$ is a subgroup of $\mathbb{C}^*$.

40. Let G consist of the 2×2 matrices of the form

$$\begin{pmatrix} \cos\theta & -\sin\theta \\ \sin\theta & \cos\theta \end{pmatrix},$$

where $\theta \in \mathbb{R}$. Prove that G is a subgroup of $SL_2(\mathbb{R})$.

41. Prove that

$$G = \{a + b\sqrt{2} : a, b \in \mathbb{Q} \text{ and } a \text{ and } b \text{ are not both zero}\}$$

is a subgroup of $\mathbb{R}^*$ under the group operation of multiplication.

42. Let G be the group of 2×2 matrices under addition and

$$H = \left\{ \begin{pmatrix} a & b \\ c & d \end{pmatrix} : a + d = 0 \right\}.$$

Prove that H is a subgroup of G.

43. Prove or disprove: $SL_2(\mathbb{Z})$, the set of 2×2 matrices with integer entries and determinant one, is a subgroup of $SL_2(\mathbb{R})$.

44. List the subgroups of the quaternion group, Q_8.

45. Prove that the intersection of two subgroups of a group G is also a subgroup of G.

46. Prove or disprove: If H and K are subgroups of a group G, then $H \cup K$ is a subgroup of G.

47. Prove or disprove: If H and K are subgroups of a group G, then $HK = \{hk : h \in H \text{ and } k \in K\}$ is a subgroup of G. What if G is abelian?

48. Let G be a group and $g \in G$. Show that

$$Z(G) = \{x \in G : gx = xg \text{ for all } g \in G\}$$

is a subgroup of G. This subgroup is called the *center* of G.

49. Let a and b be elements of a group G. If $a^4 b = ba$ and $a^3 = e$, prove that $ab = ba$.

50. Give an example of an infinite group in which every nontrivial subgroup is infinite.

51. If $xy = x^{-1}y^{-1}$ for all x and y in G, prove that G must be abelian.

52. Prove or disprove: Every proper subgroup of a nonabelian group is nonabelian.

53. Let H be a subgroup of G and

$$C(H) = \{g \in G : gh = hg \text{ for all } h \in H\}.$$

Prove $C(H)$ is a subgroup of G. This subgroup is called the *centralizer* of H in G.

54. Let H be a subgroup of G. If $g \in G$, show that $gHg^{-1} = \{ghg^{-1} : h \in H\}$ is also a subgroup of G.

3.6 Additional Exercises: Detecting Errors

1. **UPC Symbols.** Universal Product Code (UPC) symbols are found on most products in grocery and retail stores. The UPC symbol is a 12-digit code identifying the manufacturer of a product and the product itself (Figure 3.32). The first 11 digits contain information about the product; the twelfth digit is used for error detection. If $d_1 d_2 \cdots d_{12}$ is a valid UPC number, then

$$3 \cdot d_1 + 1 \cdot d_2 + 3 \cdot d_3 + \cdots + 3 \cdot d_{11} + 1 \cdot d_{12} \equiv 0 \pmod{10}.$$

(a) Show that the UPC number 0-50000-30042-6, which appears in Figure 3.32, is a valid UPC number.

(b) Show that the number 0-50000-30043-6 is not a valid UPC number.

(c) Write a formula to calculate the check digit, d_{12}, in the UPC number.

(d) The UPC error detection scheme can detect most transposition errors; that is, it can determine if two digits have been interchanged. Show that the transposition error 0-05000-30042-6 is not detected. Find a transposition error that is detected. Can you find a general rule for the types of transposition errors that can be detected?

(e) Write a program that will determine whether or not a UPC number is valid.

0 50000 30042 6

Figure 3.32. A UPC code

2. It is often useful to use an inner product notation for this type of error detection scheme; hence, we will use the notion

$$(d_1, d_2, \ldots, d_k) \cdot (w_1, w_2, \ldots, w_k) \equiv 0 \pmod{n}$$

to mean
$$d_1 w_1 + d_2 w_2 + \cdots + d_k w_k \equiv 0 \pmod{n}.$$

Suppose that $(d_1, d_2, \ldots, d_k) \cdot (w_1, w_2, \ldots, w_k) \equiv 0 \pmod{n}$ is an error detection scheme for the k-digit identification number $d_1 d_2 \cdots d_k$, where $0 \le d_i < n$. Prove that all single-digit errors are detected if and only if $\gcd(w_i, n) = 1$ for $1 \le i \le k$.

3. Let $(d_1, d_2, \ldots, d_k) \cdot (w_1, w_2, \ldots, w_k) \equiv 0 \pmod{n}$ be an error detection scheme for the k-digit identification number $d_1 d_2 \cdots d_k$, where $0 \le d_i < n$. Prove that all transposition errors of two digits d_i and d_j are detected if and only if $\gcd(w_i - w_j, n) = 1$ for i and j between 1 and k.

4. **ISBN Codes.** Every book has an International Standard Book Number (ISBN) code. This is a 10-digit code indicating the book's publisher and title. The tenth digit is a check digit satisfying

    $$(d_1, d_2, \ldots, d_{10}) \cdot (10, 9, \ldots, 1) \equiv 0 \pmod{11}.$$

One problem is that d_{10} might have to be a 10 to make the inner product zero; in this case, 11 digits would be needed to make this scheme work. Therefore, the character X is used for the eleventh digit. So ISBN 3-540-96035-X is a valid ISBN code.

 (a) Is 0-534-91500-0 a valid ISBN code? What about 0-534-91700-0 and 0-534-19500-0?

 (b) Does this method detect all single-digit errors? What about all transposition errors?

 (c) How many different ISBN codes are there?

 (d) Write a computer program that will calculate the check digit for the first nine digits of an ISBN code.

 (e) A publisher has houses in Germany and the United States. Its German prefix is 3-540. If its United States prefix will be 0-abc, find abc such that the rest of the ISBN code will be the same for a book printed in Germany and in the United States. Under the ISBN coding method the first digit identifies the language; German is 3 and English is 0. The next group of numbers identifies the publisher, and the last group identifies the specific book.

3.7 References and Suggested Readings

[1] Burnside, W. *Theory of Groups of Finite Order*. 2nd ed. Cambridge University Press, Cambridge, 1911; Dover, New York, 1953. A classic. Also available at books.google.com.

[2] Gallian, J. A. and Winters, S. "Modular Arithmetic in the Marketplace," *The American Mathematical Monthly* **95** (1988): 548–51.

[3] Gallian, J. A. *Contemporary Abstract Algebra*. 7th ed. Brooks/Cole, Belmont, CA, 2009.

[4] Hall, M. *Theory of Groups*. 2nd ed. American Mathematical Society, Providence, 1959.

[5] Kurosh, A. E. *The Theory of Groups*, vols. I and II. American Mathematical Society, Providence, 1979.

[6] Rotman, J. J. *An Introduction to the Theory of Groups*. 4th ed. Springer, New York, 1995.

$$\boxed{4}$$

Cyclic Groups

The groups $\mathbb{Z}$ and $\mathbb{Z}_n$, which are among the most familiar and easily under-stood groups, are both examples of what are called cyclic groups. In this chapter we will study the properties of cyclic groups and cyclic subgroups, which play a fundamental part in the classification of all abelian groups.

4.1 Cyclic Subgroups

Often a subgroup will depend entirely on a single element of the group; that is, knowing that particular element will allow us to compute any other element in the subgroup.

Example 4.1. Suppose that we consider $3 \in \mathbb{Z}$ and look at all multiples (both positive and negative) of 3. As a set, this is

$$3\mathbb{Z} = \{\ldots, -3, 0, 3, 6, \ldots\}.$$

It is easy to see that $3\mathbb{Z}$ is a subgroup of the integers. This subgroup is completely determined by the element 3 since we can obtain all of the other elements of the group by taking multiples of 3. Every element in the subgroup is "generated" by 3. □

Example 4.2. If $H = \{2^n : n \in \mathbb{Z}\}$, then H is a subgroup of the multiplicative group of nonzero rational numbers, $\mathbb{Q}^*$. If $a = 2^m$ and $b = 2^n$ are in H, then $ab^{-1} = 2^m 2^{-n} = 2^{m-n}$ is also in H. By Proposition 3.31 on page 49, H is a subgroup of $\mathbb{Q}^*$ determined by the element 2. □

Theorem 4.3. Let G be a group and a be any element in G. Then the set

$$\langle a \rangle = \{a^k : k \in \mathbb{Z}\}$$

is a subgroup of G. Furthermore, $\langle a \rangle$ is the smallest subgroup of G that contains a.

Proof. The identity is in $\langle a \rangle$ since $a^0 = e$. If g and h are any two elements in $\langle a \rangle$, then by the definition of $\langle a \rangle$ we can write $g = a^m$ and $h = a^n$ for some integers m and n. So $gh = a^m a^n = a^{m+n}$ is again in $\langle a \rangle$. Finally, if $g = a^n$ in $\langle a \rangle$, then the inverse $g^{-1} = a^{-n}$ is also in $\langle a \rangle$. Clearly, any subgroup H of G containing a must contain all the powers of a by closure; hence, H contains $\langle a \rangle$. Therefore, $\langle a \rangle$ is the smallest subgroup of G containing a. ∎

Remark 4.4. If we are using the "+" notation, as in the case of the integers under addition, we write $\langle a \rangle = \{na : n \in \mathbb{Z}\}$.

For $a \in G$, we call $\langle a \rangle$ the *cyclic subgroup* generated by a. If G contains some element a such that $G = \langle a \rangle$, then G is a *cyclic group*. In this case a is a ***generator*** of G. If a is an element of a group G, we define the ***order*** of a to be the smallest positive integer n such that $a^n = e$, and we write $|a| = n$. If there is no such integer n, we say that the order of a is infinite and write $|a| = \infty$ to denote the order of a.

Example 4.5. Notice that a cyclic group can have more than a single generator. Both 1 and 5 generate $\mathbb{Z}_6$; hence, $\mathbb{Z}_6$ is a cyclic group. Not every element in a cyclic group is necessarily a generator of the group. The order of $2 \in \mathbb{Z}_6$ is 3. The cyclic subgroup generated by 2 is $\langle 2 \rangle = \{0, 2, 4\}$. □

The groups $\mathbb{Z}$ and $\mathbb{Z}_n$ are cyclic groups. The elements 1 and -1 are generators for $\mathbb{Z}$. We can certainly generate $\mathbb{Z}_n$ with 1 although there may be other generators of $\mathbb{Z}_n$, as in the case of $\mathbb{Z}_6$.

Example 4.6. The group of units, $U(9)$, in $\mathbb{Z}_9$ is a cyclic group. As a set, $U(9)$ is $\{1, 2, 4, 5, 7, 8\}$. The element 2 is a generator for $U(9)$ since

$$2^1 = 2 \qquad 2^2 = 4$$
$$2^3 = 8 \qquad 2^4 = 7$$
$$2^5 = 5 \qquad 2^6 = 1.$$

□

Example 4.7. Not every group is a cyclic group. Consider the symmetry group of an equilateral triangle S_3. The multiplication table for this group is Figure 3.7 on page 41. The subgroups of S_3 are shown in Figure 4.8 on the following page. Notice that every subgroup is cyclic; however, no single element generates the entire group. □

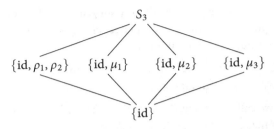

Figure 4.8. Subgroups of S_3

Theorem 4.9. Every cyclic group is abelian.

Proof. Let G be a cyclic group and $a \in G$ be a generator for G. If g and h are in G, then they can be written as powers of a, say $g = a^r$ and $h = a^s$. Since

$$gh = a^r a^s = a^{r+s} = a^{s+r} = a^s a^r = hg,$$

G is abelian.   ∎

Subgroups of Cyclic Groups

We can ask some interesting questions about cyclic subgroups of a group and subgroups of a cyclic group. If G is a group, which subgroups of G are cyclic? If G is a cyclic group, what type of subgroups does G possess?

Theorem 4.10. Every subgroup of a cyclic group is cyclic.

Proof. The main tools used in this proof are the division algorithm and the Principle of Well-Ordering. Let G be a cyclic group generated by a and suppose that H is a subgroup of G. If $H = \{e\}$, then trivially H is cyclic. Suppose that H contains some other element g distinct from the identity. Then g can be written as a^n for some integer n. Since H is a subgroup, $g^{-1} = a^{-n}$ must also be in H. Since either n or $-n$ is positive, we can assume that H contains positive powers of a and $n > 0$. Let m be the smallest natural number such that $a^m \in H$. Such an m exists by the Principle of Well-Ordering.

We claim that $h = a^m$ is a generator for H. We must show that every $h' \in H$ can be written as a power of h. Since $h' \in H$ and H is a subgroup of G, $h' = a^k$ for some integer k. Using the division algorithm, we can find numbers q and r such that $k = mq + r$ where $0 \leq r < m$; hence,

$$a^k = a^{mq+r} = (a^m)^q a^r = h^q a^r.$$

So $a^r = a^k h^{-q}$. Since a^k and h^{-q} are in H, a^r must also be in H. However, m was the smallest positive number such that a^m was in H; consequently, $r = 0$ and so

$k = mq$. Therefore,

$$h' = a^k = a^{mq} = h^q$$

and H is generated by h. ■

Corollary 4.11. The subgroups of $\mathbb{Z}$ are exactly $n\mathbb{Z}$ for $n = 0, 1, 2, \ldots$.

Proposition 4.12. Let G be a cyclic group of order n and suppose that a is a generator for G. Then $a^k = e$ if and only if n divides k.

Proof. First suppose that $a^k = e$. By the division algorithm, $k = nq + r$ where $0 \leq r < n$; hence,

$$e = a^k = a^{nq+r} = a^{nq}a^r = ea^r = a^r.$$

Since the smallest positive integer m such that $a^m = e$ is n, $r = 0$.

Conversely, if n divides k, then $k = ns$ for some integer s. Consequently,

$$a^k = a^{ns} = (a^n)^s = e^s = e.$$

■

Theorem 4.13. Let G be a cyclic group of order n and suppose that $a \in G$ is a generator of the group. If $b = a^k$, then the order of b is n/d, where $d = \gcd(k, n)$.

Proof. We wish to find the smallest integer m such that $e = b^m = a^{km}$. By Proposition 4.12, this is the smallest integer m such that n divides km or, equivalently, n/d divides $m(k/d)$. Since d is the greatest common divisor of n and k, n/d and k/d are relatively prime. Hence, for n/d to divide $m(k/d)$ it must divide m. The smallest such m is n/d. ■

Corollary 4.14. The generators of $\mathbb{Z}_n$ are the integers r such that $1 \leq r < n$ and $\gcd(r, n) = 1$.

Example 4.15. Let us examine the group $\mathbb{Z}_{16}$. The numbers 1, 3, 5, 7, 9, 11, 13, and 15 are the elements of $\mathbb{Z}_{16}$ that are relatively prime to 16. Each of these elements generates $\mathbb{Z}_{16}$. For example,

$1 \cdot 9 = 9$	$2 \cdot 9 = 2$	$3 \cdot 9 = 11$
$4 \cdot 9 = 4$	$5 \cdot 9 = 13$	$6 \cdot 9 = 6$
$7 \cdot 9 = 15$	$8 \cdot 9 = 8$	$9 \cdot 9 = 1$
$10 \cdot 9 = 10$	$11 \cdot 9 = 3$	$12 \cdot 9 = 12$
$13 \cdot 9 = 5$	$14 \cdot 9 = 14$	$15 \cdot 9 = 7.$

□

4.2 Multiplicative Group of Complex Numbers

The *complex numbers* are defined as

$$\mathbb{C} = \{a + bi : a, b \in \mathbb{R}\},$$

where $i^2 = -1$. If $z = a + bi$, then a is the *real part* of z and b is the *imaginary part* of z.

To add two complex numbers $z = a + bi$ and $w = c + di$, we just add the corresponding real and imaginary parts:

$$z + w = (a + bi) + (c + di) = (a + c) + (b + d)i.$$

Remembering that $i^2 = -1$, we multiply complex numbers just like polynomials. The product of z and w is

$$(a + bi)(c + di) = ac + bdi^2 + adi + bci = (ac - bd) + (ad + bc)i.$$

Every nonzero complex number $z = a + bi$ has a multiplicative inverse; that is, there exists a $z^{-1} \in \mathbb{C}^*$ such that $zz^{-1} = z^{-1}z = 1$. If $z = a + bi$, then

$$z^{-1} = \frac{a - bi}{a^2 + b^2}.$$

The *complex conjugate* of a complex number $z = a + bi$ is defined to be $\bar{z} = a - bi$. The *absolute value* or *modulus* of $z = a + bi$ is $|z| = \sqrt{a^2 + b^2}$.

Example 4.16. Let $z = 2 + 3i$ and $w = 1 - 2i$. Then

$$z + w = (2 + 3i) + (1 - 2i) = 3 + i$$

and

$$zw = (2 + 3i)(1 - 2i) = 8 - i.$$

Also,

$$z^{-1} = \frac{2}{13} - \frac{3}{13}i$$
$$|z| = \sqrt{13}$$
$$\bar{z} = 2 - 3i.$$

$\square$

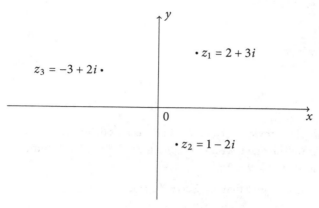

Figure 4.17. Rectangular coordinates of a complex number

There are several ways of graphically representing complex numbers. We can represent a complex number $z = a + bi$ as an ordered pair on the xy plane where a is the x (or real) coordinate and b is the y (or imaginary) coordinate. This is called the *rectangular* or *Cartesian* representation. The rectangular representations of $z_1 = 2 + 3i$, $z_2 = 1 - 2i$, and $z_3 = -3 + 2i$ are depicted in Figure 4.17.

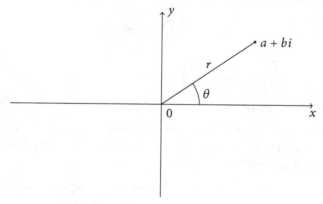

Figure 4.18. Polar coordinates of a complex number

Nonzero complex numbers can also be represented using *polar coordinates*. To specify any nonzero point on the plane, it suffices to give an angle θ from the positive x axis in the counterclockwise direction and a distance r from the origin, as in Figure 4.18. We can see that

$$z = a + bi = r(\cos \theta + i \sin \theta).$$

Hence,

$$r = |z| = \sqrt{a^2 + b^2}$$

and

$$a = r \cos \theta$$
$$b = r \sin \theta.$$

We sometimes abbreviate $r(\cos \theta + i \sin \theta)$ as $r \operatorname{cis} \theta$. To assure that the representation of z is well-defined, we also require that $0° \le \theta < 360°$. If the measurement is in radians, then $0 \le \theta < 2\pi$.

Example 4.19. Suppose that $z = 2 \operatorname{cis} 60°$. Then

$$a = 2 \cos 60° = 1$$

and

$$b = 2 \sin 60° = \sqrt{3}.$$

Hence, the rectangular representation is $z = 1 + \sqrt{3}\, i$.

Conversely, if we are given a rectangular representation of a complex number, it is often useful to know the number's polar representation. If $z = 3\sqrt{2} - 3\sqrt{2}\, i$, then

$$r = \sqrt{a^2 + b^2} = \sqrt{36} = 6$$

and

$$\theta = \arctan\left(\frac{b}{a}\right) = \arctan(-1) = 315°,$$

so $3\sqrt{2} - 3\sqrt{2}\, i = 6 \operatorname{cis} 315°$. ☐

The polar representation of a complex number makes it easy to find products and powers of complex numbers. The proof of the following proposition is straightforward and is left as an exercise.

Proposition 4.20. Let $z = r \operatorname{cis} \theta$ and $w = s \operatorname{cis} \phi$ be two nonzero complex numbers. Then

$$zw = rs \operatorname{cis}(\theta + \phi).$$

Example 4.21. If $z = 3 \operatorname{cis}(\pi/3)$ and $w = 2 \operatorname{cis}(\pi/6)$, then $zw = 6 \operatorname{cis}(\pi/2) = 6i$. ☐

Theorem 4.22. DeMoivre. Let $z = r \operatorname{cis} \theta$ be a nonzero complex number. Then

$$[r \operatorname{cis} \theta]^n = r^n \operatorname{cis}(n\theta)$$

for $n = 1, 2, \ldots$.

Proof. We will use induction on n. For $n = 1$ the theorem is trivial. Assume that the theorem is true for all k such that $1 \leq k \leq n$. Then

$$z^{n+1} = z^n z$$
$$= r^n(\cos n\theta + i \sin n\theta)r(\cos\theta + i \sin\theta)$$
$$= r^{n+1}[(\cos n\theta \cos\theta - \sin n\theta \sin\theta) + i(\sin n\theta \cos\theta + \cos n\theta \sin\theta)]$$
$$= r^{n+1}[\cos(n\theta + \theta) + i \sin(n\theta + \theta)]$$
$$= r^{n+1}[\cos(n+1)\theta + i \sin(n+1)\theta].$$

∎

Example 4.23. Suppose that $z = 1 + i$ and we wish to compute z^{10}. Rather than computing $(1 + i)^{10}$ directly, it is much easier to switch to polar coordinates and calculate z^{10} using DeMoivre's Theorem:

$$z^{10} = (1 + i)^{10}$$
$$= \left(\sqrt{2}\operatorname{cis}\left(\frac{\pi}{4}\right)\right)^{10}$$
$$= (\sqrt{2})^{10}\operatorname{cis}\left(\frac{5\pi}{2}\right)$$
$$= 32\operatorname{cis}\left(\frac{\pi}{2}\right)$$
$$= 32i.$$

□

The Circle Group and the Roots of Unity

The multiplicative group of the complex numbers, $\mathbb{C}^*$, possesses some interesting subgroups. Whereas $\mathbb{Q}^*$ and $\mathbb{R}^*$ have no interesting subgroups of finite order, $\mathbb{C}^*$ has many. We first consider the *circle group*,

$$\mathbb{T} = \{z \in \mathbb{C} : |z| = 1\}.$$

The following proposition is a direct result of Proposition 4.20.

Proposition 4.24. The circle group is a subgroup of $\mathbb{C}^*$.

Although the circle group has infinite order, it has many interesting finite subgroups. Suppose that $H = \{1, -1, i, -i\}$. Then H is a subgroup of the circle group. Also, $1, -1, i,$ and $-i$ are exactly those complex numbers that satisfy the equation $z^4 = 1$. The complex numbers satisfying the equation $z^n = 1$ are called the *nth roots of unity*.

Theorem 4.25. If $z^n = 1$, then the nth roots of unity are

$$z = \operatorname{cis}\left(\frac{2k\pi}{n}\right),$$

where $k = 0, 1, \ldots, n - 1$. Furthermore, the nth roots of unity form a cyclic subgroup of $\mathbb{T}$ of order n

Proof. By DeMoivre's Theorem,

$$z^n = \operatorname{cis}\left(n\frac{2k\pi}{n}\right) = \operatorname{cis}(2k\pi) = 1.$$

The z's are distinct since the numbers $2k\pi/n$ are all distinct and are greater than or equal to 0 but less than 2π. The fact that these are all of the roots of the equation $z^n = 1$ follows from from Corollary 17.9 on page 272, which states that a polynomial of degree n can have at most n roots. We will leave the proof that the nth roots of unity form a cyclic subgroup of $\mathbb{T}$ as an exercise. ∎

A generator for the group of the nth roots of unity is called a ***primitive nth root of unity***.

Example 4.26. The 8th roots of unity can be represented as eight equally spaced points on the unit circle (Figure 4.27). The primitive 8th roots of unity are

$$\omega = \frac{\sqrt{2}}{2} + \frac{\sqrt{2}}{2}i$$

$$\omega^3 = -\frac{\sqrt{2}}{2} + \frac{\sqrt{2}}{2}i$$

$$\omega^5 = -\frac{\sqrt{2}}{2} - \frac{\sqrt{2}}{2}i$$

$$\omega^7 = \frac{\sqrt{2}}{2} - \frac{\sqrt{2}}{2}i.$$

□

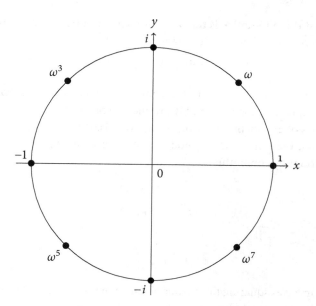

Figure 4.27. 8th roots of unity

4.3 The Method of Repeated Squares

Computing large powers can be very time-consuming. Just as anyone can compute 2^2 or 2^8, everyone knows how to compute

$$2^{2^{1,000,000}}.$$

However, such numbers are so large that we do not want to attempt the calculations; moreover, past a certain point the computations would not be feasible even if we had every computer in the world at our disposal. Even writing down the decimal representation of a very large number may not be reasonable. It could be thousands or even millions of digits long. However, if we could compute something like

$$2^{37,398,332} \pmod{46,389},$$

we could very easily write the result down since it would be a number between 0 and 46,388. If we want to compute powers modulo n quickly and efficiently, we will have to be clever.[1]

---

[1]The results in this section are needed only in Chapter 7.

The first thing to notice is that any number a can be written as the sum of distinct powers of 2; that is, we can write

$$a = 2^{k_1} + 2^{k_2} + \cdots + 2^{k_n},$$

where $k_1 < k_2 < \cdots < k_n$. This is just the binary representation of a. For example, the binary representation of 57 is 111001, since we can write $57 = 2^0 + 2^3 + 2^4 + 2^5$.

The laws of exponents still work in $\mathbb{Z}_n$; that is, if $b \equiv a^x \pmod{n}$ and $c \equiv a^y \pmod{n}$, then $bc \equiv a^{x+y} \pmod{n}$. We can compute $a^{2^k} \pmod{n}$ in k multiplications by computing

$$a^{2^0} \pmod{n}$$

$$a^{2^1} \pmod{n}$$

$$\vdots$$

$$a^{2^k} \pmod{n}.$$

Each step involves squaring the answer obtained in the previous step, dividing by n, and taking the remainder.

Example 4.28. We will compute $271^{321} \pmod{481}$. Notice that

$$321 = 2^0 + 2^6 + 2^8;$$

hence, computing $271^{321} \pmod{481}$ is the same as computing

$$271^{2^0+2^6+2^8} \equiv 271^{2^0} \cdot 271^{2^6} \cdot 271^{2^8} \pmod{481}.$$

So it will suffice to compute $271^{2^i} \pmod{481}$ where $i = 0, 6, 8$. It is very easy to see that

$$271^{2^1} = 73{,}441 \equiv 329 \pmod{481}.$$

We can square this result to obtain a value for $271^{2^2} \pmod{481}$:

$$271^{2^2} \equiv (271^{2^1})^2 \pmod{481}$$

$$\equiv (329)^2 \pmod{481}$$

$$\equiv 108{,}241 \pmod{481}$$

$$\equiv 16 \pmod{481}.$$

We are using the fact that $(a^{2^n})^2 \equiv a^{2 \cdot 2^n} \equiv a^{2^{n+1}} \pmod{n}$. Continuing, we can calculate

$$271^{2^6} \equiv 419 \pmod{481}$$

and

$$271^{2^8} \equiv 16 \pmod{481}.$$

Therefore,

$$271^{321} \equiv 271^{2^0 + 2^6 + 2^8} \pmod{481}$$
$$\equiv 271^{2^0} \cdot 271^{2^6} \cdot 271^{2^8} \pmod{481}$$
$$\equiv 271 \cdot 419 \cdot 16 \pmod{481}$$
$$\equiv 1{,}816{,}784 \pmod{481}$$
$$\equiv 47 \pmod{481}.$$

$\square$

The method of repeated squares will prove to be a very useful tool when we explore RSA cryptography in Chapter 7. To encode and decode messages in a reasonable manner under this scheme, it is necessary to be able to quickly compute large powers of integers mod n.

Sage. Sage support for cyclic groups is a little spotty — but we can still make effective use of Sage and perhaps this situation could change soon.

4.4 Reading Questions

1. What is the order of the element 3 in $U(20)$?

2. What is the order of the element 5 in $U(23)$?

3. Find three generators of $\mathbb{Z}_8$.

4. Find three generators of the 5th roots of unity.

5. Show how to compute $15^{40} \pmod{23}$ efficiently by hand. Check your answer with Sage.

4.5 Exercises

1. Prove or disprove each of the following statements.

    (a) All of the generators of $\mathbb{Z}_{60}$ are prime.

    (b) $U(8)$ is cyclic.

    (c) $\mathbb{Q}$ is cyclic.

    (d) If every proper subgroup of a group G is cyclic, then G is a cyclic group.

    (e) A group with a finite number of subgroups is finite.

2. Find the order of each of the following elements.

    (a) $5 \in \mathbb{Z}_{12}$ (d) $-i \in \mathbb{C}^*$

    (b) $\sqrt{3} \in \mathbb{R}$ (e) $72 \in \mathbb{Z}_{240}$

    (c) $\sqrt{3} \in \mathbb{R}^*$ (f) $312 \in \mathbb{Z}_{471}$

3. List all of the elements in each of the following subgroups.

    (a) The subgroup of $\mathbb{Z}$ generated by 7

    (b) The subgroup of $\mathbb{Z}_{24}$ generated by 15

    (c) All subgroups of $\mathbb{Z}_{12}$

    (d) All subgroups of $\mathbb{Z}_{60}$

    (e) All subgroups of $\mathbb{Z}_{13}$

    (f) All subgroups of $\mathbb{Z}_{48}$

    (g) The subgroup generated by 3 in $U(20)$

    (h) The subgroup generated by 5 in $U(18)$

    (i) The subgroup of $\mathbb{R}^*$ generated by 7

    (j) The subgroup of $\mathbb{C}^*$ generated by i where $i^2 = -1$

    (k) The subgroup of $\mathbb{C}^*$ generated by $2i$

    (l) The subgroup of $\mathbb{C}^*$ generated by $(1+i)/\sqrt{2}$

    (m) The subgroup of $\mathbb{C}^*$ generated by $(1+\sqrt{3}\,i)/2$

4. Find the subgroups of $GL_2(\mathbb{R})$ generated by each of the following matrices.

    (a) $\begin{pmatrix} 0 & 1 \\ -1 & 0 \end{pmatrix}$ (c) $\begin{pmatrix} 1 & -1 \\ 1 & 0 \end{pmatrix}$ (e) $\begin{pmatrix} 1 & -1 \\ -1 & 0 \end{pmatrix}$

    (b) $\begin{pmatrix} 0 & 1/3 \\ 3 & 0 \end{pmatrix}$ (d) $\begin{pmatrix} 1 & -1 \\ 0 & 1 \end{pmatrix}$ (f) $\begin{pmatrix} \sqrt{3}/2 & 1/2 \\ -1/2 & \sqrt{3}/2 \end{pmatrix}$

5. Find the order of every element in $\mathbb{Z}_{18}$.

6. Find the order of every element in the symmetry group of the square, D_4.

7. What are all of the cyclic subgroups of the quaternion group, Q_8?

8. List all of the cyclic subgroups of $U(30)$.

9. List every generator of each subgroup of order 8 in $\mathbb{Z}_{32}$.

10. Find all elements of finite order in each of the following groups. Here the "$*$" indicates the set with zero removed.

    (a) $\mathbb{Z}$ (b) $\mathbb{Q}^*$ (c) $\mathbb{R}^*$

11. If $a^{24} = e$ in a group G, what are the possible orders of a?

12. Find a cyclic group with exactly one generator. Can you find cyclic groups with exactly two generators? Four generators? How about n generators?

13. For $n \leq 20$, which groups $U(n)$ are cyclic? Make a conjecture as to what is true in general. Can you prove your conjecture?

14. Let

$$A = \begin{pmatrix} 0 & 1 \\ -1 & 0 \end{pmatrix} \quad \text{and} \quad B = \begin{pmatrix} 0 & -1 \\ 1 & -1 \end{pmatrix}$$

be elements in $GL_2(\mathbb{R})$. Show that A and B have finite orders but AB does not.

15. Evaluate each of the following.
    (a) $(3 - 2i) + (5i - 6)$
    (b) $(4 - 5i) - \overline{(4i - 4)}$
    (c) $(5 - 4i)(7 + 2i)$
    (d) $(9 - i)\overline{(9 - i)}$
    (e) i^{45}
    (f) $(1 + i) + \overline{(1 + i)}$

16. Convert the following complex numbers to the form $a + bi$.
    (a) $2\operatorname{cis}(\pi/6)$
    (b) $5\operatorname{cis}(9\pi/4)$
    (c) $3\operatorname{cis}(\pi)$
    (d) $\operatorname{cis}(7\pi/4)/2$

17. Change the following complex numbers to polar representation.
    (a) $1 - i$
    (b) -5
    (c) $2 + 2i$
    (d) $\sqrt{3} + i$
    (e) $-3i$
    (f) $2i + 2\sqrt{3}$

18. Calculate each of the following expressions.
    (a) $(1 + i)^{-1}$
    (b) $(1 - i)^6$
    (c) $(\sqrt{3} + i)^5$
    (d) $(-i)^{10}$
    (e) $((1 - i)/2)^4$
    (f) $(-\sqrt{2} - \sqrt{2}\,i)^{12}$
    (g) $(-2 + 2i)^{-5}$

19. Prove each of the following statements.
    (a) $|z| = |\bar{z}|$
    (b) $z\bar{z} = |z|^2$
    (c) $z^{-1} = \bar{z}/|z|^2$
    (d) $|z + w| \leq |z| + |w|$
    (e) $|z - w| \geq ||z| - |w||$
    (f) $|zw| = |z||w|$

20. List and graph the 6th roots of unity. What are the generators of this group? What are the primitive 6th roots of unity?

21. List and graph the 5th roots of unity. What are the generators of this group? What are the primitive 5th roots of unity?

22. Calculate each of the following.

    (a) $292^{3171} \pmod{582}$ (c) $2071^{9521} \pmod{4724}$

    (b) $2557^{341} \pmod{5681}$ (d) $971^{321} \pmod{765}$

23. Let $a, b \in G$. Prove the following statements.

    (a) The order of a is the same as the order of a^{-1}.

    (b) For all $g \in G$, $|a| = |g^{-1}ag|$.

    (c) The order of ab is the same as the order of ba.

24. Let p and q be distinct primes. How many generators does $\mathbb{Z}_{pq}$ have?

25. Let p be prime and r be a positive integer. How many generators does $\mathbb{Z}_{p^r}$ have?

26. Prove that $\mathbb{Z}_p$ has no nontrivial subgroups if p is prime.

27. If g and h have orders 15 and 16 respectively in a group G, what is the order of $\langle g \rangle \cap \langle h \rangle$?

28. Let a be an element in a group G. What is a generator for the subgroup $\langle a^m \rangle \cap \langle a^n \rangle$?

29. Prove that $\mathbb{Z}_n$ has an even number of generators for $n > 2$.

30. Suppose that G is a group and let $a, b \in G$. Prove that if $|a| = m$ and $|b| = n$ with $\gcd(m, n) = 1$, then $\langle a \rangle \cap \langle b \rangle = \{e\}$.

31. Let G be an abelian group. Show that the elements of finite order in G form a subgroup. This subgroup is called the **torsion subgroup** of G.

32. Let G be a finite cyclic group of order n generated by x. Show that if $y = x^k$ where $\gcd(k, n) = 1$, then y must be a generator of G.

33. If G is an abelian group that contains a pair of cyclic subgroups of order 2, show that G must contain a subgroup of order 4. Does this subgroup have to be cyclic?

34. Let G be an abelian group of order pq where $\gcd(p, q) = 1$. If G contains elements a and b of order p and q respectively, then show that G is cyclic.

35. Prove that the subgroups of $\mathbb{Z}$ are exactly $n\mathbb{Z}$ for $n = 0, 1, 2, \ldots$.

36. Prove that the generators of $\mathbb{Z}_n$ are the integers r such that $1 \le r < n$ and $\gcd(r, n) = 1$.

37. Prove that if G has no proper nontrivial subgroups, then G is a cyclic group.

38. Prove that the order of an element in a cyclic group G must divide the order of the group.

39. Prove that if G is a cyclic group of order m and $d \mid m$, then G must have a subgroup of order d.

40. For what integers n is -1 an nth root of unity?

41. If $z = r(\cos\theta + i\sin\theta)$ and $w = s(\cos\phi + i\sin\phi)$ are two nonzero complex numbers, show that

$$zw = rs[\cos(\theta + \phi) + i\sin(\theta + \phi)].$$

42. Prove that the circle group is a subgroup of $\mathbb{C}^*$.

43. Prove that the nth roots of unity form a cyclic subgroup of $\mathbb{T}$ of order n.

44. Let $\alpha \in \mathbb{T}$. Prove that $\alpha^m = 1$ and $\alpha^n = 1$ if and only if $\alpha^d = 1$ for $d = \gcd(m, n)$.

45. Let $z \in \mathbb{C}^*$. If $|z| \neq 1$, prove that the order of z is infinite.

46. Let $z = \cos\theta + i\sin\theta$ be in $\mathbb{T}$ where $\theta \in \mathbb{Q}$. Prove that the order of z is infinite.

4.6 Programming Exercises

1. Write a computer program that will write any decimal number as the sum of distinct powers of 2. What is the largest integer that your program will handle?

2. Write a computer program to calculate $a^x \pmod{n}$ by the method of repeated squares. What are the largest values of n and x that your program will accept?

4.7 References and Suggested Readings

[1]   Koblitz, N. *A Course in Number Theory and Cryptography*. 2nd ed. Springer, New York, 1994.

[2]   Pomerance, C. "Cryptology and Computational Number Theory—An Introduction," in *Cryptology and Computational Number Theory*, Pomerance, C., ed. Proceedings of Symposia in Applied Mathematics, vol. 42, American Mathematical Society, Providence, RI, 1990. This book gives an excellent account of how the method of repeated squares is used in cryptography.

5

Permutation Groups

$\mathcal{P}$ermutation groups are central to the study of geometric symmetries and to Galois theory, the study of finding solutions of polynomial equations. They also provide abundant examples of nonabelian groups.

Let us recall for a moment the symmetries of the equilateral triangle $\triangle ABC$ from Chapter 3. The symmetries actually consist of permutations of the three vertices, where a **permutation** of the set $S = \{A, B, C\}$ is a one-to-one and onto map $\pi : S \to S$. The three vertices have the following six permutations.

$$\begin{pmatrix} A & B & C \\ A & B & C \end{pmatrix} \qquad \begin{pmatrix} A & B & C \\ C & A & B \end{pmatrix} \qquad \begin{pmatrix} A & B & C \\ B & C & A \end{pmatrix}$$

$$\begin{pmatrix} A & B & C \\ A & C & B \end{pmatrix} \qquad \begin{pmatrix} A & B & C \\ C & B & A \end{pmatrix} \qquad \begin{pmatrix} A & B & C \\ B & A & C \end{pmatrix}$$

We have used the array

$$\begin{pmatrix} A & B & C \\ B & C & A \end{pmatrix}$$

to denote the permutation that sends A to B, B to C, and C to A. That is,

$$A \mapsto B$$
$$B \mapsto C$$
$$C \mapsto A.$$

The symmetries of a triangle form a group. In this chapter we will study groups of this type.

5.1 Definitions and Notation

In general, the permutations of a set X form a group S_X. If X is a finite set, we can assume $X = \{1, 2, \ldots, n\}$. In this case we write S_n instead of S_X. The following theorem says that S_n is a group. We call this group the **symmetric group** on n letters.

Theorem 5.1. The symmetric group on n letters, S_n, is a group with $n!$ elements, where the binary operation is the composition of maps.

Proof. The identity of S_n is just the identity map that sends 1 to 1, 2 to 2, ..., n to n. If $f : S_n \to S_n$ is a permutation, then f^{-1} exists, since f is one-to-one and onto; hence, every permutation has an inverse. Composition of maps is associative, which makes the group operation associative. We leave the proof that $|S_n| = n!$ as an exercise. ∎

A subgroup of S_n is called a ***permutation group***.

Example 5.2. Consider the subgroup G of S_5 consisting of the identity permutation id and the permutations

$$\sigma = \begin{pmatrix} 1 & 2 & 3 & 4 & 5 \\ 1 & 2 & 3 & 5 & 4 \end{pmatrix}$$

$$\tau = \begin{pmatrix} 1 & 2 & 3 & 4 & 5 \\ 3 & 2 & 1 & 4 & 5 \end{pmatrix}$$

$$\mu = \begin{pmatrix} 1 & 2 & 3 & 4 & 5 \\ 3 & 2 & 1 & 5 & 4 \end{pmatrix}.$$

The following table tells us how to multiply elements in the permutation group G.

$\circ$	id	σ	τ	μ
id	id	σ	τ	μ
σ	σ	id	μ	τ
τ	τ	μ	id	σ
μ	μ	τ	σ	id

□

Remark 5.3. Though it is natural to multiply elements in a group from left to right, functions are composed from right to left. Let σ and τ be permutations on a set X. To compose σ and τ as functions, we calculate $(\sigma \circ \tau)(x) = \sigma(\tau(x))$. That is, we do τ first, then σ. There are several ways to approach this inconsistency. *We will adopt the convention of multiplying permutations right to left. To compute $\sigma\tau$, do τ first and then σ.* That is, by $\sigma\tau(x)$ we mean $\sigma(\tau(x))$. (Another way of solving this problem would be to write functions on the right; that is, instead of writing $\sigma(x)$, we could write $(x)\sigma$. We could also multiply permutations left to right to agree with the usual way of multiplying elements in a group. Certainly all of these methods have been used.

Example 5.4. Permutation multiplication is not usually commutative. Let

$$\sigma = \begin{pmatrix} 1 & 2 & 3 & 4 \\ 4 & 1 & 2 & 3 \end{pmatrix}$$

$$\tau = \begin{pmatrix} 1 & 2 & 3 & 4 \\ 2 & 1 & 4 & 3 \end{pmatrix}.$$

Then

$$\sigma\tau = \begin{pmatrix} 1 & 2 & 3 & 4 \\ 1 & 4 & 3 & 2 \end{pmatrix},$$

but

$$\tau\sigma = \begin{pmatrix} 1 & 2 & 3 & 4 \\ 3 & 2 & 1 & 4 \end{pmatrix}.$$

□

Cycle Notation

The notation that we have used to represent permutations up to this point is cumbersome, to say the least. To work effectively with permutation groups, we need a more streamlined method of writing down and manipulating permutations.

A permutation $\sigma \in S_X$ is a *cycle of length* k if there exist $a_1, a_2, \ldots, a_k \in X$ such that

$$\sigma(a_1) = a_2$$
$$\sigma(a_2) = a_3$$
$$\vdots$$
$$\sigma(a_k) = a_1$$

and $\sigma(x) = x$ for all other elements $x \in X$. We will write $(a_1, a_2, \ldots, a_k)$ to denote the cycle σ. Cycles are the building blocks of all permutations.

Example 5.5. The permutation

$$\sigma = \begin{pmatrix} 1 & 2 & 3 & 4 & 5 & 6 & 7 \\ 6 & 3 & 5 & 1 & 4 & 2 & 7 \end{pmatrix} = (1\,6\,2\,3\,5\,4)$$

is a cycle of length 6, whereas

$$\tau = \begin{pmatrix} 1 & 2 & 3 & 4 & 5 & 6 \\ 1 & 4 & 2 & 3 & 5 & 6 \end{pmatrix} = (2\,4\,3)$$

is a cycle of length 3.

Not every permutation is a cycle. Consider the permutation

$$\begin{pmatrix} 1 & 2 & 3 & 4 & 5 & 6 \\ 2 & 4 & 1 & 3 & 6 & 5 \end{pmatrix} = (1\,2\,4\,3)(5\,6).$$

This permutation actually contains a cycle of length 2 and a cycle of length 4. □

Example 5.6. It is very easy to compute products of cycles. Suppose that

$$\sigma = (1\,3\,5\,2) \quad \text{and} \quad \tau = (2\,5\,6).$$

If we think of σ as

$$1 \mapsto 3, \qquad 3 \mapsto 5, \qquad 5 \mapsto 2, \qquad 2 \mapsto 1,$$

and τ as

$$2 \mapsto 5, \qquad 5 \mapsto 6, \qquad 6 \mapsto 2,$$

then for $\sigma\tau$ remembering that we apply τ first and then σ, it must be the case that

$$1 \mapsto 3, \qquad 3 \mapsto 5, \qquad 5 \mapsto 6, \qquad 6 \mapsto 2 \mapsto 1,$$

or $\sigma\tau = (1\,3\,5\,6)$. If $\mu = (1\,6\,3\,4)$, then $\sigma\mu = (1\,6\,5\,2)(3\,4)$. □

Two cycles in S_X, $\sigma = (a_1, a_2, \ldots, a_k)$ and $\tau = (b_1, b_2, \ldots, b_l)$, are **disjoint** if $a_i \neq b_j$ for all i and j.

Example 5.7. The cycles $(1\,3\,5)$ and $(2\,7)$ are disjoint; however, the cycles $(1\,3\,5)$ and $(3\,4\,7)$ are not. Calculating their products, we find that

$$(1\,3\,5)(2\,7) = (1\,3\,5)(2\,7)$$
$$(1\,3\,5)(3\,4\,7) = (1\,3\,4\,7\,5).$$

The product of two cycles that are not disjoint may reduce to something less complicated; the product of disjoint cycles cannot be simplified. □

Proposition 5.8. *Let σ and τ be two disjoint cycles in S_X. Then $\sigma\tau = \tau\sigma$.*

Proof. Let $\sigma = (a_1, a_2, \ldots, a_k)$ and $\tau = (b_1, b_2, \ldots, b_l)$. We must show that $\sigma\tau(x) = \tau\sigma(x)$ for all $x \in X$. If x is neither in $\{a_1, a_2, \ldots, a_k\}$ nor $\{b_1, b_2, \ldots, b_l\}$, then both σ and τ fix x. That is, $\sigma(x) = x$ and $\tau(x) = x$. Hence,

$$\sigma\tau(x) = \sigma(\tau(x)) = \sigma(x) = x = \tau(x) = \tau(\sigma(x)) = \tau\sigma(x).$$

Do not forget that we are multiplying permutations right to left, which is the opposite of the order in which we usually multiply group elements. Now suppose that $x \in$

$\{a_1, a_2, \ldots, a_k\}$. Then $\sigma(a_i) = a_{(i \bmod k)+1}$; that is,

$$a_1 \mapsto a_2$$
$$a_2 \mapsto a_3$$
$$\vdots$$
$$a_{k-1} \mapsto a_k$$
$$a_k \mapsto a_1.$$

However, $\tau(a_i) = a_i$ since σ and τ are disjoint. Therefore,

$$\begin{aligned}
\sigma\tau(a_i) &= \sigma(\tau(a_i)) \\
&= \sigma(a_i) \\
&= a_{(i \bmod k)+1} \\
&= \tau(a_{(i \bmod k)+1}) \\
&= \tau(\sigma(a_i)) \\
&= \tau\sigma(a_i).
\end{aligned}$$

Similarly, if $x \in \{b_1, b_2, \ldots, b_l\}$, then σ and τ also commute. ∎

Theorem 5.9. Every permutation in S_n can be written as the product of disjoint cycles.

Proof. We can assume that $X = \{1, 2, \ldots, n\}$. If $\sigma \in S_n$ and we define X_1 to be $\{\sigma(1), \sigma^2(1), \ldots\}$, then the set X_1 is finite since X is finite. Now let i be the first integer in X that is not in X_1 and define X_2 by $\{\sigma(i), \sigma^2(i), \ldots\}$. Again, X_2 is a finite set. Continuing in this manner, we can define finite disjoint sets $X_3, X_4, \ldots$. Since X is a finite set, we are guaranteed that this process will end and there will be only a finite number of these sets, say r. If σ_i is the cycle defined by

$$\sigma_i(x) = \begin{cases} \sigma(x) & x \in X_i \\ x & x \notin X_i \end{cases},$$

then $\sigma = \sigma_1 \sigma_2 \cdots \sigma_r$. Since sets $X_1, X_2, \ldots, X_r$ are disjoint, the cycles $\sigma_1, \sigma_2, \ldots, \sigma_r$ must also be disjoint. ∎

Example 5.10. Let

$$\sigma = \begin{pmatrix} 1 & 2 & 3 & 4 & 5 & 6 \\ 6 & 4 & 3 & 1 & 5 & 2 \end{pmatrix}$$

$$\tau = \begin{pmatrix} 1 & 2 & 3 & 4 & 5 & 6 \\ 3 & 2 & 1 & 5 & 6 & 4 \end{pmatrix}.$$

Using cycle notation, we can write

$$\sigma = (1\,6\,2\,4)$$
$$\tau = (1\,3)(4\,5\,6)$$
$$\sigma\tau = (1\,3\,6)(2\,4\,5)$$
$$\tau\sigma = (1\,4\,3)(2\,5\,6).$$

□

Remark 5.11. From this point forward we will find it convenient to use cycle notation to represent permutations. When using cycle notation, we often denote the identity permutation by (1).

Transpositions

The simplest permutation is a cycle of length 2. Such cycles are called *transpositions*. Since

$$(a_1, a_2, \ldots, a_n) = (a_1, a_n)(a_1, a_{n-1})\cdots(a_1, a_3)(a_1, a_2),$$

any cycle can be written as the product of transpositions, leading to the following proposition.

Proposition 5.12. Any permutation of a finite set containing at least two elements can be written as the product of transpositions.

Example 5.13. Consider the permutation

$$(1\,6)(2\,5\,3) = (1\,6)(2\,3)(2\,5) = (1\,6)(4\,5)(2\,3)(4\,5)(2\,5).$$

As we can see, there is no unique way to represent permutation as the product of transpositions. For instance, we can write the identity permutation as $(1\,2)(1\,2)$, as $(1\,3)(2\,4)(1\,3)(2\,4)$, and in many other ways. However, as it turns out, no permutation can be written as the product of both an even number of transpositions and an odd number of transpositions. For instance, we could represent the permutation $(1\,6)$ by

$$(2\,3)(1\,6)(2\,3)$$

or by

$$(3\,5)(1\,6)(1\,3)(1\,6)(1\,3)(3\,5)(5\,6),$$

but $(1\,6)$ will always be the product of an odd number of transpositions.   □

Lemma 5.14. If the identity is written as the product of r transpositions,

$$\text{id} = \tau_1 \tau_2 \cdots \tau_r,$$

then r is an even number.

Proof. We will employ induction on r. A transposition cannot be the identity; hence, $r > 1$. If $r = 2$, then we are done. Suppose that $r > 2$. In this case the product of the last two transpositions, $\tau_{r-1}\tau_r$, must be one of the following cases:

$$(a,b)(a,b) = \text{id}$$
$$(b,c)(a,b) = (a,c)(b,c)$$
$$(c,d)(a,b) = (a,b)(c,d)$$
$$(a,c)(a,b) = (a,b)(b,c),$$

where a, b, c, and d are distinct.

The first equation simply says that a transposition is its own inverse. If this case occurs, delete $\tau_{r-1}\tau_r$ from the product to obtain

$$\text{id} = \tau_1 \tau_2 \cdots \tau_{r-3}\tau_{r-2}.$$

By induction $r - 2$ is even; hence, r must be even.

In each of the other three cases, we can replace $\tau_{r-1}\tau_r$ with the right-hand side of the corresponding equation to obtain a new product of r transpositions for the identity. In this new product the last occurrence of a will be in the next-to-the-last transposition. We can continue this process with $\tau_{r-2}\tau_{r-1}$ to obtain either a product of $r - 2$ transpositions or a new product of r transpositions where the last occurrence of a is in τ_{r-2}. If the identity is the product of $r - 2$ transpositions, then again we are done, by our induction hypothesis; otherwise, we will repeat the procedure with $\tau_{r-3}\tau_{r-2}$.

At some point either we will have two adjacent, identical transpositions canceling each other out or a will be shuffled so that it will appear only in the first transposition. However, the latter case cannot occur, because the identity would not fix a in this instance. Therefore, the identity permutation must be the product of $r - 2$ transpositions and, again by our induction hypothesis, we are done. ∎

Theorem 5.15. If a permutation σ can be expressed as the product of an even number of transpositions, then any other product of transpositions equaling σ must also contain an even number of transpositions. Similarly, if σ can be expressed as the product of an odd number of transpositions, then any other product of transpositions equaling σ must also contain an odd number of transpositions.

Proof. Suppose that

$$\sigma = \sigma_1\sigma_2\cdots\sigma_m = \tau_1\tau_2\cdots\tau_n,$$

where m is even. We must show that n is also an even number. The inverse of σ is $\sigma_m\cdots\sigma_1$. Since

$$\text{id} = \sigma\sigma_m\cdots\sigma_1 = \tau_1\cdots\tau_n\sigma_m\cdots\sigma_1,$$

n must be even by Lemma 5.14. The proof for the case in which σ can be expressed as an odd number of transpositions is left as an exercise. ∎

In light of Theorem 5.15, we define a permutation to be *even* if it can be expressed as an even number of transpositions and *odd* if it can be expressed as an odd number of transpositions.

The Alternating Groups

One of the most important subgroups of S_n is the set of all even permutations, A_n. The group A_n is called the **alternating group on n letters**.

Theorem 5.16. *The set A_n is a subgroup of S_n.*

Proof. Since the product of two even permutations must also be an even permutation, A_n is closed. The identity is an even permutation and therefore is in A_n. If σ is an even permutation, then

$$\sigma = \sigma_1\sigma_2\cdots\sigma_r,$$

where σ_i is a transposition and r is even. Since the inverse of any transposition is itself,

$$\sigma^{-1} = \sigma_r\sigma_{r-1}\cdots\sigma_1$$

is also in A_n. ∎

Proposition 5.17. *The number of even permutations in S_n, $n \geq 2$, is equal to the number of odd permutations; hence, the order of A_n is $n!/2$.*

Proof. Let A_n be the set of even permutations in S_n and B_n be the set of odd permutations. If we can show that there is a bijection between these sets, they must contain the same number of elements. Fix a transposition σ in S_n. Since $n \geq 2$, such a σ exists. Define

$$\lambda_\sigma : A_n \to B_n$$

by

$$\lambda_\sigma(\tau) = \sigma\tau.$$

Suppose that $\lambda_\sigma(\tau) = \lambda_\sigma(\mu)$. Then $\sigma\tau = \sigma\mu$ and so

$$\tau = \sigma^{-1}\sigma\tau = \sigma^{-1}\sigma\mu = \mu.$$

Therefore, λ_σ is one-to-one. We will leave the proof that λ_σ is surjective to the reader. ∎

Example 5.18. The group A_4 is the subgroup of S_4 consisting of even permutations. There are twelve elements in A_4:

(1)	(12)(34)	(13)(24)	(14)(23)
(123)	(132)	(124)	(142)
(134)	(143)	(234)	(243).

One of the end-of-chapter exercises will be to write down all the subgroups of A_4. You will find that there is no subgroup of order 6. Does this surprise you? □

↫   Historical Note   ↬

Lagrange first thought of permutations as functions from a set to itself, but it was Cauchy who developed the basic theorems and notation for permutations. He was the first to use cycle notation. Augustin-Louis Cauchy (1789–1857) was born in Paris at the height of the French Revolution. His family soon left Paris for the village of Arcueil to escape the Reign of Terror. One of the family's neighbors there was Pierre-Simon Laplace (1749–1827), who encouraged him to seek a career in mathematics. Cauchy began his career as a mathematician by solving a problem in geometry given to him by Lagrange. Cauchy wrote over 800 papers on such diverse topics as differential equations, finite groups, applied mathematics, and complex analysis. He was one of the mathematicians responsible for making calculus rigorous. Perhaps more theorems and concepts in mathematics have the name Cauchy attached to them than that of any other mathematician.

5.2 Dihedral Groups

Another special type of permutation group is the dihedral group. Recall the symmetry group of an equilateral triangle in Chapter 3. Such groups consist of the rigid motions of a regular n-sided polygon or n-gon. For $n = 3, 4, \ldots$, we define the **nth dihedral group** to be the group of rigid motions of a regular n-gon. We will denote this group by D_n. We can number the vertices of a regular n-gon by $1, 2, \ldots, n$ (Figure 5.19). Notice that there are exactly n choices to replace the first vertex. If we replace the first vertex by k, then the second vertex must be replaced either by vertex $k + 1$ or by vertex $k - 1$; hence, there are $2n$ possible rigid motions of the n-gon. We summarize these results in the following theorem.

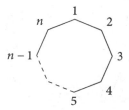

Figure 5.19. A regular n-gon

Theorem 5.20. The dihedral group, D_n, is a subgroup of S_n of order $2n$.

Theorem 5.21. The group D_n, $n \geq 3$, consists of all products of the two elements r and s, satisfying the relations

$$r^n = 1$$
$$s^2 = 1$$
$$srs = r^{-1}.$$

Proof. The possible motions of a regular n-gon are either reflections or rotations (Figure 5.22 on the following page). There are exactly n possible rotations:

$$\text{id}, \frac{360°}{n}, 2 \cdot \frac{360°}{n}, \ldots, (n-1) \cdot \frac{360°}{n}.$$

We will denote the rotation $360°/n$ by r. The rotation r generates all of the other rotations. That is,

$$r^k = k \cdot \frac{360°}{n}.$$

Label the n reflections $s_1, s_2, \ldots, s_n$, where s_k is the reflection that leaves vertex k fixed. There are two cases of reflections, depending on whether n is even or odd. If there are an even number of vertices, then two vertices are left fixed by a reflection, and $s_1 = s_{n/2+1}, s_2 = s_{n/2+2}, \ldots, s_{n/2} = s_n$. If there are an odd number of vertices, then only a single vertex is left fixed by a reflection and $s_1, s_2, \ldots, s_n$ are distinct (Figure 5.23 on the next page). In either case, the order of each s_k is two. Let $s = s_1$. Then $s^2 = 1$ and $r^n = 1$. Since any rigid motion t of the n-gon replaces the first vertex by the vertex k, the second vertex must be replaced by either $k + 1$ or by $k - 1$. If the second vertex is replaced by $k + 1$, then $t = r^k$. If the second vertex is replaced by $k - 1$, then $t = r^k s$ (since we are in an abstract group, we will adopt the convention that group elements are multiplied left to right). Hence, r and s generate D_n. That is, D_n consists of all finite products of r and s,

$$D_n = \{1, r, r^2, \ldots, r^{n-1}, s, rs, r^2 s, \ldots, r^{n-1} s\}.$$

We will leave the proof that $srs = r^{-1}$ as an exercise.   ■

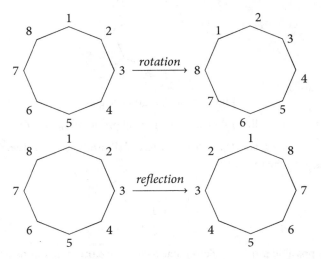

Figure 5.22. Rotations and reflections of a regular n-gon

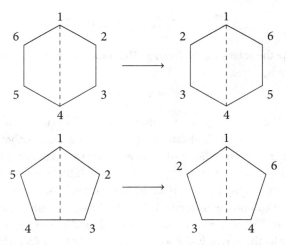

Figure 5.23. Types of reflections of a regular n-gon

Example 5.24. The group of rigid motions of a square, D_4, consists of eight elements. With the vertices numbered 1, 2, 3, 4 (Figure 5.25), the rotations are

$$r = (1\,2\,3\,4)$$
$$r^2 = (13)(24)$$
$$r^3 = (1\,4\,3\,2)$$
$$r^4 = (1)$$

and the reflections are

$$s_1 = (2\,4)$$
$$s_2 = (1\,3).$$

The order of D_4 is 8. The remaining two elements are

$$rs_1 = (1\,2)(3\,4)$$
$$r^3 s_1 = (1\,4)(2\,3).$$

$\square$

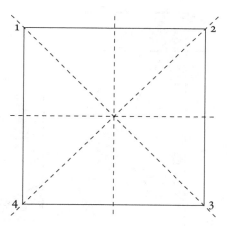

Figure 5.25. The group D_4

The Motion Group of a Cube

We can investigate the groups of rigid motions of geometric objects other than a regular n-sided polygon to obtain interesting examples of permutation groups. Let us consider the group of rigid motions of a cube. By rigid motion, we mean a rotation with the axis of rotation about opposing faces, edges, or vertices. One of the first questions that we can ask about this group is "what is its order?" A cube has 6 sides. If a particular side is facing upward, then there are four possible rotations of the cube that will preserve the upward-facing side. Hence, the order of the group is $6 \cdot 4 = 24$. We have just proved the following proposition.

Proposition 5.26. The group of rigid motions of a cube contains 24 elements.

Theorem 5.27. The group of rigid motions of a cube is S_4.

Proof. From Proposition 5.26, we already know that the motion group of the cube has 24 elements, the same number of elements as there are in S_4. There are exactly four diagonals in the cube. If we label these diagonals 1, 2, 3, and 4, we must show that the motion group of the cube will give us any permutation of the diagonals (Figure 5.28). If we can obtain all of these permutations, then S_4 and the group of rigid motions of the cube must be the same. To obtain a transposition we can rotate the cube 180° about the axis joining the midpoints of opposite edges (Figure 5.29). There are six such axes, giving all transpositions in S_4. Since every element in S_4 is the product of a finite number of transpositions, the motion group of a cube must be S_4.   ■

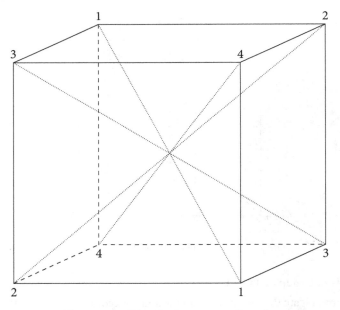

Figure 5.28. The motion group of a cube

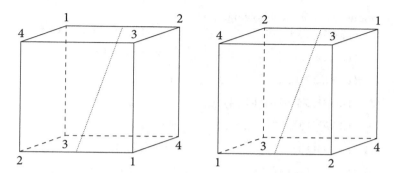

Figure 5.29. Transpositions in the motion group of a cube

Sage.   A permutation group is a very concrete representation of a group, and Sage support for permutations groups is very good — making Sage a natural place for beginners to learn about group theory.

5.3 Reading Questions

1. Express $(1\,3\,4)(3\,5\,4)$ as a cycle, or a product of disjoint cycles. (Interpret the composition of functions in the order used by Sage, which is the reverse of the order used in the book.)

2. What is a transposition?

3. What does it mean for a permutation to be even or odd?

4. Describe another group that is fundamentally the same as A_3.

5. Write the elements of the symmetry group of a pentagon using permutations in cycle notation. Do this exercise by hand, and without the assistance of Sage.

5.4 Exercises

1. Write the following permutations in cycle notation.

    (a)
    $$\begin{pmatrix} 1 & 2 & 3 & 4 & 5 \\ 2 & 4 & 1 & 5 & 3 \end{pmatrix}$$

    (c)
    $$\begin{pmatrix} 1 & 2 & 3 & 4 & 5 \\ 3 & 5 & 1 & 4 & 2 \end{pmatrix}$$

    (b)
    $$\begin{pmatrix} 1 & 2 & 3 & 4 & 5 \\ 4 & 2 & 5 & 1 & 3 \end{pmatrix}$$

    (d)
    $$\begin{pmatrix} 1 & 2 & 3 & 4 & 5 \\ 1 & 4 & 3 & 2 & 5 \end{pmatrix}$$

2. Compute each of the following.

   (a) $(1345)(234)$

   (b) $(12)(1253)$

   (c) $(143)(23)(24)$

   (d) $(1423)(34)(56)(1324)$

   (e) $(1254)(13)(25)$

   (f) $(1254)(13)(25)^2$

   (g) $(1254)^{-1}(123)(45)(1254)$

   (h) $(1254)^2(123)(45)$

   (i) $(123)(45)(1254)^{-2}$

   (j) $(1254)^{100}$

   (k) $|(1254)|$

   (l) $|(1254)^2|$

   (m) $(12)^{-1}$

   (n) $(12537)^{-1}$

   (o) $[(12)(34)(12)(47)]^{-1}$

   (p) $[(1235)(467)]^{-1}$

3. Express the following permutations as products of transpositions and identify them as even or odd.

   (a) (14356)

   (b) $(156)(234)$

   (c) $(1426)(142)$

   (d) $(17254)(1423)(154632)$

   (e) (142637)

4. Find $(a_1, a_2, \ldots, a_n)^{-1}$.

5. List all of the subgroups of S_4. Find each of the following sets:

   (a) $\{\sigma \in S_4 : \sigma(1) = 3\}$

   (b) $\{\sigma \in S_4 : \sigma(2) = 2\}$

   (c) $\{\sigma \in S_4 : \sigma(1) = 3 \text{ and } \sigma(2) = 2\}$.

   Are any of these sets subgroups of S_4?

6. Find all of the subgroups in A_4. What is the order of each subgroup?

7. Find all possible orders of elements in S_7 and A_7.

8. Show that A_{10} contains an element of order 15.

9. Does A_8 contain an element of order 26?

10. Find an element of largest order in S_n for $n = 3, \ldots, 10$.

11. What are the possible cycle structures of elements of A_5? What about A_6?

12. Let $\sigma \in S_n$ have order n. Show that for all integers i and j, $\sigma^i = \sigma^j$ if and only if $i \equiv j \pmod{n}$.

13. Let $\sigma = \sigma_1 \cdots \sigma_m \in S_n$ be the product of disjoint cycles. Prove that the order of σ is the least common multiple of the lengths of the cycles $\sigma_1, \ldots, \sigma_m$.

14. Using cycle notation, list the elements in D_5. What are r and s? Write every element as a product of r and s.

15. If the diagonals of a cube are labeled as Figure 5.28 on page 86, to which motion of the cube does the permutation $(12)(34)$ correspond? What about the other permutations of the diagonals?

16. Find the group of rigid motions of a tetrahedron. Show that this is the same group as A_4.

17. Prove that S_n is nonabelian for $n \geq 3$.

18. Show that A_n is nonabelian for $n \geq 4$.

19. Prove that D_n is nonabelian for $n \geq 3$.

20. Let $\sigma \in S_n$ be a cycle. Prove that σ can be written as the product of at most $n - 1$ transpositions.

21. Let $\sigma \in S_n$. If σ is not a cycle, prove that σ can be written as the product of at most $n - 2$ transpositions.

22. If σ can be expressed as an odd number of transpositions, show that any other product of transpositions equaling σ must also be odd.

23. If σ is a cycle of odd length, prove that σ^2 is also a cycle.

24. Show that a 3-cycle is an even permutation.

25. Prove that in A_n with $n \geq 3$, any permutation is a product of cycles of length 3.

26. Prove that any element in S_n can be written as a finite product of the following permutations.

   (a) $(12), (13), \ldots, (1n)$

   (b) $(12), (23), \ldots, (n - 1, n)$

   (c) $(12), (12 \ldots n)$

27. Let G be a group and define a map $\lambda_g : G \to G$ by $\lambda_g(a) = ga$. Prove that λ_g is a permutation of G.

28. Prove that there exist $n!$ permutations of a set containing n elements.

29. Recall that the *center* of a group G is

$$Z(G) = \{g \in G : gx = xg \text{ for all } x \in G\}.$$

Find the center of D_8. What about the center of D_{10}? What is the center of D_n?

30. Let $\tau = (a_1, a_2, \ldots, a_k)$ be a cycle of length k.

(a) Prove that if σ is any permutation, then

$$\sigma \tau \sigma^{-1} = (\sigma(a_1), \sigma(a_2), \ldots, \sigma(a_k))$$

is a cycle of length k.

(b) Let μ be a cycle of length k. Prove that there is a permutation σ such that $\sigma \tau \sigma^{-1} = \mu$.

31. For α and β in S_n, define $\alpha \sim \beta$ if there exists an $\sigma \in S_n$ such that $\sigma \alpha \sigma^{-1} = \beta$. Show that $\sim$ is an equivalence relation on S_n.

32. Let $\sigma \in S_X$. If $\sigma^n(x) = y$ for some $n \in \mathbb{Z}$, we will say that $x \sim y$.

(a) Show that $\sim$ is an equivalence relation on X.

(b) Define the *orbit* of $x \in X$ under $\sigma \in S_X$ to be the set $\mathcal{O}_{x,\sigma} = \{y : x \sim y\}$. Compute the orbits of each element in $\{1, 2, 3, 4, 5\}$ under each of the following elements in S_5:

$$\alpha = (1\,2\,5\,4); \qquad \beta = (1\,2\,3)(4\,5); \qquad \gamma = (1\,3)(2\,5).$$

(c) If $\mathcal{O}_{x,\sigma} \cap \mathcal{O}_{y,\sigma} \neq \varnothing$, prove that $\mathcal{O}_{x,\sigma} = \mathcal{O}_{y,\sigma}$. The orbits under a permutation σ are the equivalence classes corresponding to the equivalence relation $\sim$.

(d) A subgroup H of S_X is *transitive* if for every $x, y \in X$, there exists a $\sigma \in H$ such that $\sigma(x) = y$. Prove that $\langle \sigma \rangle$ is transitive if and only if $\mathcal{O}_{x,\sigma} = X$ for some $x \in X$.

33. Let $\alpha \in S_n$ for $n \geq 3$. If $\alpha\beta = \beta\alpha$ for all $\beta \in S_n$, prove that α must be the identity permutation; hence, the center of S_n is the trivial subgroup.

34. If α is even, prove that α^{-1} is also even. Does a corresponding result hold if α is odd?

35. If $\sigma \in A_n$ and $\tau \in S_n$, show that $\tau^{-1}\sigma\tau \in A_n$.

36. Show that $\alpha^{-1}\beta^{-1}\alpha\beta$ is even for $\alpha, \beta \in S_n$.

37. Let r and s be the elements in D_n described in Theorem 5.21 on page 83.

(a) Show that $srs = r^{-1}$.

(b) Show that $r^k s = s r^{-k}$ in D_n.

(c) Prove that the order of $r^k \in D_n$ is $n/\gcd(k, n)$.

6

Cosets and Lagrange's Theorem

Lagrange's Theorem, one of the most important results in finite group theory, states that the order of a subgroup must divide the order of the group. This theorem provides a powerful tool for analyzing finite groups; it gives us an idea of exactly what type of subgroups we might expect a finite group to possess. Central to understanding Lagranges's Theorem is the notion of a coset.

6.1 Cosets

Let G be a group and H a subgroup of G. Define a *left coset* of H with *representative* $g \in G$ to be the set

$$gH = \{gh : h \in H\}.$$

Right cosets can be defined similarly by

$$Hg = \{hg : h \in H\}.$$

If left and right cosets coincide or if it is clear from the context to which type of coset that we are referring, we will use the word *coset* without specifying left or right.

Example 6.1. Let H be the subgroup of $\mathbb{Z}_6$ consisting of the elements 0 and 3. The cosets are

$$0 + H = 3 + H = \{0, 3\}$$
$$1 + H = 4 + H = \{1, 4\}$$
$$2 + H = 5 + H = \{2, 5\}.$$

We will always write the cosets of subgroups of $\mathbb{Z}$ and $\mathbb{Z}_n$ with the additive notation we have used for cosets here. In a commutative group, left and right cosets are always identical. □

Example 6.2. Let H be the subgroup of S_3 defined by the permutations

$$\{(1), (123), (132)\}.$$

The left cosets of H are

$$(1)H = (123)H = (132)H = \{(1), (123), (132)\}$$
$$(12)H = (13)H = (23)H = \{(12), (13), (23)\}.$$

The right cosets of H are exactly the same as the left cosets:

$$H(1) = H(123) = H(132) = \{(1), (123), (132)\}$$
$$H(12) = H(13) = H(23) = \{(12), (13), (23)\}.$$

It is not always the case that a left coset is the same as a right coset. Let K be the subgroup of S_3 defined by the permutations $\{(1), (12)\}$. Then the left cosets of K are

$$(1)K = (12)K = \{(1), (12)\}$$
$$(13)K = (123)K = \{(13), (123)\}$$
$$(23)K = (132)K = \{(23), (132)\};$$

however, the right cosets of K are

$$K(1) = K(12) = \{(1), (12)\}$$
$$K(13) = K(132) = \{(13), (132)\}$$
$$K(23) = K(123) = \{(23), (123)\}.$$

□

The following lemma is quite useful when dealing with cosets. (We leave its proof as an exercise.)

Lemma 6.3. Let H be a subgroup of a group G and suppose that $g_1, g_2 \in G$. The following conditions are equivalent.

1. $g_1 H = g_2 H$;

2. $H g_1^{-1} = H g_2^{-1}$;

3. $g_1 H \subset g_2 H$;

4. $g_2 \in g_1 H$;

5. $g_1^{-1} g_2 \in H$.

In all of our examples the cosets of a subgroup H partition the larger group G. The following theorem proclaims that this will always be the case.

Theorem 6.4. Let H be a subgroup of a group G. Then the left cosets of H in G partition G. That is, the group G is the disjoint union of the left cosets of H in G.

Proof. Let g_1H and g_2H be two cosets of H in G. We must show that either $g_1H \cap g_2H = \varnothing$ or $g_1H = g_2H$. Suppose that $g_1H \cap g_2H \neq \varnothing$ and $a \in g_1H \cap g_2H$. Then by the definition of a left coset, $a = g_1h_1 = g_2h_2$ for some elements h_1 and h_2 in H. Hence, $g_1 = g_2h_2h_1^{-1}$ or $g_1 \in g_2H$. By Lemma 6.3, $g_1H = g_2H$. ∎

Remark 6.5. There is nothing special in this theorem about left cosets. Right cosets also partition G; the proof of this fact is exactly the same as the proof for left cosets except that all group multiplications are done on the opposite side of H.

Let G be a group and H be a subgroup of G. Define the **index** of H in G to be the number of left cosets of H in G. We will denote the index by $[G : H]$.

Example 6.6. Let $G = \mathbb{Z}_6$ and $H = \{0, 3\}$. Then $[G : H] = 3$. □

Example 6.7. Suppose that $G = S_3$, $H = \{(1), (123), (132)\}$, and $K = \{(1), (12)\}$. Then $[G : H] = 2$ and $[G : K] = 3$. □

Theorem 6.8. Let H be a subgroup of a group G. The number of left cosets of H in G is the same as the number of right cosets of H in G.

Proof. Let $\mathcal{L}_H$ and $\mathcal{R}_H$ denote the set of left and right cosets of H in G, respectively. If we can define a bijective map $\phi : \mathcal{L}_H \to \mathcal{R}_H$, then the theorem will be proved. If $gH \in \mathcal{L}_H$, let $\phi(gH) = Hg^{-1}$. By Lemma 6.3, the map ϕ is well-defined; that is, if $g_1H = g_2H$, then $Hg_1^{-1} = Hg_2^{-1}$. To show that ϕ is one-to-one, suppose that

$$Hg_1^{-1} = \phi(g_1H) = \phi(g_2H) = Hg_2^{-1}.$$

Again by Lemma 6.3, $g_1H = g_2H$. The map ϕ is onto since $\phi(g^{-1}H) = Hg$. ∎

6.2 Lagrange's Theorem

Proposition 6.9. Let H be a subgroup of G with $g \in G$ and define a map $\phi : H \to gH$ by $\phi(h) = gh$. The map ϕ is bijective; hence, the number of elements in H is the same as the number of elements in gH.

Proof. We first show that the map ϕ is one-to-one. Suppose that $\phi(h_1) = \phi(h_2)$ for elements $h_1, h_2 \in H$. We must show that $h_1 = h_2$, but $\phi(h_1) = gh_1$ and $\phi(h_2) = gh_2$. So $gh_1 = gh_2$, and by left cancellation $h_1 = h_2$. To show that ϕ is onto is easy. By definition every element of gH is of the form gh for some $h \in H$ and $\phi(h) = gh$. ∎

Theorem 6.10. Lagrange. Let G be a finite group and let H be a subgroup of G. Then $|G|/|H| = [G : H]$ is the number of distinct left cosets of H in G. In particular, the number of elements in H must divide the number of elements in G.

Proof. The group G is partitioned into $[G : H]$ distinct left cosets. Each left coset has $|H|$ elements; therefore, $|G| = [G : H]|H|$. ∎

Corollary 6.11. Suppose that G is a finite group and $g \in G$. Then the order of g must divide the number of elements in G.

Corollary 6.12. Let $|G| = p$ with p a prime number. Then G is cyclic and any $g \in G$ such that $g \neq e$ is a generator.

Proof. Let g be in G such that $g \neq e$. Then by Corollary 6.11, the order of g must divide the order of the group. Since $|\langle g \rangle| > 1$, it must be p. Hence, g generates G.

∎

Corollary 6.12 suggests that groups of prime order p must somehow look like $\mathbb{Z}_p$.

Corollary 6.13. Let H and K be subgroups of a finite group G such that $G \supset H \supset K$. Then

$$[G : K] = [G : H][H : K].$$

Proof. Observe that

$$[G : K] = \frac{|G|}{|K|} = \frac{|G|}{|H|} \cdot \frac{|H|}{|K|} = [G : H][H : K].$$

∎

Remark 6.14. The converse of Lagrange's Theorem is false. The group A_4 has order 12; however, it can be shown that it does not possess a subgroup of order 6. According to Lagrange's Theorem, subgroups of a group of order 12 can have orders of either 1, 2, 3, 4, or 6. However, we are not guaranteed that subgroups of every possible order exist. To prove that A_4 has no subgroup of order 6, we will assume that it does have such a subgroup H and show that a contradiction must occur. Since A_4 contains eight 3-cycles, we know that H must contain a 3-cycle. We will show that if H contains one 3-cycle, then it must contain more than 6 elements.

Proposition 6.15. The group A_4 has no subgroup of order 6.

Proof. Since $[A_4 : H] = 2$, there are only two cosets of H in A_4. Inasmuch as one of the cosets is H itself, right and left cosets must coincide; therefore, $gH = Hg$ or $gHg^{-1} = H$ for every $g \in A_4$. Since there are eight 3-cycles in A_4, at least one 3-cycle must be in H. Without loss of generality, assume that (123) is in H. Then $(123)^{-1} = (132)$ must also be in H. Since $ghg^{-1} \in H$ for all $g \in A_4$ and all $h \in H$ and

$$(124)(123)(124)^{-1} = (124)(123)(142) = (243)$$
$$(243)(123)(243)^{-1} = (243)(123)(234) = (142)$$

we can conclude that H must have at least seven elements

$$(1), (123), (132), (243), (243)^{-1} = (234), (142), (142)^{-1} = (124).$$

Therefore, A_4 has no subgroup of order 6. ∎

In fact, we can say more about when two cycles have the same length.

Theorem 6.16. Two cycles τ and μ in S_n have the same length if and only if there exists a $\sigma \in S_n$ such that $\mu = \sigma \tau \sigma^{-1}$.

Proof. Suppose that

$$\tau = (a_1, a_2, \ldots, a_k)$$
$$\mu = (b_1, b_2, \ldots, b_k).$$

Define σ to be the permutation

$$\sigma(a_1) = b_1$$
$$\sigma(a_2) = b_2$$
$$\vdots$$
$$\sigma(a_k) = b_k.$$

Then $\mu = \sigma \tau \sigma^{-1}$.

Conversely, suppose that $\tau = (a_1, a_2, \ldots, a_k)$ is a k-cycle and $\sigma \in S_n$. If $\sigma(a_i) = b$ and $\sigma(a_{(i \bmod k)+1}) = b'$, then $\mu(b) = b'$. Hence,

$$\mu = (\sigma(a_1), \sigma(a_2), \ldots, \sigma(a_k)).$$

Since σ is one-to-one and onto, μ is a cycle of the same length as τ. ∎

6.3 Fermat's and Euler's Theorems

The *Euler ϕ-function* is the map $\phi : \mathbb{N} \to \mathbb{N}$ defined by $\phi(n) = 1$ for $n = 1$, and, for $n > 1$, $\phi(n)$ is the number of positive integers m with $1 \leq m < n$ and $\gcd(m, n) = 1$.

From Proposition 3.4 on page 37, we know that the order of $U(n)$, the group of units in $\mathbb{Z}_n$, is $\phi(n)$. For example, $|U(12)| = \phi(12) = 4$ since the numbers that are relatively prime to 12 are 1, 5, 7, and 11. For any prime p, $\phi(p) = p - 1$. We state these results in the following theorem.

Theorem 6.17. Let $U(n)$ be the group of units in $\mathbb{Z}_n$. Then $|U(n)| = \phi(n)$.

The following theorem is an important result in number theory, due to Leonhard Euler.

Theorem 6.18. Euler's Theorem. Let a and n be integers such that $n > 0$ and $\gcd(a, n) = 1$. Then $a^{\phi(n)} \equiv 1 \pmod{n}$.

Proof. By Theorem 6.17 on the preceding page the order of $U(n)$ is $\phi(n)$. Consequently, $a^{\phi(n)} = 1$ for all $a \in U(n)$; or $a^{\phi(n)} - 1$ is divisible by n. Therefore, $a^{\phi(n)} \equiv 1 \pmod{n}$. ■

If we consider the special case of Euler's Theorem in which $n = p$ is prime and recall that $\phi(p) = p - 1$, we obtain the following result, due to Pierre de Fermat.

Theorem 6.19. Fermat's Little Theorem. Let p be any prime number and suppose that $p \nmid a$ (p does not divide a). Then

$$a^{p-1} \equiv 1 \pmod{p}.$$

Furthermore, for any integer b, $b^p \equiv b \pmod{p}$.

Sage.   Sage can create all the subgroups of a group, so long as the group is not too large. It can also create the cosets of a subgroup.

↪   Historical Note   ↩

Joseph-Louis Lagrange (1736–1813), born in Turin, Italy, was of French and Italian descent. His talent for mathematics became apparent at an early age. Leonhard Euler recognized Lagrange's abilities when Lagrange, who was only 19, communicated to Euler some work that he had done in the calculus of variations. That year he was also named a professor at the Royal Artillery School in Turin. At the age of 23 he joined the Berlin Academy. Frederick the Great had written to Lagrange proclaiming that the "greatest king in Europe" should have the "greatest mathematician in Europe" at his court. For 20 years Lagrange held the position vacated by his mentor, Euler. His works include contributions to number theory, group theory, physics and mechanics, the calculus of variations, the theory of equations, and differential equations. Along with Laplace and Lavoisier, Lagrange was one of the people responsible for designing the metric system. During his life Lagrange profoundly influenced the development of mathematics, leaving much to the next generation of mathematicians in the form of examples and new problems to be solved.

6.4 Reading Questions

1. State Lagrange's Theorem in your own words.

2. Determine the left cosets of $\langle 3 \rangle$ in $\mathbb{Z}_9$.

3. The set $\{(), (12)(34), (13)(24), (14)(23)\}$ is a subgroup of S_4. What is its index in S_4?

4. Suppose G is a group of order 29. Describe G.

5. The number $p = 137909$ is prime. Explain how to compute

$$57^{137909} \pmod{137909}$$

without a calculator.

6.5 Exercises

1. Suppose that G is a finite group with an element g of order 5 and an element h of order 7. Why must $|G| \geq 35$?

2. Suppose that G is a finite group with 60 elements. What are the orders of possible subgroups of G?

3. Prove or disprove: Every subgroup of the integers has finite index.

4. Prove or disprove: Every subgroup of the integers has finite order.

5. List the left and right cosets of the subgroups in each of the following.
   (a) $\langle 8 \rangle$ in $\mathbb{Z}_{24}$ (e) A_n in S_n

   (b) $\langle 3 \rangle$ in $U(8)$ (f) D_4 in S_4

   (c) $3\mathbb{Z}$ in $\mathbb{Z}$ (g) $\mathbb{T}$ in $\mathbb{C}^*$

   (d) A_4 in S_4 (h) $H = \{(1), (123), (132)\}$ in S_4

6. Describe the left cosets of $SL_2(\mathbb{R})$ in $GL_2(\mathbb{R})$. What is the index of $SL_2(\mathbb{R})$ in $GL_2(\mathbb{R})$?

7. Verify Euler's Theorem for $n = 15$ and $a = 4$.

8. Use Fermat's Little Theorem to show that if $p = 4n + 3$ is prime, there is no solution to the equation $x^2 \equiv -1 \pmod{p}$.

9. Show that the integers have infinite index in the additive group of rational numbers.

10. Show that the additive group of real numbers has infinite index in the additive group of the complex numbers.

11. Let H be a subgroup of a group G and suppose that $g_1, g_2 \in G$. Prove that the following conditions are equivalent.

  (a) $g_1 H = g_2 H$

  (b) $H g_1^{-1} = H g_2^{-1}$

  (c) $g_1 H \subset g_2 H$

  (d) $g_2 \in g_1 H$

  (e) $g_1^{-1} g_2 \in H$

12. If $ghg^{-1} \in H$ for all $g \in G$ and $h \in H$, show that right cosets are identical to left cosets. That is, show that $gH = Hg$ for all $g \in G$.

13. What fails in the proof of Theorem 6.8 on page 93 if $\phi : \mathcal{L}_H \to \mathcal{R}_H$ is defined by $\phi(gH) = Hg$?

14. Suppose that $g^n = e$. Show that the order of g divides n.

15. The *cycle structure* of a permutation σ is defined as the unordered list of the sizes of the cycles in the cycle decomposition σ. For example, the permutation $\sigma = (1\,2)(3\,4\,5)(7\,8)(9)$ has cycle structure $(2, 3, 2, 1)$ which can also be written as $(1, 2, 2, 3)$.

    Show that any two permutations $\alpha, \beta \in S_n$ have the same cycle structure if and only if there exists a permutation γ such that $\beta = \gamma \alpha \gamma^{-1}$. If $\beta = \gamma \alpha \gamma^{-1}$ for some $\gamma \in S_n$, then α and β are *conjugate*.

16. If $|G| = 2n$, prove that the number of elements of order 2 is odd. Use this result to show that G must contain a subgroup of order 2.

17. Suppose that $[G : H] = 2$. If a and b are not in H, show that $ab \in H$.

18. If $[G : H] = 2$, prove that $gH = Hg$.

19. Let H and K be subgroups of a group G. Prove that $gH \cap gK$ is a coset of $H \cap K$ in G.

20. Let H and K be subgroups of a group G. Define a relation $\sim$ on G by $a \sim b$ if there exists an $h \in H$ and a $k \in K$ such that $hak = b$. Show that this relation is an equivalence relation. The corresponding equivalence classes are called *double cosets*. Compute the double cosets of $H = \{(1), (1\,2\,3), (1\,3\,2)\}$ in A_4.

21. Let G be a cyclic group of order n. Show that there are exactly $\phi(n)$ generators for G.

22. Let $n = p_1^{e_1} p_2^{e_2} \cdots p_k^{e_k}$, where $p_1, p_2, \ldots, p_k$ are distinct primes. Prove that

$$\phi(n) = n\left(1 - \frac{1}{p_1}\right)\left(1 - \frac{1}{p_2}\right)\cdots\left(1 - \frac{1}{p_k}\right).$$

23. Show that $n = \sum_{d|n} \phi(d)$ for all positive integers n.

7
Introduction to Cryptography

*C*ryptography is the study of sending and receiving secret messages. The aim of cryptography is to send messages across a channel so that only the intended recipient of the message can read it. In addition, when a message is received, the recipient usually requires some assurance that the message is authentic; that is, that it has not been sent by someone who is trying to deceive the recipient. Modern cryptography is heavily dependent on abstract algebra and number theory.

The message to be sent is called the *plaintext* message. The disguised message is called the *ciphertext*. The plaintext and the ciphertext are both written in an *alphabet*, consisting of *letters* or *characters*. Characters can include not only the familiar alphabetic characters A, . . ., Z and a, . . ., z but also digits, punctuation marks, and blanks. A *cryptosystem*, or *cipher*, has two parts: *encryption*, the process of transforming a plaintext message to a ciphertext message, and *decryption*, the reverse transformation of changing a ciphertext message into a plaintext message.

There are many different families of cryptosystems, each distinguished by a particular encryption algorithm. Cryptosystems in a specified cryptographic family are distinguished from one another by a parameter to the encryption function called a *key*. A classical cryptosystem has a single key, which must be kept secret, known only to the sender and the receiver of the message. If person A wishes to send secret messages to two different people B and C, and does not wish to have B understand C's messages or vice versa, A must use two separate keys, so one cryptosystem is used for exchanging messages with B, and another is used for exchanging messages with C.

Systems that use two separate keys, one for encoding and another for decoding, are called *public key cryptosystems*. Since knowledge of the encoding key does not allow anyone to guess at the decoding key, the encoding key can be made public. A public key cryptosystem allows A and B to send messages to C using the same encoding key. Anyone is capable of encoding a message to be sent to C, but only C knows how to decode such a message.

7.1 Private Key Cryptography

In *single* or *private key cryptosystems* the same key is used for both encrypting and decrypting messages. To encrypt a plaintext message, we apply to the message some function which is kept secret, say f. This function will yield an encrypted message. Given the encrypted form of the message, we can recover the original message by applying the inverse transformation f^{-1}. The transformation f must be relatively easy to compute, as must f^{-1}; however, f must be extremely difficult to guess from available examples of coded messages.

Example 7.1. One of the first and most famous private key cryptosystems was the shift code used by Julius Caesar. We first digitize the alphabet by letting $A = 00, B = 01, \ldots, Z = 25$. The encoding function will be

$$f(p) = p + 3 \bmod 26;$$

that is, $A \mapsto D, B \mapsto E, \ldots, Z \mapsto C$. The decoding function is then

$$f^{-1}(p) = p - 3 \bmod 26 = p + 23 \bmod 26.$$

Suppose we receive the encoded message DOJHEUD. To decode this message, we first digitize it:

$$3, 14, 9, 7, 4, 20, 3.$$

Next we apply the inverse transformation to get

$$0, 11, 6, 4, 1, 17, 0,$$

or ALGEBRA. Notice here that there is nothing special about either of the numbers 3 or 26. We could have used a larger alphabet or a different shift. □

Cryptanalysis is concerned with deciphering a received or intercepted message. Methods from probability and statistics are great aids in deciphering an intercepted message; for example, the frequency analysis of the characters appearing in the intercepted message often makes its decryption possible.

Example 7.2. Suppose we receive a message that we know was encrypted by using a shift transformation on single letters of the 26-letter alphabet. To find out exactly what the shift transformation was, we must compute b in the equation $f(p) = p + b \bmod 26$. We can do this using frequency analysis. The letter $E = 04$ is the most commonly occurring letter in the English language. Suppose that $S = 18$ is the most commonly occurring letter in the ciphertext. Then we have good reason to suspect that $18 = 4 + b \bmod 26$, or $b = 14$. Therefore, the most likely encrypting function is

$$f(p) = p + 14 \bmod 26.$$

The corresponding decrypting function is

$$f^{-1}(p) = p + 12 \bmod 26.$$

It is now easy to determine whether or not our guess is correct.   □

Simple shift codes are examples of *monoalphabetic cryptosystems*. In these ciphers a character in the enciphered message represents exactly one character in the original message. Such cryptosystems are not very sophisticated and are quite easy to break. In fact, in a simple shift as described in Example 7.1, there are only 26 possible keys. It would be quite easy to try them all rather than to use frequency analysis.

Let us investigate a slightly more sophisticated cryptosystem. Suppose that the encoding function is given by

$$f(p) = ap + b \bmod 26.$$

We first need to find out when a decoding function f^{-1} exists. Such a decoding function exists when we can solve the equation

$$c = ap + b \bmod 26$$

for p. By Proposition 3.4 on page 37, this is possible exactly when a has an inverse or, equivalently, when $\gcd(a, 26) = 1$. In this case

$$f^{-1}(p) = a^{-1}p - a^{-1}b \bmod 26.$$

Such a cryptosystem is called an *affine cryptosystem*.

Example 7.3. Let us consider the affine cryptosystem $f(p) = ap + b \bmod 26$. For this cryptosystem to work we must choose an $a \in \mathbb{Z}_{26}$ that is invertible. This is only possible if $\gcd(a, 26) = 1$. Recognizing this fact, we will let $a = 5$ since $\gcd(5, 26) = 1$. It is easy to see that $a^{-1} = 21$. Therefore, we can take our encryption function to be $f(p) = 5p + 3 \bmod 26$. Thus, ALGEBRA is encoded as $3, 6, 7, 23, 8, 10, 3$, or DGHXIKD. The decryption function will be

$$f^{-1}(p) = 21p - 21 \cdot 3 \bmod 26 = 21p + 15 \bmod 26.$$

                                                                                      □

A cryptosystem would be more secure if a ciphertext letter could represent more than one plaintext letter. To give an example of this type of cryptosystem, called a *polyalphabetic cryptosystem*, we will generalize affine codes by using matrices. The idea works roughly the same as before; however, instead of encrypting one letter at a time we will encrypt pairs of letters. We can store a pair of letters p_1 and p_2 in a vector

$$\mathbf{p} = \begin{pmatrix} p_1 \\ p_2 \end{pmatrix}.$$

Let A be a 2×2 invertible matrix with entries in $\mathbb{Z}_{26}$. We can define an encoding function by

$$f(\mathbf{p}) = A\mathbf{p} + \mathbf{b},$$

where $\mathbf{b}$ is a fixed column vector and matrix operations are performed in $\mathbb{Z}_{26}$. The decoding function must be

$$f^{-1}(\mathbf{p}) = A^{-1}\mathbf{p} - A^{-1}\mathbf{b}.$$

Example 7.4. Suppose that we wish to encode the word HELP. The corresponding digit string is $7, 4, 11, 15$. If

$$A = \begin{pmatrix} 3 & 5 \\ 1 & 2 \end{pmatrix},$$

then

$$A^{-1} = \begin{pmatrix} 2 & 21 \\ 25 & 3 \end{pmatrix}.$$

If $\mathbf{b} = (2,2)^t$, then our message is encrypted as RRGR. The encrypted letter R represents more than one plaintext letter. $\square$

Frequency analysis can still be performed on a polyalphabetic cryptosystem, because we have a good understanding of how pairs of letters appear in the English language. The pair *th* appears quite often; the pair *qz* never appears. To avoid decryption by a third party, we must use a larger matrix than the one we used in Example 7.4.

7.2 Public Key Cryptography

If traditional cryptosystems are used, anyone who knows enough to encode a message will also know enough to decode an intercepted message. In 1976, W. Diffie and M. Hellman proposed public key cryptography, which is based on the observation that the encryption and decryption procedures need not have the same key. This removes the requirement that the encoding key be kept secret. The encoding function f must be relatively easy to compute, but f^{-1} must be extremely difficult to compute without some additional information, so that someone who knows only the encrypting key cannot find the decrypting key without prohibitive computation. It is interesting to note that to date, no system has been proposed that has been proven to be "one-way;" that is, for any existing public key cryptosystem, it has never been shown to be computationally prohibitive to decode messages with only knowledge of the encoding key.

The RSA Cryptosystem

The RSA cryptosystem introduced by R. Rivest, A. Shamir, and L. Adleman in 1978, is based on the difficulty of factoring large numbers. Though it is not a difficult task to find two large random primes and multiply them together, factoring a 150-digit number that is the product of two large primes would take 100 million computers operating at 10 million instructions per second about 50 million years under the fastest algorithms available in the early 1990s. Although the algorithms have improved, factoring a number that is a product of two large primes is still computationally prohibitive.

The RSA cryptosystem works as follows. Suppose that we choose two random 150-digit prime numbers p and q. Next, we compute the product $n = pq$ and also compute $\phi(n) = m = (p-1)(q-1)$, where ϕ is the Euler ϕ-function. Now we start choosing random integers E until we find one that is relatively prime to m; that is, we choose E such that $\gcd(E, m) = 1$. Using the Euclidean algorithm, we can find a number D such that $DE \equiv 1 \pmod{m}$. The numbers n and E are now made public.

Suppose now that person B (Bob) wishes to send person A (Alice) a message over a public line. Since E and n are known to everyone, anyone can encode messages. Bob first digitizes the message according to some scheme, say A = $00, B = 02, \ldots, Z = 25$. If necessary, he will break the message into pieces such that each piece is a positive integer less than n. Suppose x is one of the pieces. Bob forms the number $y = x^E$ mod n and sends y to Alice. For Alice to recover x, she need only compute $x = y^D$ mod n. Only Alice knows D.

Example 7.5. Before exploring the theory behind the RSA cryptosystem or attempting to use large integers, we will use some small integers just to see that the system does indeed work. Suppose that we wish to send some message, which when digitized is 25. Let $p = 23$ and $q = 29$. Then

$$n = pq = 667$$

and

$$\phi(n) = m = (p-1)(q-1) = 616.$$

We can let $E = 487$, since $\gcd(616, 487) = 1$. The encoded message is computed to be

$$25^{487} \bmod 667 = 169.$$

This computation can be reasonably done by using the method of repeated squares as described in Chapter 4. Using the Euclidean algorithm, we determine that $191E = 1 + 151m$; therefore, the decrypting key is $(n, D) = (667, 191)$. We can recover the original message by calculating

$$169^{191} \bmod 667 = 25.$$

$\square$

Now let us examine why the RSA cryptosystem works. We know that $DE \equiv 1 \pmod{m}$; hence, there exists a k such that

$$DE = km + 1 = k\phi(n) + 1.$$

There are two cases to consider. In the first case assume that $\gcd(x, n) = 1$. Then by Theorem 6.18 on page 96,

$$y^D = (x^E)^D = x^{DE} = x^{km+1} = (x^{\phi(n)})^k x = (1)^k x = x \bmod n.$$

So we see that Alice recovers the original message x when she computes $y^D \bmod n$.

For the other case, assume that $\gcd(x, n) \neq 1$. Since $n = pq$ and $x < n$, we know x is a multiple of p or a multiple of q, but not both. We will describe the first possibility only, since the second is entirely similar. There is then an integer r, with $r < q$ and $x = rp$. Note that we have $\gcd(x, q) = 1$ and that $m = \phi(n) = (p-1)(q-1) = \phi(p)\phi(q)$. Then, using Theorem 6.18 on page 96, but now mod q,

$$x^{km} = x^{k\phi(p)\phi(q)} = (x^{\phi(q)})^{k\phi(p)} = (1)^{k\phi(p)} = 1 \bmod q.$$

So there is an integer t such that $x^{km} = 1 + tq$. Thus, Alice also recovers the message in this case,

$$y^D = x^{km+1} = x^{km}x = (1 + tq)x = x + tq(rp) = x + trn = x \bmod n.$$

We can now ask how one would go about breaking the RSA cryptosystem. To find D given n and E, we simply need to factor n and solve for D by using the Euclidean algorithm. If we had known that $667 = 23 \cdot 29$ in Example 7.5 on the previous page, we could have recovered D.

Message Verification

There is a problem of message verification in public key cryptosystems. Since the encoding key is public knowledge, anyone has the ability to send an encoded message. If Alice receives a message from Bob, she would like to be able to verify that it was Bob who actually sent the message. Suppose that Bob's encrypting key is (n', E') and his decrypting key is (n', D'). Also, suppose that Alice's encrypting key is (n, E) and her decrypting key is (n, D). Since encryption keys are public information, they can exchange coded messages at their convenience. Bob wishes to assure Alice that the message he is sending is authentic. Before Bob sends the message x to Alice, he decrypts x with his own key:

$$x' = x^{D'} \bmod n'.$$

Anyone can change x' back to x just by encryption, but only Bob has the ability to form x'. Now Bob encrypts x' with Alice's encryption key to form

$$y' = x'^{E} \bmod n,$$

a message that only Alice can decode. Alice decodes the message and then encodes the result with Bob's key to read the original message, a message that could have only been sent by Bob.

⤙ Historical Note ⤚

Encrypting secret messages goes as far back as ancient Greece and Rome. As we know, Julius Caesar used a simple shift code to send and receive messages. However, the formal study of encoding and decoding messages probably began with the Arabs in the 1400s. In the fifteenth and six-teenth centuries mathematicians such as Alberti and Viete discovered that monoalphabetic cryptosystems offered no real security. In the 1800s, F. W. Kasiski established methods for breaking ciphers in which a ciphertext letter can represent more than one plaintext letter, if the same key was used several times. This discovery led to the use of cryptosystems with keys that were used only a single time. Cryptography was placed on firm mathematical foundations by such people as W. Friedman and L. Hill in the early part of the twentieth century.

The period after World War I saw the development of special-purpose machines for encrypting and decrypting messages, and mathematicians were very active in cryptography during World War II. Efforts to penetrate the cryptosystems of the Axis nations were organized in England and in the United States by such notable mathematicians as Alan Turing and A. A. Albert. The Allies gained a tremendous advantage in World War II by breaking the ciphers produced by the German Enigma machine and the Japanese Purple ciphers.

By the 1970s, interest in commercial cryptography had begun to take hold. There was a growing need to protect banking transactions, computer data, and electronic mail. In the early 1970s, IBM developed and implemented LUZIFER, the forerunner of the National Bureau of Standards' Data Encryption Standard (DES).

The concept of a public key cryptosystem, due to Diffie and Hellman, is very recent (1976). It was further developed by Rivest, Shamir, and Adleman with the RSA cryptosystem (1978). It is not known how secure any of these systems are. The trapdoor knapsack cryptosystem, developed by Merkle and Hellman, has been broken. It is still an open question whether or not the RSA system can be broken. In 1991, RSA Laboratories published a list of semiprimes (numbers with exactly two prime factors) with a cash prize for whoever was able to provide a factorization (http://www.emc.com/emc-plus/rsa-labs/historical/the-rsa-challenge-numbers.htm). Although the challenge ended in 2007, many of these numbers have not yet been factored.

There been a great deal of controversy about research in cryptography and cryptography itself. In 1929, when Henry Stimson, Secretary of State under Herbert Hoover, dismissed the Black Chamber (the State Department's cryptography division) on the ethical grounds that "gentlemen do not read each other's mail." During the last two decades of the twentieth century, the National Security Agency wanted to keep information about cryptography secret, whereas the academic community fought for the right to publish basic research. Currently, research in mathematical cryptography and computational number theory is very active, and mathematicians are free to publish their results in these areas.

Sage. Sage's early development featured powerful routines for number theory, and later included significant support for algebraic structures and other areas of discrete mathematics. So it is a natural tool for the study of cryptology, including topics like RSA, elliptic curve cryptography, and AES (Advanced Encryption Standard).

7.3 Reading Questions

1. Use the `euler_phi()` function in Sage to compute $\phi(893\,456\,123)$.

2. Use the `power_mod()` function in Sage to compute $7^{324} \pmod{895}$.

3. Explain the mathematical basis for saying: encrypting a message using an RSA public key is very simple computationally, while decrypting a communication without the private key is very hard computationally.

4. Explain how in RSA message encoding differs from message verification.

5. Explain how one could be justified in saying that Diffie and Hellman's proposal in 1976 was "revolutionary."

7.4 Exercises

1. Encode IXLOVEXMATH using the cryptosystem in Example 7.1 on page 100.

2. Decode ZLOOA WKLVA EHARQ WKHA ILQDO, which was encoded using the cryptosystem in Example 7.1 on page 100.

3. Assuming that monoalphabetic code was used to encode the following secret message, what was the original message?

    APHUO EGEHP PEXOV FKEUH CKVUE CHKVE APHUO
    EGEHU EXOVL EXDKT VGEFT EHFKE UHCKF TZEXO
    VEZDT TVKUE XOVKV ENOHK ZFTEH TEHKQ LEROF
    PVEHP PEXOV ERYKP GERYT GVKEG XDRTE RGAGA

    What is the significance of this message in the history of cryptography?

4. What is the total number of possible monoalphabetic cryptosystems? How secure are such cryptosystems?

5. Prove that a 2×2 matrix A with entries in $\mathbb{Z}_{26}$ is invertible if and only if $\gcd(\det(A), 26) = 1$.

6. Given the matrix

$$A = \begin{pmatrix} 3 & 4 \\ 2 & 3 \end{pmatrix},$$

    use the encryption function $f(\mathbf{p}) = A\mathbf{p} + \mathbf{b}$ to encode the message CRYPTOLOGY, where $\mathbf{b} = (2, 5)^t$. What is the decoding function?

7. Encrypt each of the following RSA messages x so that x is divided into blocks of integers of length 2; that is, if $x = 142528$, encode 14, 25, and 28 separately.

    (a) $n = 3551, E = 629, x = 31$

    (b) $n = 2257, E = 47, x = 23$

    (c) $n = 120979, E = 13251, x = 142371$

    (d) $n = 45629, E = 781, x = 231561$

8. Compute the decoding key D for each of the encoding keys in Exercise 7.4.7.

9. Decrypt each of the following RSA messages y.

    (a) $n = 3551, D = 1997, y = 2791$

    (b) $n = 5893, D = 81, y = 34$

    (c) $n = 120979, D = 27331, y = 112135$

    (d) $n = 79403, D = 671, y = 129381$

10. For each of the following encryption keys (n, E) in the RSA cryptosystem, compute D.

    (a) $(n, E) = (451, 231)$

    (b) $(n, E) = (3053, 1921)$

    (c) $(n, E) = (37986733, 12371)$

    (d) $(n, E) = (16394854313, 34578451)$

11. Encrypted messages are often divided into blocks of n letters. A message such as THE WORLD WONDERS WHY might be encrypted as JIW OCFRJ LPOEVYQ IOC but sent as JIW OCF RJL POE VYQ IOC. What are the advantages of using blocks of n letters?

12. Find integers n, E, and X such that $X^E \equiv X \pmod{n}$. Is this a potential problem in the RSA cryptosystem?

13. Every person in the class should construct an RSA cryptosystem using primes that are 10 to 15 digits long. Hand in (n, E) and an encoded message. Keep D secret. See if you can break one another's codes.

7.5 Additional Exercises: Primality and Factoring

In the RSA cryptosystem it is important to be able to find large prime numbers easily. Also, this cryptosystem is not secure if we can factor a composite number that is the product of two large primes. The solutions to both of these problems are quite easy. To find out if a number n is prime or to factor n, we can use trial division. We simply divide n by $d = 2, 3, \ldots, \sqrt{n}$. Either a factorization will be obtained, or n is prime if no d divides n. The problem is that such a computation is prohibitively time-consuming if n is very large.

1. A better algorithm for factoring odd positive integers is *Fermat's factorization algorithm*.

    (a) Let $n = ab$ be an odd composite number. Prove that n can be written as the difference of two perfect squares:

    $$n = x^2 - y^2 = (x - y)(x + y).$$

    Consequently, a positive odd integer can be factored exactly when we can find integers x and y such that $n = x^2 - y^2$.

    (b) Write a program to implement the following factorization algorithm based on the observation in part (a). The expression ceiling(sqrt(n)) means the smallest integer greater than or equal to the square root of n. Write another program to do factorization using trial division and compare the speed of the two algorithms. Which algorithm is faster and why?

```
x := ceiling(sqrt(n))
y := 1

1 : while x^2 - y^2 > n do
      y := y + 1

if x^2 - y^2 < n then
    x := x + 1
    y := 1
    goto 1
else if x^2 - y^2 = 0 then
    a := x - y
    b := x + y
    write n = a * b
```

2. **Primality Testing.** Recall Fermat's Little Theorem from Chapter 6. Let p be prime with $\gcd(a, p) = 1$. Then $a^{p-1} \equiv 1 \pmod{p}$. We can use Fermat's Little Theorem as a screening test for primes. For example, 15 cannot be prime since

$$2^{15-1} \equiv 2^{14} \equiv 4 \pmod{15}.$$

However, 17 is a potential prime since

$$2^{17-1} \equiv 2^{16} \equiv 1 \pmod{17}.$$

We say that an odd composite number n is a *pseudoprime* if

$$2^{n-1} \equiv 1 \pmod{n}.$$

Which of the following numbers are primes and which are pseudoprimes?

   (a) 342   (c) 601   (e) 771

   (b) 811   (d) 561   (f) 631

3. Let n be an odd composite number and b be a positive integer such that $\gcd(b, n) = 1$. If $b^{n-1} \equiv 1 \pmod{n}$, then n is a *pseudoprime base b*. Show that 341 is a pseudoprime base 2 but not a pseudoprime base 3.

4. Write a program to determine all primes less than 2000 using trial division. Write a second program that will determine all numbers less than 2000 that are either primes or pseudoprimes. Compare the speed of the two programs. How many pseudoprimes are there below 2000?

    There exist composite numbers that are pseudoprimes for all bases to which they are relatively prime. These numbers are called *Carmichael numbers*. The first Carmichael number is $561 = 3 \cdot 11 \cdot 17$. In 1992, Alford, Granville, and Pomerance proved that there are an infinite number of Carmichael numbers [4]. However, Carmichael numbers are very rare. There are only 2163 Carmichael numbers less than 25×10^9. For more sophisticated primality tests, see [1], [6], or [7].

7.6 References and Suggested Readings

[1] Bressoud, D. M. *Factorization and Primality Testing.* Springer-Verlag, New York, 1989.

[2] Diffie, W. and Hellman, M. E. "New Directions in Cryptography," *IEEE Trans. Inform. Theory* 22 (1976), 644–54.

[3] Gardner, M. "Mathematical games: A new kind of cipher that would take millions of years to break," *Scientific American* 237 (1977), 120–24.

[4] Granville, A. "Primality Testing and Carmichael Numbers," *Notices of the American Mathematical Society* 39 (1992), 696–700.

[5] Hellman, M. E. "The Mathematics of Public Key Cryptography," *Scientific American* 241 (1979), 130–39.

[6] Koblitz, N. *A Course in Number Theory and Cryptography.* 2nd ed. Springer, New York, 1994.

[7]   Pomerance, C., ed. "Cryptology and Computational Number Theory", *Proceedings of Symposia in Applied Mathematics* **42** (1990) American Mathematical Society, Providence, RI.

[8]   Rivest, R. L., Shamir, A., and Adleman, L., "A Method for Obtaining Signatures and Public-key Cryptosystems," *Comm.* ACM **21** (1978), 120–26.

8

Algebraic Coding Theory

*C*oding theory is an application of algebra that has become increasingly important over the last several decades. When we transmit data, we are concerned about sending a message over a channel that could be affected by "noise." We wish to be able to encode and decode the information in a manner that will allow the detection, and possibly the correction, of errors caused by noise. This situation arises in many areas of communications, including radio, telephone, television, computer communications, and digital media technology. Probability, combinatorics, group theory, linear algebra, and polynomial rings over finite fields all play important roles in coding theory.

8.1 Error-Detecting and Correcting Codes

Let us examine a simple model of a communications system for transmitting and receiving coded messages (Figure 8.1).

Uncoded messages may be composed of letters or characters, but typically they consist of binary m-tuples. These messages are encoded into codewords, consisting of binary n-tuples, by a device called an *encoder*. The message is transmitted and then decoded. We will consider the occurrence of errors during transmission. An *error* occurs if there is a change in one or more bits in the codeword. A *decoding scheme* is a method that either converts an arbitrarily received n-tuple into a meaningful decoded message or gives an error message for that n-tuple. If the received message is a codeword (one of the special n-tuples allowed to be transmitted), then the decoded message must be the unique message that was encoded into the codeword. For received non-codewords, the decoding scheme will give an error indication, or, if we are more clever, will actually try to correct the error and reconstruct the original message. Our goal is to transmit error-free messages as cheaply and quickly as possible.

m-digit message

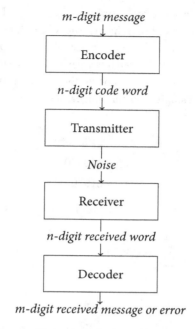

Figure 8.1. Encoding and decoding messages

Example 8.2. One possible coding scheme would be to send a message several times and to compare the received copies with one another. Suppose that the message to be encoded is a binary n-tuple $(x_1, x_2, \ldots, x_n)$. The message is encoded into a binary $3n$-tuple by simply repeating the message three times:

$$(x_1, x_2, \ldots, x_n) \mapsto (x_1, x_2, \ldots, x_n, x_1, x_2, \ldots, x_n, x_1, x_2, \ldots, x_n).$$

To decode the message, we choose as the ith digit the one that appears in the ith place in at least two of the three transmissions. For example, if the original message is (0110), then the transmitted message will be $(0110\ 0110\ 0110)$. If there is a transmission error in the fifth digit, then the received codeword will be $(0110\ 1110\ 0110)$, which will be correctly decoded as (0110).[2] This triple-repetition method will automatically detect and correct all single errors, but it is slow and inefficient: to send a message consisting of n bits, $2n$ extra bits are required, and we can only detect and correct single errors. We will see that it is possible to find an encoding scheme that will encode a message of n bits into m bits with m much smaller than $3n$. □

---

[2]We will adopt the convention that bits are numbered left to right in binary n-tuples.

Example 8.3.   *Even parity*, a commonly used coding scheme, is much more efficient than the simple repetition scheme. The ASCII (American Standard Code for Information Interchange) coding system uses binary 8-tuples, yielding $2^8 = 256$ possible 8-tuples. However, only seven bits are needed since there are only $2^7 = 128$ ASCII characters. What can or should be done with the extra bit? Using the full eight bits, we can detect single transmission errors. For example, the ASCII codes for A, B, and C are

$$A = 65_{10} = 01000001_2,$$
$$B = 66_{10} = 01000010_2,$$
$$C = 67_{10} = 01000011_2.$$

Notice that the leftmost bit is always set to 0; that is, the 128 ASCII characters have codes

$$00000000_2 = 0_{10},$$
$$\vdots$$
$$01111111_2 = 127_{10}.$$

The bit can be used for error checking on the other seven bits. It is set to either 0 or 1 so that the total number of 1 bits in the representation of a character is even. Using even parity, the codes for A, B, and C now become

$$A = 01000001_2,$$
$$B = 01000010_2,$$
$$C = 11000011_2.$$

Suppose an A is sent and a transmission error in the sixth bit is caused by noise over the communication channel so that $(0100\ 0101)$ is received. We know an error has occurred since the received word has an odd number of 1s, and we can now request that the codeword be transmitted again. When used for error checking, the leftmost bit is called a *parity check bit*.

By far the most common error-detecting codes used in computers are based on the addition of a parity bit. Typically, a computer stores information in m-tuples called *words*. Common word lengths are 8, 16, and 32 bits. One bit in the word is set aside as the parity check bit, and is not used to store information. This bit is set to either 0 or 1, depending on the number of 1s in the word.

Adding a parity check bit allows the detection of all single errors because changing a single bit either increases or decreases the number of 1s by one, and in either case the parity has been changed from even to odd, so the new word is not a codeword. (We could also construct an error detection scheme based on *odd parity*; that is, we could set the parity check bit so that a codeword always has an odd number of 1s.)   □

The even parity system is easy to implement, but has two drawbacks. First, multiple errors are not detectable. Suppose an A is sent and the first and seventh bits are changed from 0 to 1. The received word is a codeword, but will be decoded into a C instead of an A. Second, we do not have the ability to correct errors. If the 8-tuple (1001 1000) is received, we know that an error has occurred, but we have no idea which bit has been changed. We will now investigate a coding scheme that will not only allow us to detect transmission errors but will actually correct the errors.

Example 8.4. Suppose that our original message is either a 0 or a 1, and that 0 encodes to (000) and 1 encodes to (111). If only a single error occurs during transmission, we can detect and correct the error. For example, if a (101) is received, then the second bit must have been changed from a 1 to a 0. The originally transmitted codeword must have been (111). This method will detect and correct all single errors.

Table 8.5. A repetition code

Transmitted	Received Word							
Codeword	000	001	010	011	100	101	110	111
000	0	1	1	2	1	2	2	3
111	3	2	2	1	2	1	1	0

In Table 8.5, we present all possible words that might be received for the transmitted codewords (000) and (111). Table 8.5 also shows the number of bits by which each received 3-tuple differs from each original codeword.   □

Maximum-Likelihood Decoding

The coding scheme presented in Example 8.4 is not a complete solution to the problem because it does not account for the possibility of multiple errors. For example, either a (000) or a (111) could be sent and a (001) received. We have no means of deciding from the received word whether there was a single error in the third bit or two errors, one in the first bit and one in the second. No matter what coding scheme is used, an incorrect message could be received. We could transmit a (000), have errors in all three bits, and receive the codeword (111). It is important to make explicit assumptions about the likelihood and distribution of transmission errors so that, in a particular application, it will be known whether a given error detection scheme is appropriate. We will assume that transmission errors are rare, and, that when they do occur, they occur independently in each bit; that is, if p is the probability of an error in one bit and q is the probability of an error in a different bit, then the probability of errors occurring in both of these bits at the same time is pq. We will also assume that a received n-tuple is decoded into a codeword that is closest to it; that is, we assume that the receiver uses *maximum-likelihood decoding*. [3]

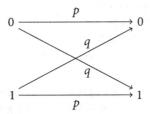

Figure 8.6. Binary symmetric channel

A *binary symmetric channel* is a model that consists of a transmitter capable of sending a binary signal, either a 0 or a 1, together with a receiver. Let p be the probability that the signal is correctly received. Then $q = 1 - p$ is the probability of an incorrect reception. If a 1 is sent, then the probability that a 1 is received is p and the probability that a 0 is received is q (Figure 8.6). The probability that no errors occur during the transmission of a binary codeword of length n is p^n. For example, if $p = 0.999$ and a message consisting of 10,000 bits is sent, then the probability of a perfect transmission is

$$(0.999)^{10,000} \approx 0.00005.$$

Theorem 8.7.   If a binary n-tuple $(x_1, \ldots, x_n)$ is transmitted across a binary symmetric channel with probability p that no error will occur in each coordinate, then the probability that there are errors in exactly k coordinates is

$$\binom{n}{k} q^k p^{n-k}.$$

Proof. Fix k different coordinates. We first compute the probability that an error has occurred in this fixed set of coordinates. The probability of an error occurring in a particular one of these k coordinates is q; the probability that an error will not occur in any of the remaining $n - k$ coordinates is p. The probability of each of these n independent events is $q^k p^{n-k}$. The number of possible error patterns

---

[3]This section requires a knowledge of probability, but can be skipped without loss of continuity.

with exactly k errors occurring is equal to

$$\binom{n}{k} = \frac{n!}{k!(n-k)!},$$

the number of combinations of n things taken k at a time. Each of these error patterns has probability $q^k p^{n-k}$ of occurring; hence, the probability of all of these error patterns is

$$\binom{n}{k} q^k p^{n-k}.$$

■

Example 8.8. Suppose that $p = 0.995$ and a 500-bit message is sent. The probability that the message was sent error-free is

$$p^n = (0.995)^{500} \approx 0.082.$$

The probability of exactly one error occurring is

$$\binom{n}{1} q p^{n-1} = 500(0.005)(0.995)^{499} \approx 0.204.$$

The probability of exactly two errors is

$$\binom{n}{2} q^2 p^{n-2} = \frac{500 \cdot 499}{2}(0.005)^2(0.995)^{498} \approx 0.257.$$

The probability of more than two errors is approximately

$$1 - 0.082 - 0.204 - 0.257 = 0.457.$$

□

Block Codes

If we are to develop efficient error-detecting and error-correcting codes, we will need more sophisticated mathematical tools. Group theory will allow faster methods of encoding and decoding messages. A code is an (n, m)-*block code* if the information that is to be coded can be divided into blocks of m binary digits, each of which can be encoded into n binary digits. More specifically, an (n, m)-block code consists of an *encoding function*

$$E : \mathbb{Z}_2^m \rightarrow \mathbb{Z}_2^n$$

and a *decoding function*

$$D : \mathbb{Z}_2^n \rightarrow \mathbb{Z}_2^m.$$

A *codeword* is any element in the image of E. We also require that E be one-to-one so that two information blocks will not be encoded into the same codeword. If our code is to be error-correcting, then D must be onto.

Example 8.9. The even-parity coding system developed to detect single errors in ASCII characters is an $(8,7)$-block code. The encoding function is

$$E(x_7, x_6, \ldots, x_1) = (x_8, x_7, \ldots, x_1),$$

where $x_8 = x_7 + x_6 + \cdots + x_1$ with addition in $\mathbb{Z}_2$.   □

Let $\mathbf{x} = (x_1, \ldots, x_n)$ and $\mathbf{y} = (y_1, \ldots, y_n)$ be binary n-tuples. The *Hamming distance* or *distance*, $d(\mathbf{x}, \mathbf{y})$, between $\mathbf{x}$ and $\mathbf{y}$ is the number of bits in which $\mathbf{x}$ and $\mathbf{y}$ differ. The distance between two codewords is the minimum number of transmission errors required to change one codeword into the other. The *minimum distance* for a code, $d_{\min}$, is the minimum of all distances $d(\mathbf{x}, \mathbf{y})$, where $\mathbf{x}$ and $\mathbf{y}$ are distinct codewords. The *weight*, $w(\mathbf{x})$, of a binary codeword $\mathbf{x}$ is the number of 1s in $\mathbf{x}$. Clearly, $w(\mathbf{x}) = d(\mathbf{x}, \mathbf{0})$, where $\mathbf{0} = (00\cdots0)$.

Example 8.10. Let $\mathbf{x} = (10101)$, $\mathbf{y} = (11010)$, and $\mathbf{z} = (00011)$ be all of the codewords in some code C. Then we have the following Hamming distances:

$$d(\mathbf{x}, \mathbf{y}) = 4, \qquad d(\mathbf{x}, \mathbf{z}) = 3, \qquad d(\mathbf{y}, \mathbf{z}) = 3.$$

The minimum distance for this code is 3. We also have the following weights:

$$w(\mathbf{x}) = 3, \qquad w(\mathbf{y}) = 3, \qquad w(\mathbf{z}) = 2.$$

□

The following proposition lists some basic properties about the weight of a codeword and the distance between two codewords. The proof is left as an exercise.

Proposition 8.11. Let $\mathbf{x}$, $\mathbf{y}$, and $\mathbf{z}$ be binary n-tuples. Then

1. $w(\mathbf{x}) = d(\mathbf{x}, \mathbf{0})$;

2. $d(\mathbf{x}, \mathbf{y}) \geq 0$;

3. $d(\mathbf{x}, \mathbf{y}) = 0$ exactly when $\mathbf{x} = \mathbf{y}$;

4. $d(\mathbf{x}, \mathbf{y}) = d(\mathbf{y}, \mathbf{x})$;

5. $d(\mathbf{x}, \mathbf{y}) \leq d(\mathbf{x}, \mathbf{z}) + d(\mathbf{z}, \mathbf{y})$.

The weights in a particular code are usually much easier to compute than the Hamming distances between all codewords in the code. If a code is set up carefully, we can use this fact to our advantage.

Suppose that $\mathbf{x} = (1101)$ and $\mathbf{y} = (1100)$ are codewords in some code. If we transmit (1101) and an error occurs in the rightmost bit, then (1100) will be received. Since (1100) is a codeword, the decoder will decode (1100) as the transmitted message. This code is clearly not very appropriate for error detection. The problem is that $d(\mathbf{x}, \mathbf{y}) = 1$. If $\mathbf{x} = (1100)$ and $\mathbf{y} = (1010)$ are codewords, then $d(\mathbf{x}, \mathbf{y}) = 2$. If $\mathbf{x}$ is transmitted and a single error occurs, then $\mathbf{y}$ can never be received. Table 8.12 gives the distances between all 4-bit codewords in which the first three bits carry information and the fourth is an even parity check bit. We can see that the minimum distance here is 2; hence, the code is suitable as a single error-detecting code.

Table 8.12. Distances between 4-bit codewords

	0000	0011	0101	0110	1001	1010	1100	1111
0000	0	2	2	2	2	2	2	4
0011	2	0	2	2	2	2	4	2
0101	2	2	0	2	2	4	2	2
0110	2	2	2	0	4	2	2	2
1001	2	2	2	4	0	2	2	2
1010	2	2	4	2	2	0	2	2
1100	2	4	2	2	2	2	0	2
1111	4	2	2	2	2	2	2	0

To determine exactly what the error-detecting and error-correcting capabilities for a code are, we need to analyze the minimum distance for the code. Let $\mathbf{x}$ and $\mathbf{y}$ be codewords. If $d(\mathbf{x}, \mathbf{y}) = 1$ and an error occurs where $\mathbf{x}$ and $\mathbf{y}$ differ, then $\mathbf{x}$ is changed to $\mathbf{y}$. The received codeword is $\mathbf{y}$ and no error message is given. Now suppose $d(\mathbf{x}, \mathbf{y}) = 2$. Then a single error cannot change $\mathbf{x}$ to $\mathbf{y}$. Therefore, if $d_{\min} = 2$, we have the ability to detect single errors. However, suppose that $d(\mathbf{x}, \mathbf{y}) = 2$, $\mathbf{y}$ is sent, and a noncodeword $\mathbf{z}$ is received such that

$$d(\mathbf{x}, \mathbf{z}) = d(\mathbf{y}, \mathbf{z}) = 1.$$

Then the decoder cannot decide between $\mathbf{x}$ and $\mathbf{y}$. Even though we are aware that an error has occurred, we do not know what the error is.

Suppose $d_{\min} \geq 3$. Then the maximum-likelihood decoding scheme corrects all single errors. Starting with a codeword $\mathbf{x}$, an error in the transmission of a single bit gives $\mathbf{y}$ with $d(\mathbf{x}, \mathbf{y}) = 1$, but $d(\mathbf{z}, \mathbf{y}) \geq 2$ for any other codeword $\mathbf{z} \neq \mathbf{x}$. If we do not require the correction of errors, then we can detect multiple errors when a code has a minimum distance that is greater than or equal to 3.

Theorem 8.13. Let C be a code with $d_{\min} = 2n + 1$. Then C can correct any n or fewer errors. Furthermore, any $2n$ or fewer errors can be detected in C.

Proof. Suppose that a codeword $\mathbf{x}$ is sent and the word $\mathbf{y}$ is received with at most n errors. Then $d(\mathbf{x}, \mathbf{y}) \leq n$. If $\mathbf{z}$ is any codeword other than $\mathbf{x}$, then

$$2n + 1 \leq d(\mathbf{x}, \mathbf{z}) \leq d(\mathbf{x}, \mathbf{y}) + d(\mathbf{y}, \mathbf{z}) \leq n + d(\mathbf{y}, \mathbf{z}).$$

Hence, $d(\mathbf{y}, \mathbf{z}) \geq n + 1$ and $\mathbf{y}$ will be correctly decoded as $\mathbf{x}$. Now suppose that $\mathbf{x}$ is transmitted and $\mathbf{y}$ is received and that at least one error has occurred, but not more than $2n$ errors. Then $1 \leq d(\mathbf{x}, \mathbf{y}) \leq 2n$. Since the minimum distance between codewords is $2n + 1$, $\mathbf{y}$ cannot be a codeword. Consequently, the code can detect between 1 and $2n$ errors. ∎

Example 8.14. In Table 8.15, the codewords $\mathbf{c}_1 = (00000)$, $\mathbf{c}_2 = (00111)$, $\mathbf{c}_3 = (11100)$, and $\mathbf{c}_4 = (11011)$ determine a single error-correcting code. □

Table 8.15. Hamming distances for an error-correcting code

	00000	00111	11100	11011
00000	0	3	3	4
00111	3	0	4	3
11100	3	4	0	3
11011	4	3	3	0

⟿   Historical Note   ⟾

Modern coding theory began in 1948 with C. Shannon's paper, "A Mathematical Theory of Information" [7]. This paper offered an example of an algebraic code, and Shannon's Theorem proclaimed exactly how good codes could be expected to be. Richard Hamming began working with linear codes at Bell Labs in the late 1940s and early 1950s after becoming frustrated because the programs that he was running could not recover from simple errors generated by noise. Coding theory has grown tremendously in the past several decades. *The Theory of Error-Correcting Codes*, by MacWilliams and Sloane [5], published in 1977, already contained over 1500 references. Linear codes (Reed-Muller $(32, 6)$-block codes) were used on NASA's Mariner space probes. More recent space probes such as Voyager have used what are called convolution codes. Currently, very active research is being done with Goppa codes, which are heavily dependent on algebraic geometry.

8.2 Linear Codes

To gain more knowledge of a particular code and develop more efficient techniques of encoding, decoding, and error detection, we need to add additional structure to our codes. One way to accomplish this is to require that the code also be a group. A *group code* is a code that is also a subgroup of $\mathbb{Z}_2^n$.

To check that a code is a group code, we need only verify one thing. If we add any two elements in the code, the result must be an n-tuple that is again in the code. It is not necessary to check that the inverse of the n-tuple is in the code, since every codeword is its own inverse, nor is it necessary to check that 0 is a codeword. For instance,

$$(11000101) + (11000101) = (00000000).$$

Example 8.16. Suppose that we have a code that consists of the following 7-tuples:

(0000000)	(0001111)	(0010101)	(0011010)
(0100110)	(0101001)	(0110011)	(0111100)
(1000011)	(1001100)	(1010110)	(1011001)
(1100101)	(1101010)	(1110000)	(1111111).

It is a straightforward though tedious task to verify that this code is also a subgroup of $\mathbb{Z}_2^7$ and, therefore, a group code. This code is a single error-detecting and single error-correcting code, but it is a long and tedious process to compute all of the distances between pairs of codewords to determine that $d_{\min} = 3$. It is much easier to see that the minimum weight of all the nonzero codewords is 3. As we will soon see, this is no coincidence. However, the relationship between weights and distances in a particular code is heavily dependent on the fact that the code is a group.   □

Lemma 8.17. Let $\mathbf{x}$ and $\mathbf{y}$ be binary n-tuples. Then $w(\mathbf{x} + \mathbf{y}) = d(\mathbf{x}, \mathbf{y})$.

Proof. Suppose that $\mathbf{x}$ and $\mathbf{y}$ are binary n-tuples. Then the distance between $\mathbf{x}$ and $\mathbf{y}$ is exactly the number of places in which $\mathbf{x}$ and $\mathbf{y}$ differ. But $\mathbf{x}$ and $\mathbf{y}$ differ in a particular coordinate exactly when the sum in the coordinate is 1, since

$$1 + 1 = 0$$
$$0 + 0 = 0$$
$$1 + 0 = 1$$
$$0 + 1 = 1.$$

Consequently, the weight of the sum must be the distance between the two codewords.   ∎

Theorem 8.18. Let $d_{\min}$ be the minimum distance for a group code C. Then $d_{\min}$ is the minimum weight of all the nonzero codewords in C. That is,

$$d_{\min} = \min\{w(\mathbf{x}) : \mathbf{x} \ne \mathbf{0}\}.$$

Proof. Observe that

$$
\begin{aligned}
d_{\min} &= \min\{d(\mathbf{x}, \mathbf{y}) : \mathbf{x} \ne \mathbf{y}\} \\
&= \min\{d(\mathbf{x}, \mathbf{y}) : \mathbf{x} + \mathbf{y} \ne \mathbf{0}\} \\
&= \min\{w(\mathbf{x} + \mathbf{y}) : \mathbf{x} + \mathbf{y} \ne \mathbf{0}\} \\
&= \min\{w(\mathbf{z}) : \mathbf{z} \ne \mathbf{0}\}.
\end{aligned}
$$

∎

Linear Codes

From Example 8.16 on the preceding page, it is now easy to check that the minimum nonzero weight is 3; hence, the code does indeed detect and correct all single errors. We have now reduced the problem of finding "good" codes to that of generating group codes. One easy way to generate group codes is to employ a bit of matrix theory.

Define the ***inner product*** of two binary n-tuples to be

$$\mathbf{x} \cdot \mathbf{y} = x_1 y_1 + \cdots + x_n y_n,$$

where $\mathbf{x} = (x_1, x_2, \ldots, x_n)^{\text{t}}$ and $\mathbf{y} = (y_1, y_2, \ldots, y_n)^{\text{t}}$ are column vectors.[4] For example, if $\mathbf{x} = (011001)^{\text{t}}$ and $\mathbf{y} = (110101)^{\text{t}}$, then $\mathbf{x} \cdot \mathbf{y} = 0$. We can also look at an inner product as the product of a row matrix with a column matrix; that is,

$$\mathbf{x} \cdot \mathbf{y} = \mathbf{x}^{\text{t}} \mathbf{y}$$

---

[4]Since we will be working with matrices, we will write binary n-tuples as column vectors for the remainder of this chapter.

$$= \begin{pmatrix} x_1 & x_2 & \cdots & x_n \end{pmatrix} \begin{pmatrix} y_1 \\ y_2 \\ \vdots \\ y_n \end{pmatrix}$$

$$= x_1 y_1 + x_2 y_2 + \cdots + x_n y_n.$$

Example 8.19. Suppose that the words to be encoded consist of all binary 3-tuples and that our encoding scheme is even-parity. To encode an arbitrary 3-tuple, we add a fourth bit to obtain an even number of 1s. Notice that an arbitrary n-tuple $\mathbf{x} = (x_1, x_2, \ldots, x_n)^t$ has an even number of 1s exactly when $x_1 + x_2 + \cdots + x_n = 0$; hence, a 4-tuple $\mathbf{x} = (x_1, x_2, x_3, x_4)^t$ has an even number of 1s if $x_1 + x_2 + x_3 + x_4 = 0$, or

$$\mathbf{x} \cdot \mathbf{1} = \mathbf{x}^t \mathbf{1} = \begin{pmatrix} x_1 & x_2 & x_3 & x_4 \end{pmatrix} \begin{pmatrix} 1 \\ 1 \\ 1 \\ 1 \end{pmatrix} = 0.$$

This example leads us to hope that there is a connection between matrices and coding theory. □

Let $\mathbb{M}_{m \times n}(\mathbb{Z}_2)$ denote the set of all $m \times n$ matrices with entries in $\mathbb{Z}_2$. We do matrix operations as usual except that all our addition and multiplication operations occur in $\mathbb{Z}_2$. Define the ***null space*** of a matrix $H \in \mathbb{M}_{m \times n}(\mathbb{Z}_2)$ to be the set of all binary n-tuples $\mathbf{x}$ such that $H\mathbf{x} = \mathbf{0}$. We denote the null space of a matrix H by $\text{Null}(H)$.

Example 8.20. Suppose that

$$H = \begin{pmatrix} 0 & 1 & 0 & 1 & 0 \\ 1 & 1 & 1 & 1 & 0 \\ 0 & 0 & 1 & 1 & 1 \end{pmatrix}.$$

For a 5-tuple $\mathbf{x} = (x_1, x_2, x_3, x_4, x_5)^t$ to be in the null space of H, $H\mathbf{x} = \mathbf{0}$. Equivalently, the following system of equations must be satisfied:

$$x_2 + x_4 = 0$$
$$x_1 + x_2 + x_3 + x_4 = 0$$
$$x_3 + x_4 + x_5 = 0.$$

The set of binary 5-tuples satisfying these equations is

$$(00000) \quad (11110) \quad (10101) \quad (01011).$$

This code is easily determined to be a group code.   □

Theorem 8.21. Let H be in $\mathbb{M}_{m \times n}(\mathbb{Z}_2)$. Then the null space of H is a group code.

Proof. Since each element of $\mathbb{Z}_2^n$ is its own inverse, the only thing that really needs to be checked here is closure. Let $\mathbf{x}, \mathbf{y} \in \text{Null}(H)$ for some matrix H in $\mathbb{M}_{m \times n}(\mathbb{Z}_2)$. Then $H\mathbf{x} = \mathbf{0}$ and $H\mathbf{y} = \mathbf{0}$. So

$$H(\mathbf{x} + \mathbf{y}) = H\mathbf{x} + H\mathbf{y} = \mathbf{0} + \mathbf{0} = \mathbf{0}.$$

Hence, $\mathbf{x} + \mathbf{y}$ is in the null space of H and therefore must be a codeword.   ■

A code is a *linear code* if it is determined by the null space of some matrix $H \in \mathbb{M}_{m \times n}(\mathbb{Z}_2)$.

Example 8.22. Let C be the code given by the matrix

$$H = \begin{pmatrix} 0 & 0 & 0 & 1 & 1 & 1 \\ 0 & 1 & 1 & 0 & 1 & 1 \\ 1 & 0 & 1 & 0 & 0 & 1 \end{pmatrix}.$$

Suppose that the 6-tuple $\mathbf{x} = (010011)^t$ is received. It is a simple matter of matrix multiplication to determine whether or not $\mathbf{x}$ is a codeword. Since

$$H\mathbf{x} = \begin{pmatrix} 0 \\ 1 \\ 1 \end{pmatrix},$$

the received word is not a codeword. We must either attempt to correct the word or request that it be transmitted again.   □

8.3 Parity-Check and Generator Matrices

We need to find a systematic way of generating linear codes as well as fast methods of decoding. By examining the properties of a matrix H and by carefully choosing H, it is possible to develop very efficient methods of encoding and decoding messages. To this end, we will introduce standard generator and canonical parity-check matrices.

Suppose that H is an $m \times n$ matrix with entries in $\mathbb{Z}_2$ and $n > m$. If the last m columns of the matrix form the $m \times m$ identity matrix, I_m, then the matrix is a *canonical parity-check matrix*. More specifically, $H = (A \mid I_m)$, where A is the

$m \times (n - m)$ matrix

$$\begin{pmatrix} a_{11} & a_{12} & \cdots & a_{1,n-m} \\ a_{21} & a_{22} & \cdots & a_{2,n-m} \\ \vdots & \vdots & \ddots & \vdots \\ a_{m1} & a_{m2} & \cdots & a_{m,n-m} \end{pmatrix}$$

and I_m is the $m \times m$ identity matrix

$$\begin{pmatrix} 1 & 0 & \cdots & 0 \\ 0 & 1 & \cdots & 0 \\ \vdots & \vdots & \ddots & \vdots \\ 0 & 0 & \cdots & 1 \end{pmatrix}.$$

With each canonical parity-check matrix we can associate an $n \times (n - m)$ **standard generator matrix**

$$G = \left(\frac{I_{n-m}}{A} \right).$$

Our goal will be to show that an **x** satisfying $G\mathbf{x} = \mathbf{y}$ exists if and only if $H\mathbf{y} = \mathbf{0}$. Given a message block **x** to be encoded, the matrix G will allow us to quickly encode it into a linear codeword **y**.

Example 8.23. Suppose that we have the following eight words to be encoded:

$$(000), (001), (010), \ldots, (111).$$

For

$$A = \begin{pmatrix} 0 & 1 & 1 \\ 1 & 1 & 0 \\ 1 & 0 & 1 \end{pmatrix},$$

the associated standard generator and canonical parity-check matrices are

$$G = \begin{pmatrix} 1 & 0 & 0 \\ 0 & 1 & 0 \\ 0 & 0 & 1 \\ 0 & 1 & 1 \\ 1 & 1 & 0 \\ 1 & 0 & 1 \end{pmatrix}$$

and

$$H = \begin{pmatrix} 0 & 1 & 1 & 1 & 0 & 0 \\ 1 & 1 & 0 & 0 & 1 & 0 \\ 1 & 0 & 1 & 0 & 0 & 1 \end{pmatrix},$$

respectively.

Observe that the rows in H represent the parity checks on certain bit positions in a 6-tuple. The 1s in the identity matrix serve as parity checks for the 1s in the same row. If $\mathbf{x} = (x_1, x_2, x_3, x_4, x_5, x_6)$, then

$$\mathbf{0} = H\mathbf{x} = \begin{pmatrix} x_2 + x_3 + x_4 \\ x_1 + x_2 + x_5 \\ x_1 + x_3 + x_6 \end{pmatrix},$$

which yields a system of equations:

$$x_2 + x_3 + x_4 = 0$$
$$x_1 + x_2 + x_5 = 0$$
$$x_1 + x_3 + x_6 = 0.$$

Here x_4 serves as a check bit for x_2 and x_3; x_5 is a check bit for x_1 and x_2; and x_6 is a check bit for x_1 and x_3. The identity matrix keeps x_4, x_5, and x_6 from having to check on each other. Hence, x_1, x_2, and x_3 can be arbitrary but x_4, x_5, and x_6 must be chosen to ensure parity. The null space of H is easily computed to be

(000000)   (001101)   (010110)   (011011)
(100011)   (101110)   (110101)   (111000).

An even easier way to compute the null space is with the generator matrix G (Table 8.24). □

Table 8.24. A matrix-generated code

Message Word x	Codeword Gx
000	000000
001	001101
010	010110
011	011011
100	100011
101	101110
110	110101
111	111000

Theorem 8.25. If $H \in \mathbb{M}_{m \times n}(\mathbb{Z}_2)$ is a canonical parity-check matrix, then $\text{Null}(H)$ consists of all $\mathbf{x} \in \mathbb{Z}_2^n$ whose first $n - m$ bits are arbitrary but whose last m bits are determined by $H\mathbf{x} = \mathbf{0}$. Each of the last m bits serves as an even parity check bit for some of the first $n - m$ bits. Hence, H gives rise to an $(n, n - m)$-block code.

We leave the proof of this theorem as an exercise. In light of the theorem, the first $n - m$ bits in $\mathbf{x}$ are called *information bits* and the last m bits are called *check bits*. In Example 8.23 on page 125, the first three bits are the information bits and the last three are the check bits.

Theorem 8.26. Suppose that G is an $n \times k$ standard generator matrix. Then $C = \{\mathbf{y} : G\mathbf{x} = \mathbf{y} \text{ for } \mathbf{x} \in \mathbb{Z}_2^k\}$ is an (n, k)-block code. More specifically, C is a group code.

Proof. Let $G\mathbf{x}_1 = \mathbf{y}_1$ and $G\mathbf{x}_2 = \mathbf{y}_2$ be two codewords. Then $\mathbf{y}_1 + \mathbf{y}_2$ is in C since

$$G(\mathbf{x}_1 + \mathbf{x}_2) = G\mathbf{x}_1 + G\mathbf{x}_2 = \mathbf{y}_1 + \mathbf{y}_2.$$

We must also show that two message blocks cannot be encoded into the same codeword. That is, we must show that if $G\mathbf{x} = G\mathbf{y}$, then $\mathbf{x} = \mathbf{y}$. Suppose that $G\mathbf{x} = G\mathbf{y}$. Then

$$G\mathbf{x} - G\mathbf{y} = G(\mathbf{x} - \mathbf{y}) = \mathbf{0}.$$

However, the first k coordinates in $G(\mathbf{x} - \mathbf{y})$ are exactly $x_1 - y_1, \ldots, x_k - y_k$, since they are determined by the identity matrix, I_k, part of G. Hence, $G(\mathbf{x} - \mathbf{y}) = \mathbf{0}$ exactly when $\mathbf{x} = \mathbf{y}$. ∎

Before we can prove the relationship between canonical parity-check matrices and standard generating matrices, we need to prove a lemma.

Lemma 8.27. Let $H = (A \mid I_m)$ be an $m \times n$ canonical parity-check matrix and $G = \left(\frac{I_{n-m}}{A}\right)$ be the corresponding $n \times (n - m)$ standard generator matrix. Then $HG = \mathbf{0}$.

Proof. Let $C = HG$. The ijth entry in C is

$$c_{ij} = \sum_{k=1}^{n} h_{ik} g_{kj}$$

$$= \sum_{k=1}^{n-m} h_{ik} g_{kj} + \sum_{k=n-m+1}^{n} h_{ik} g_{kj}$$

$$= \sum_{k=1}^{n-m} a_{ik} \delta_{kj} + \sum_{k=n-m+1}^{n} \delta_{i-(m-n),k} a_{kj}$$

$$= a_{ij} + a_{ij}$$

$$= 0,$$

where

$$\delta_{ij} = \begin{cases} 1 & i = j \\ 0 & i \neq j \end{cases}$$

is the Kronecker delta.   ∎

Theorem 8.28. Let $H = (A \mid I_m)$ be an $m \times n$ canonical parity-check matrix and let $G = \left(\frac{I_{n-m}}{A} \right)$ be the $n \times (n - m)$ standard generator matrix associated with H. Let C be the code generated by G. Then $\mathbf{y}$ is in C if and only if $H\mathbf{y} = \mathbf{0}$. In particular, C is a linear code with canonical parity-check matrix H.

Proof. First suppose that $\mathbf{y} \in C$. Then $G\mathbf{x} = \mathbf{y}$ for some $\mathbf{x} \in \mathbb{Z}_2^m$. By Lemma 8.27 on the previous page, $H\mathbf{y} = HG\mathbf{x} = \mathbf{0}$.

Conversely, suppose that $\mathbf{y} = (y_1, \ldots, y_n)^t$ is in the null space of H. We need to find an $\mathbf{x}$ in $\mathbb{Z}_2^{n-m}$ such that $G\mathbf{x}^t = \mathbf{y}$. Since $H\mathbf{y} = \mathbf{0}$, the following set of equations must be satisfied:

$$a_{11}y_1 + a_{12}y_2 + \cdots + a_{1,n-m}y_{n-m} + y_{n-m+1} = 0$$
$$a_{21}y_1 + a_{22}y_2 + \cdots + a_{2,n-m}y_{n-m} + y_{n-m+2} = 0$$
$$\vdots$$
$$a_{m1}y_1 + a_{m2}y_2 + \cdots + a_{m,n-m}y_{n-m} + y_{n-m+m} = 0.$$

Equivalently, $y_{n-m+1}, \ldots, y_n$ are determined by $y_1, \ldots, y_{n-m}$:

$$y_{n-m+1} = a_{11}y_1 + a_{12}y_2 + \cdots + a_{1,n-m}y_{n-m}$$
$$y_{n-m+2} = a_{21}y_1 + a_{22}y_2 + \cdots + a_{2,n-m}y_{n-m}$$
$$\vdots$$
$$y_n = a_{m1}y_1 + a_{m2}y_2 + \cdots + a_{m,n-m}y_{n-m}.$$

Consequently, we can let $x_i = y_i$ for $i = 1, \ldots, n - m$.   ∎

It would be helpful if we could compute the minimum distance of a linear code directly from its matrix H in order to determine the error-detecting and error-correcting capabilities of the code. Suppose that

$$\mathbf{e}_1 = (100\cdots00)^t$$
$$\mathbf{e}_2 = (010\cdots00)^t$$
$$\vdots$$
$$\mathbf{e}_n = (000\cdots01)^t$$

are the n-tuples in $\mathbb{Z}_2^n$ of weight 1. For an $m \times n$ binary matrix H, $H\mathbf{e}_i$ is exactly the ith column of the matrix H.

Example 8.29. Observe that

$$
\begin{pmatrix} 1 & 1 & 1 & 0 & 0 \\ 1 & 0 & 0 & 1 & 0 \\ 1 & 1 & 0 & 0 & 1 \end{pmatrix}
\begin{pmatrix} 0 \\ 1 \\ 0 \\ 0 \\ 0 \end{pmatrix}
= \begin{pmatrix} 1 \\ 0 \\ 1 \end{pmatrix}.
$$

□

We state this result in the following proposition and leave the proof as an exercise.

Proposition 8.30. Let $\mathbf{e}_i$ be the binary n-tuple with a 1 in the ith coordinate and 0's elsewhere and suppose that $H \in \mathbb{M}_{m \times n}(\mathbb{Z}_2)$. Then $H\mathbf{e}_i$ is the ith column of the matrix H.

Theorem 8.31. Let H be an $m \times n$ binary matrix. Then the null space of H is a single error-detecting code if and only if no column of H consists entirely of zeros.

Proof. Suppose that $\mathrm{Null}(H)$ is a single error-detecting code. Then the minimum distance of the code must be at least 2. Since the null space is a group code, it is sufficient to require that the code contain no codewords of less than weight 2 other than the zero codeword. That is, $\mathbf{e}_i$ must not be a codeword for $i = 1, \ldots, n$. Since $H\mathbf{e}_i$ is the ith column of H, the only way in which $\mathbf{e}_i$ could be in the null space of H would be if the ith column were all zeros, which is impossible; hence, the code must have the capability to detect at least single errors.

Conversely, suppose that no column of H is the zero column. By Proposition 8.30, $H\mathbf{e}_i \neq \mathbf{0}$. ∎

Example 8.32. If we consider the matrices

$$
H_1 = \begin{pmatrix} 1 & 1 & 1 & 0 & 0 \\ 1 & 0 & 0 & 1 & 0 \\ 1 & 1 & 0 & 0 & 1 \end{pmatrix}
$$

and

$$
H_2 = \begin{pmatrix} 1 & 1 & 1 & 0 & 0 \\ 1 & 0 & 0 & 0 & 0 \\ 1 & 1 & 0 & 0 & 1 \end{pmatrix},
$$

then the null space of H_1 is a single error-detecting code and the null space of H_2 is not.   □

We can even do better than Theorem 8.31 on the previous page. This theorem gives us conditions on a matrix H that tell us when the minimum weight of the code formed by the null space of H is 2. We can also determine when the minimum distance of a linear code is 3 by examining the corresponding matrix.

Example 8.33. If we let

$$H = \begin{pmatrix} 1 & 1 & 1 & 0 \\ 1 & 0 & 0 & 1 \\ 1 & 1 & 0 & 0 \end{pmatrix}$$

and want to determine whether or not H is the canonical parity-check matrix for an error-correcting code, it is necessary to make certain that Null(H) does not contain any 4-tuples of weight 2. That is, (1100), (1010), (1001), (0110), (0101), and (0011) must not be in Null(H). The next theorem states that we can indeed determine that the code generated by H is error-correcting by examining the columns of H. Notice in this example that not only does H have no zero columns, but also that no two columns are the same.   □

Theorem 8.34. Let H be a binary matrix. The null space of H is a single error-correcting code if and only if H does not contain any zero columns and no two columns of H are identical.

Proof. The n-tuple $\mathbf{e}_i + \mathbf{e}_j$ has 1s in the ith and jth entries and 0s elsewhere, and $w(\mathbf{e}_i + \mathbf{e}_j) = 2$ for $i \neq j$. Since

$$\mathbf{0} = H(\mathbf{e}_i + \mathbf{e}_j) = H\mathbf{e}_i + H\mathbf{e}_j$$

can only occur if the ith and jth columns are identical, the null space of H is a single error-correcting code.   ■

Suppose now that we have a canonical parity-check matrix H with three rows. Then we might ask how many more columns we can add to the matrix and still have a null space that is a single error-detecting and single error-correcting code. Since each column has three entries, there are $2^3 = 8$ possible distinct columns. We cannot add the columns

$$\begin{pmatrix} 0 \\ 0 \\ 0 \end{pmatrix}, \begin{pmatrix} 1 \\ 0 \\ 0 \end{pmatrix}, \begin{pmatrix} 0 \\ 1 \\ 0 \end{pmatrix}, \begin{pmatrix} 0 \\ 0 \\ 1 \end{pmatrix}.$$

So we can add as many as four columns and still maintain a minimum distance of 3.

In general, if H is an $m \times n$ canonical parity-check matrix, then there are $n - m$ information positions in each codeword. Each column has m bits, so there are 2^m possible distinct columns. It is necessary that the columns $\mathbf{0}, \mathbf{e}_1, \ldots, \mathbf{e}_m$ be excluded, leaving $2^m - (1 + m)$ remaining columns for information if we are still to maintain the ability not only to detect but also to correct single errors.

8.4 Efficient Decoding

We are now at the stage where we are able to generate linear codes that detect and correct errors fairly easily, but it is still a time-consuming process to decode a received n-tuple and determine which is the closest codeword, because the received n-tuple must be compared to each possible codeword to determine the proper decoding. This can be a serious impediment if the code is very large.

Example 8.35. Given the binary matrix

$$H = \begin{pmatrix} 1 & 1 & 1 & 0 & 0 \\ 0 & 1 & 0 & 1 & 0 \\ 1 & 0 & 0 & 0 & 1 \end{pmatrix}$$

and the 5-tuples $\mathbf{x} = (11011)^t$ and $\mathbf{y} = (01011)^t$, we can compute

$$H\mathbf{x} = \begin{pmatrix} 0 \\ 0 \\ 0 \end{pmatrix} \quad \text{and} \quad H\mathbf{y} = \begin{pmatrix} 1 \\ 0 \\ 1 \end{pmatrix}.$$

Hence, $\mathbf{x}$ is a codeword and $\mathbf{y}$ is not, since $\mathbf{x}$ is in the null space and $\mathbf{y}$ is not. Notice that $H\mathbf{y}$ is identical to the first column of H. In fact, this is where the error occurred. If we flip the first bit in $\mathbf{y}$ from 0 to 1, then we obtain $\mathbf{x}$. $\quad\square$

If H is an $m \times n$ matrix and $\mathbf{x} \in \mathbb{Z}_2^n$, then we say that the *syndrome* of $\mathbf{x}$ is $H\mathbf{x}$. The following proposition allows the quick detection and correction of errors.

Proposition 8.36. Let the $m \times n$ binary matrix H determine a linear code and let $\mathbf{x}$ be the received n-tuple. Write $\mathbf{x}$ as $\mathbf{x} = \mathbf{c} + \mathbf{e}$, where $\mathbf{c}$ is the transmitted codeword and $\mathbf{e}$ is the transmission error. Then the syndrome $H\mathbf{x}$ of the received codeword $\mathbf{x}$ is also the syndrome of the error $\mathbf{e}$.

Proof. The proof follows from the fact that

$$H\mathbf{x} = H(\mathbf{c} + \mathbf{e}) = H\mathbf{c} + H\mathbf{e} = \mathbf{0} + H\mathbf{e} = H\mathbf{e}.$$

$\blacksquare$

This proposition tells us that the syndrome of a received word depends solely on the error and not on the transmitted codeword. The proof of the following theorem follows immediately from Proposition 8.36 on the preceding page and from the fact that $H\mathbf{e}$ is the ith column of the matrix H.

Theorem 8.37. Let $H \in \mathbb{M}_{m \times n}(\mathbb{Z}_2)$ and suppose that the linear code corresponding to H is single error-correcting. Let $\mathbf{r}$ be a received n-tuple that was transmitted with at most one error. If the syndrome of $\mathbf{r}$ is $\mathbf{0}$, then no error has occurred; otherwise, if the syndrome of $\mathbf{r}$ is equal to some column of H, say the ith column, then the error has occurred in the ith bit.

Example 8.38. Consider the matrix

$$H = \begin{pmatrix} 1 & 0 & 1 & 1 & 0 & 0 \\ 0 & 1 & 1 & 0 & 1 & 0 \\ 1 & 1 & 1 & 0 & 0 & 1 \end{pmatrix}$$

and suppose that the 6-tuples $\mathbf{x} = (111110)^t$, $\mathbf{y} = (111111)^t$, and $\mathbf{z} = (010111)^t$ have been received. Then

$$H\mathbf{x} = \begin{pmatrix} 1 \\ 1 \\ 1 \end{pmatrix}, H\mathbf{y} = \begin{pmatrix} 1 \\ 1 \\ 0 \end{pmatrix}, H\mathbf{z} = \begin{pmatrix} 1 \\ 0 \\ 0 \end{pmatrix}.$$

Hence, $\mathbf{x}$ has an error in the third bit and $\mathbf{z}$ has an error in the fourth bit. The transmitted codewords for $\mathbf{x}$ and $\mathbf{z}$ must have been (110110) and (010011), respectively. The syndrome of $\mathbf{y}$ does not occur in any of the columns of the matrix H, so multiple errors must have occurred to produce $\mathbf{y}$. □

Coset Decoding

We can use group theory to obtain another way of decoding messages. A linear code C is a subgroup of $\mathbb{Z}_2^n$. *Coset* or *standard decoding* uses the cosets of C in $\mathbb{Z}_2^n$ to implement maximum-likelihood decoding. Suppose that C is an (n, m)-linear code. A coset of C in $\mathbb{Z}_2^n$ is written in the form $\mathbf{x} + C$, where $\mathbf{x} \in \mathbb{Z}_2^n$. By Lagrange's Theorem (Theorem 6.10 on page 93), there are 2^{n-m} distinct cosets of C in $\mathbb{Z}_2^n$.

Example 8.39. Let C be the $(5, 3)$-linear code given by the parity-check matrix

$$H = \begin{pmatrix} 0 & 1 & 1 & 0 & 0 \\ 1 & 0 & 0 & 1 & 0 \\ 1 & 1 & 0 & 0 & 1 \end{pmatrix}.$$

The code consists of the codewords

$$(00000) \quad (01101) \quad (10011) \quad (11110).$$

There are $2^{5-2} = 2^3$ cosets of C in $\mathbb{Z}_2^5$, each with order $2^2 = 4$. These cosets are listed in Table 8.40.   □

Table 8.40. Cosets of C

Coset Representative	Coset
C	$(00000)(01101)(10011)(11110)$
$(10000) + C$	$(10000)(11101)(00011)(01110)$
$(01000) + C$	$(01000)(00101)(11011)(10110)$
$(00100) + C$	$(00100)(01001)(10111)(11010)$
$(00010) + C$	$(00010)(01111)(10001)(11100)$
$(00001) + C$	$(00001)(01100)(10010)(11111)$
$(10100) + C$	$(00111)(01010)(10100)(11001)$
$(00110) + C$	$(00110)(01011)(10101)(11000)$

Our task is to find out how knowing the cosets might help us to decode a message. Suppose that $\mathbf{x}$ was the original codeword sent and that $\mathbf{r}$ is the n-tuple received. If $\mathbf{e}$ is the transmission error, then $\mathbf{r} = \mathbf{e} + \mathbf{x}$ or, equivalently, $\mathbf{x} = \mathbf{e} + \mathbf{r}$. However, this is exactly the statement that $\mathbf{r}$ is an element in the coset $\mathbf{e} + C$. In maximum-likelihood decoding we expect the error $\mathbf{e}$ to be as small as possible; that is, $\mathbf{e}$ will have the least weight. An n-tuple of least weight in a coset is called a *coset leader*. Once we have determined a coset leader for each coset, the decoding process becomes a task of calculating $\mathbf{r} + \mathbf{e}$ to obtain $\mathbf{x}$.

Example 8.41. In Table 8.40, notice that we have chosen a representative of the least possible weight for each coset. These representatives are coset leaders. Now suppose that $\mathbf{r} = (01111)$ is the received word. To decode $\mathbf{r}$, we find that it is in the coset $(00010) + C$; hence, the originally transmitted codeword must have been $(01101) = (01111) + (00010)$.   □

A potential problem with this method of decoding is that we might have to examine every coset for the received codeword. The following proposition gives a method of implementing coset decoding. It states that we can associate a syndrome with each coset; hence, we can make a table that designates a coset leader corresponding to each syndrome. Such a list is called a *decoding table*.

Table 8.42. Syndromes for each coset

Syndrome	Coset Leader
(000)	(00000)
(001)	(00001)
(010)	(00010)
(011)	(10000)
(100)	(00100)
(101)	(01000)
(110)	(00110)
(111)	(10100)

Proposition 8.43. Let C be an (n, k)-linear code given by the matrix H and suppose that $\mathbf{x}$ and $\mathbf{y}$ are in $\mathbb{Z}_2^n$. Then $\mathbf{x}$ and $\mathbf{y}$ are in the same coset of C if and only if $H\mathbf{x} = H\mathbf{y}$. That is, two n-tuples are in the same coset if and only if their syndromes are the same.

Proof. Two n-tuples $\mathbf{x}$ and $\mathbf{y}$ are in the same coset of C exactly when $\mathbf{x} - \mathbf{y} \in C$; however, this is equivalent to $H(\mathbf{x} - \mathbf{y}) = 0$ or $H\mathbf{x} = H\mathbf{y}$. ∎

Example 8.44. Table 8.42 is a decoding table for the code C given in Example 8.39 on page 132. If $\mathbf{x} = (01111)$ is received, then its syndrome can be computed to be

$$H\mathbf{x} = \begin{pmatrix} 0 \\ 1 \\ 0 \end{pmatrix}.$$

Examining the decoding table, we determine that the coset leader is (00010). It is now easy to decode the received codeword. □

Given an (n, k)-block code, the question arises of whether or not coset decoding is a manageable scheme. A decoding table requires a list of cosets and syndromes, one for each of the 2^{n-k} cosets of C. Suppose that we have a $(32, 24)$-block code. We have a huge number of codewords, 2^{24}, yet there are only $2^{32-24} = 2^8 = 256$ cosets.

Sage. Sage has a substantial repertoire of commands for coding theory, including the ability to build many different families of codes.

8.5 Reading Questions

1. Suppose a binary code has minimum distance $d = 6$. How many errors can be detected? How many errors can be corrected?

2. Explain why it is impossible for the 8-bit string with decimal value 56_{10} to be an ASCII code for a character. Assume the leftmost bit of the string is being used as a parity-check bit.

3. Suppose we receive the 8-bit string with decimal value 56_{10} when we are expecting ASCII characters with a parity-check bit in the first bit (leftmost). We know an error has occurred in transmission. Give one of the probable guesses for the character which was actually sent (other than '8'), under the assumption that any individual bit is rarely sent in error. Explain the logic of your answer. (You may need to consult a table of ASCII values online.)

4. Suppose a linear code C is created as the null space of the parity-check matrix

$$H = \begin{bmatrix} 0 & 1 & 0 & 1 & 0 \\ 1 & 1 & 1 & 1 & 0 \\ 0 & 0 & 1 & 1 & 1 \end{bmatrix}$$

Then $x = 11100$ is not a codeword. Describe a computation, and give the result of that computation, which verifies that x is not a codeword of the code C.

5. For H and x as in the previous question, suppose that x is received as a message. Give a maximum likelihood decoding of the received message.

8.6 Exercises

1. Why is the following encoding scheme not acceptable?

Information	0	1	2	3	4	5	6	7	8
Codeword	000	001	010	011	101	110	111	000	001

2. Without doing any addition, explain why the following set of 4-tuples in $\mathbb{Z}_2^4$ cannot be a group code.

$$(0110) \quad (1001) \quad (1010) \quad (1100)$$

3. Compute the Hamming distances between the following pairs of n-tuples.
   (a) $(011010), (011100)$
   (b) $(11110101), (01010100)$
   (c) $(00110), (01111)$
   (d) $(1001), (0111)$

4. Compute the weights of the following n-tuples.
   (a) (011010)   (c) (01111)

   (b) (11110101)   (d) (1011)

5. Suppose that a linear code C has a minimum weight of 7. What are the error-detection and error-correction capabilities of C?

6. In each of the following codes, what is the minimum distance for the code? What is the best situation we might hope for in connection with error detection and error correction?

   (a) (011010) (011100) (110111) (110000)

   (b) (011100) (011011) (111011) (100011)
       (000000) (010101) (110100) (110011)

   (c) (000000) (011100) (110101) (110001)

   (d) (0110110) (0111100) (1110000) (1111111)
       (1001001) (1000011) (0001111) (0000000)

7. Compute the null space of each of the following matrices. What type of (n, k)-block codes are the null spaces? Can you find a matrix (not necessarily a standard generator matrix) that generates each code? Are your generator matrices unique?

   (a)
   $$\begin{pmatrix} 0 & 1 & 0 & 0 & 0 \\ 1 & 0 & 1 & 0 & 1 \\ 1 & 0 & 0 & 1 & 0 \end{pmatrix}$$

   (c)
   $$\begin{pmatrix} 1 & 0 & 0 & 1 & 1 \\ 0 & 1 & 0 & 1 & 1 \end{pmatrix}$$

   (d)

   (b)
   $$\begin{pmatrix} 1 & 0 & 1 & 0 & 0 & 0 \\ 1 & 1 & 0 & 1 & 0 & 0 \\ 0 & 1 & 0 & 0 & 1 & 0 \\ 1 & 1 & 0 & 0 & 0 & 1 \end{pmatrix}$$

   (d)
   $$\begin{pmatrix} 0 & 0 & 0 & 1 & 1 & 1 & 1 \\ 0 & 1 & 1 & 0 & 0 & 1 & 1 \\ 1 & 0 & 1 & 0 & 1 & 0 & 1 \\ 0 & 1 & 1 & 0 & 0 & 1 & 1 \end{pmatrix}$$

8. Construct a $(5, 2)$-block code. Discuss both the error-detection and error-correction capabilities of your code.

9. Let C be the code obtained from the null space of the matrix

   $$H = \begin{pmatrix} 0 & 1 & 0 & 0 & 1 \\ 1 & 0 & 1 & 0 & 1 \\ 0 & 0 & 1 & 1 & 1 \end{pmatrix}.$$

Decode the message

$$01111 \quad 10101 \quad 01110 \quad 00011$$

if possible.

10. Suppose that a 1000-bit binary message is transmitted. Assume that the probability of a single error is p and that the errors occurring in different bits are independent of one another. If $p = 0.01$, what is the probability of more than one error occurring? What is the probability of exactly two errors occurring? Repeat this problem for $p = 0.0001$.

11. Which matrices are canonical parity-check matrices? For those matrices that are canonical parity-check matrices, what are the corresponding standard generator matrices? What are the error-detection and error-correction capabilities of the code generated by each of these matrices?

(a)

$$\begin{pmatrix} 1 & 1 & 0 & 0 & 0 \\ 0 & 0 & 1 & 0 & 0 \\ 0 & 0 & 0 & 1 & 0 \\ 1 & 0 & 0 & 0 & 1 \end{pmatrix}$$

(b)

$$\begin{pmatrix} 0 & 1 & 1 & 0 & 0 & 0 \\ 1 & 1 & 0 & 1 & 0 & 0 \\ 0 & 1 & 0 & 0 & 1 & 0 \\ 1 & 1 & 0 & 0 & 0 & 1 \end{pmatrix}$$

(c)

$$\begin{pmatrix} 1 & 1 & 1 & 0 \\ 1 & 0 & 0 & 1 \end{pmatrix}$$

(d)

$$\begin{pmatrix} 0 & 0 & 0 & 1 & 0 & 0 & 0 \\ 0 & 1 & 1 & 0 & 1 & 0 & 0 \\ 1 & 0 & 1 & 0 & 0 & 1 & 0 \\ 0 & 1 & 1 & 0 & 0 & 0 & 1 \end{pmatrix}$$

12. List all possible syndromes for the codes generated by each of the matrices in Exercise 8.6.11.

13. Let

$$H = \begin{pmatrix} 0 & 1 & 1 & 1 & 1 \\ 0 & 0 & 0 & 1 & 1 \\ 1 & 0 & 1 & 0 & 1 \end{pmatrix}.$$

Compute the syndrome caused by each of the following transmission errors.

(a) An error in the first bit.

(b) An error in the third bit.

(c) An error in the last bit.

(d) Errors in the third and fourth bits.

14. Let C be the group code in $\mathbb{Z}_2^3$ defined by the codewords (000) and (111). Compute the cosets of C in $\mathbb{Z}_2^3$. Why was there no need to specify right or left cosets? Give the single transmission error, if any, to which each coset corresponds.

15. For each of the following matrices, find the cosets of the corresponding code C. Give a decoding table for each code if possible.

(a)
$$\begin{pmatrix} 0 & 1 & 0 & 0 & 0 \\ 1 & 0 & 1 & 0 & 1 \\ 1 & 0 & 0 & 1 & 0 \end{pmatrix}$$

(c)
$$\begin{pmatrix} 1 & 0 & 0 & 1 & 1 \\ 0 & 1 & 0 & 1 & 1 \end{pmatrix}$$

(d)

(b)
$$\begin{pmatrix} 0 & 0 & 1 & 0 & 0 \\ 1 & 1 & 0 & 1 & 0 \\ 0 & 1 & 0 & 1 & 0 \\ 1 & 1 & 0 & 0 & 1 \end{pmatrix}$$

$$\begin{pmatrix} 1 & 0 & 0 & 1 & 1 & 1 & 1 \\ 1 & 1 & 1 & 0 & 0 & 1 & 1 \\ 1 & 0 & 1 & 0 & 1 & 0 & 1 \\ 1 & 1 & 1 & 0 & 0 & 1 & 0 \end{pmatrix}$$

16. Let $\mathbf{x}$, $\mathbf{y}$, and $\mathbf{z}$ be binary n-tuples. Prove each of the following statements.

(a) $w(\mathbf{x}) = d(\mathbf{x}, \mathbf{0})$

(b) $d(\mathbf{x}, \mathbf{y}) = d(\mathbf{x} + \mathbf{z}, \mathbf{y} + \mathbf{z})$

(c) $d(\mathbf{x}, \mathbf{y}) = w(\mathbf{x} - \mathbf{y})$

17. A *metric* on a set X is a map $d : X \times X \to \mathbb{R}$ satisfying the following conditions.

(a) $d(\mathbf{x}, \mathbf{y}) \geq 0$ for all $\mathbf{x}, \mathbf{y} \in X$;

(b) $d(\mathbf{x}, \mathbf{y}) = 0$ exactly when $\mathbf{x} = \mathbf{y}$;

(c) $d(\mathbf{x}, \mathbf{y}) = d(\mathbf{y}, \mathbf{x})$;

(d) $d(\mathbf{x}, \mathbf{y}) \leq d(\mathbf{x}, \mathbf{z}) + d(\mathbf{z}, \mathbf{y})$.

In other words, a metric is simply a generalization of the notion of distance. Prove that Hamming distance is a metric on $\mathbb{Z}_2^n$. Decoding a message actually reduces to deciding which is the closest codeword in terms of distance.

18. Let C be a linear code. Show that either the ith coordinates in the codewords of C are all zeros or exactly half of them are zeros.

19. Let C be a linear code. Show that either every codeword has even weight or exactly half of the codewords have even weight.

20. Show that the codewords of even weight in a linear code C are also a linear code.

21. If we are to use an error-correcting linear code to transmit the 128 ASCII characters, what size matrix must be used? What size matrix must be used to transmit the extended ASCII character set of 256 characters? What if we require only error detection in both cases?

22. Find the canonical parity-check matrix that gives the even parity check bit code with three information positions. What is the matrix for seven information positions? What are the corresponding standard generator matrices?

23. How many check positions are needed for a single error-correcting code with 20 information positions? With 32 information positions?

24. Let e_i be the binary n-tuple with a 1 in the ith coordinate and 0's elsewhere and suppose that $H \in \mathbb{M}_{m \times n}(\mathbb{Z}_2)$. Show that He_i is the ith column of the matrix H.

25. Let C be an (n, k)-linear code. Define the *dual* or *orthogonal code* of C to be

$$C^{\perp} = \{\mathbf{x} \in \mathbb{Z}_2^n : \mathbf{x} \cdot \mathbf{y} = 0 \text{ for all } \mathbf{y} \in C\}.$$

   (a) Find the dual code of the linear code C where C is given by the matrix

$$\begin{pmatrix} 1 & 1 & 1 & 0 & 0 \\ 0 & 0 & 1 & 0 & 1 \\ 1 & 0 & 0 & 1 & 0 \end{pmatrix}.$$

   (b) Show that $C^{\perp}$ is an $(n, n - k)$-linear code.

   (c) Find the standard generator and parity-check matrices of C and $C^{\perp}$. What happens in general? Prove your conjecture.

26. Let H be an $m \times n$ matrix over $\mathbb{Z}_2$, where the ith column is the number i written in binary with m bits. The null space of such a matrix is called a *Hamming code*.

   (a) Show that the matrix

$$H = \begin{pmatrix} 0 & 0 & 0 & 1 & 1 & 1 \\ 0 & 1 & 1 & 0 & 0 & 1 \\ 1 & 0 & 1 & 0 & 1 & 0 \end{pmatrix}$$

   generates a Hamming code. What are the error-correcting properties of a Hamming code?

(b) The column corresponding to the syndrome also marks the bit that was in error; that is, the ith column of the matrix is i written as a binary number, and the syndrome immediately tells us which bit is in error. If the received word is (101011), compute the syndrome. In which bit did the error occur in this case, and what codeword was originally transmitted?

(c) Give a binary matrix H for the Hamming code with six information positions and four check positions. What are the check positions and what are the information positions? Encode the messages (101101) and (001001). Decode the received words (0010000101) and (0000101100). What are the possible syndromes for this code?

(d) What is the number of check bits and the number of information bits in an (m, n)-block Hamming code? Give both an upper and a lower bound on the number of information bits in terms of the number of check bits. Hamming codes having the maximum possible number of information bits with k check bits are called **perfect**. Every possible syndrome except **0** occurs as a column. If the number of information bits is less than the maximum, then the code is called **shortened**. In this case, give an example showing that some syndromes can represent multiple errors.

8.7 Programming Exercises

1. Write a program to implement a $(16, 12)$-linear code. Your program should be able to encode and decode messages using coset decoding. Once your program is written, write a program to simulate a binary symmetric channel with transmission noise. Compare the results of your simulation with the theoretically predicted error probability.

8.8 References and Suggested Readings

[1]   Blake, I. F. "Codes and Designs," *Mathematics Magazine* **52** (1979), 81–95.

[2]   Hill, R. *A First Course in Coding Theory*. Oxford University Press, Oxford, 1990.

[3]   Levinson, N. "Coding Theory: A Counterexample to G. H. Hardy's Conception of Applied Mathematics," *American Mathematical Monthly* **77** (1970), 249–58.

[4]   Lidl, R. and Pilz, G. *Applied Abstract Algebra*. 2nd ed. Springer, New York, 1998.

[5] MacWilliams, F. J. and Sloane, N. J. A. *The Theory of Error-Correcting Codes*. North-Holland Mathematical Library, 16, Elsevier, Amsterdam, 1983.

[6] Roman, S. *Coding and Information Theory*. Springer-Verlag, New York, 1992.

[7] Shannon, C. E. "A Mathematical Theory of Communication," *Bell System Technical Journal* 27 (1948), 379–423, 623–56.

[8] Thompson, T. M. *From Error-Correcting Codes through Sphere Packing to Simple Groups*. Carus Monograph Series, No. 21. Mathematical Association of America, Washington, DC, 1983.

[9] van Lint, J. H. *Introduction to Coding Theory*. Springer, New York, 1999.

Isomorphisms

*M*any groups may appear to be different at first glance, but can be shown to be the same by a simple renaming of the group elements. For example, $\mathbb{Z}_4$ and the subgroup of the circle group $\mathbb{T}$ generated by i can be shown to be the same by demonstrating a one-to-one correspondence between the elements of the two groups and between the group operations. In such a case we say that the groups are isomorphic.

9.1 Definition and Examples

Two groups $(G, \cdot)$ and $(H, \circ)$ are *isomorphic* if there exists a one-to-one and onto map $\phi : G \to H$ such that the group operation is preserved; that is,

$$\phi(a \cdot b) = \phi(a) \circ \phi(b)$$

for all a and b in G. If G is isomorphic to H, we write $G \cong H$. The map ϕ is called an *isomorphism*.

Example 9.1. To show that $\mathbb{Z}_4 \cong \langle i \rangle$, define a map $\phi : \mathbb{Z}_4 \to \langle i \rangle$ by $\phi(n) = i^n$. We must show that ϕ is bijective and preserves the group operation. The map ϕ is one-to-one and onto because

$$\phi(0) = 1$$
$$\phi(1) = i$$
$$\phi(2) = -1$$
$$\phi(3) = -i.$$

Since

$$\phi(m + n) = i^{m+n} = i^m i^n = \phi(m)\phi(n),$$

the group operation is preserved. $\qquad\qquad\square$

Example 9.2. We can define an isomorphism ϕ from the additive group of real numbers $(\mathbb{R}, +)$ to the multiplicative group of positive real numbers $(\mathbb{R}^+, \cdot)$ with the exponential map; that is,

$$\phi(x + y) = e^{x+y} = e^x e^y = \phi(x)\phi(y).$$

Of course, we must still show that ϕ is one-to-one and onto, but this can be determined using calculus. □

Example 9.3. The integers are isomorphic to the subgroup of $\mathbb{Q}^*$ consisting of elements of the form 2^n. Define a map $\phi : \mathbb{Z} \to \mathbb{Q}^*$ by $\phi(n) = 2^n$. Then

$$\phi(m + n) = 2^{m+n} = 2^m 2^n = \phi(m)\phi(n).$$

By definition the map ϕ is onto the subset $\{2^n : n \in \mathbb{Z}\}$ of $\mathbb{Q}^*$. To show that the map is injective, assume that $m \neq n$. If we can show that $\phi(m) \neq \phi(n)$, then we are done. Suppose that $m > n$ and assume that $\phi(m) = \phi(n)$. Then $2^m = 2^n$ or $2^{m-n} = 1$, which is impossible since $m - n > 0$. □

Example 9.4. The groups $\mathbb{Z}_8$ and $\mathbb{Z}_{12}$ cannot be isomorphic since they have different orders; however, it is true that $U(8) \cong U(12)$. We know that

$$U(8) = \{1, 3, 5, 7\}$$
$$U(12) = \{1, 5, 7, 11\}.$$

An isomorphism $\phi : U(8) \to U(12)$ is then given by

$$1 \mapsto 1$$
$$3 \mapsto 5$$
$$5 \mapsto 7$$
$$7 \mapsto 11.$$

The map ϕ is not the only possible isomorphism between these two groups. We could define another isomorphism ψ by $\psi(1) = 1$, $\psi(3) = 11$, $\psi(5) = 5$, $\psi(7) = 7$. In fact, both of these groups are isomorphic to $\mathbb{Z}_2 \times \mathbb{Z}_2$ (see Example 3.28 on page 48). □

Example 9.5. Even though S_3 and $\mathbb{Z}_6$ possess the same number of elements, we would suspect that they are not isomorphic, because $\mathbb{Z}_6$ is abelian and S_3 is nonabelian. To demonstrate that this is indeed the case, suppose that $\phi : \mathbb{Z}_6 \to S_3$ is an isomorphism. Let $a, b \in S_3$ be two elements such that $ab \neq ba$. Since ϕ is an isomorphism, there exist elements m and n in $\mathbb{Z}_6$ such that

$$\phi(m) = a \quad \text{and} \quad \phi(n) = b.$$

However,

$$ab = \phi(m)\phi(n) = \phi(m+n) = \phi(n+m) = \phi(n)\phi(m) = ba,$$

which contradicts the fact that a and b do not commute.   □

Theorem 9.6. Let $\phi : G \to H$ be an isomorphism of two groups. Then the following statements are true.

1. $\phi^{-1} : H \to G$ is an isomorphism.

2. $|G| = |H|$.

3. If G is abelian, then H is abelian.

4. If G is cyclic, then H is cyclic.

5. If G has a subgroup of order n, then H has a subgroup of order n.

Proof. Assertions (1) and (2) follow from the fact that ϕ is a bijection. We will prove (3) here and leave the remainder of the theorem to be proved in the exercises.

(3) Suppose that h_1 and h_2 are elements of H. Since ϕ is onto, there exist elements $g_1, g_2 \in G$ such that $\phi(g_1) = h_1$ and $\phi(g_2) = h_2$. Therefore,

$$h_1 h_2 = \phi(g_1)\phi(g_2) = \phi(g_1 g_2) = \phi(g_2 g_1) = \phi(g_2)\phi(g_1) = h_2 h_1.$$

■

We are now in a position to characterize all cyclic groups.

Theorem 9.7. All cyclic groups of infinite order are isomorphic to $\mathbb{Z}$.

Proof. Let G be a cyclic group with infinite order and suppose that a is a generator of G. Define a map $\phi : \mathbb{Z} \to G$ by $\phi : n \mapsto a^n$. Then

$$\phi(m+n) = a^{m+n} = a^m a^n = \phi(m)\phi(n).$$

To show that ϕ is injective, suppose that m and n are two elements in $\mathbb{Z}$, where $m \neq n$. We can assume that $m > n$. We must show that $a^m \neq a^n$. Let us suppose the contrary; that is, $a^m = a^n$. In this case $a^{m-n} = e$, where $m - n > 0$, which contradicts the fact that a has infinite order. Our map is onto since any element in G can be written as a^n for some integer n and $\phi(n) = a^n$.   ■

Theorem 9.8. If G is a cyclic group of order n, then G is isomorphic to $\mathbb{Z}_n$.

Proof. Let G be a cyclic group of order n generated by a and define a map $\phi : \mathbb{Z}_n \to G$ by $\phi : k \mapsto a^k$, where $0 \leq k < n$. The proof that ϕ is an isomorphism is one of the end-of-chapter exercises.   ■

Corollary 9.9. If G is a group of order p, where p is a prime number, then G is isomorphic to $\mathbb{Z}_p$.

Proof. The proof is a direct result of Corollary 6.12 on page 94. ∎

The main goal in group theory is to classify all groups; however, it makes sense to consider two groups to be the same if they are isomorphic. We state this result in the following theorem, whose proof is left as an exercise.

Theorem 9.10. The isomorphism of groups determines an equivalence relation on the class of all groups.

Hence, we can modify our goal of classifying all groups to classifying all groups *up to isomorphism*; that is, we will consider two groups to be the same if they are isomorphic.

Cayley's Theorem

Cayley proved that if G is a group, it is isomorphic to a group of permutations on some set; hence, every group is a permutation group. Cayley's Theorem is what we call a representation theorem. The aim of representation theory is to find an isomorphism of some group G that we wish to study into a group that we know a great deal about, such as a group of permutations or matrices.

Example 9.11. Consider the group $\mathbb{Z}_3$. The Cayley table for $\mathbb{Z}_3$ is as follows.

+	0	1	2
0	0	1	2
1	1	2	0
2	2	0	1

The addition table of $\mathbb{Z}_3$ suggests that it is the same as the permutation group $G = \{(0), (012), (021)\}$. The isomorphism here is

$$0 \mapsto \begin{pmatrix} 0 & 1 & 2 \\ 0 & 1 & 2 \end{pmatrix} = (0)$$

$$1 \mapsto \begin{pmatrix} 0 & 1 & 2 \\ 1 & 2 & 0 \end{pmatrix} = (012)$$

$$2 \mapsto \begin{pmatrix} 0 & 1 & 2 \\ 2 & 0 & 1 \end{pmatrix} = (021).$$

□

Theorem 9.12. Cayley. Every group is isomorphic to a group of permutations.

Proof. Let G be a group. We must find a group of permutations $\overline{G}$ that is isomorphic to G. For any $g \in G$, define a function $\lambda_g : G \to G$ by $\lambda_g(a) = ga$. We claim that λ_g is a permutation of G. To show that λ_g is one-to-one, suppose that $\lambda_g(a) = \lambda_g(b)$. Then

$$ga = \lambda_g(a) = \lambda_g(b) = gb.$$

Hence, $a = b$. To show that λ_g is onto, we must prove that for each $a \in G$, there is a b such that $\lambda_g(b) = a$. Let $b = g^{-1}a$.

Now we are ready to define our group $\overline{G}$. Let

$$\overline{G} = \{\lambda_g : g \in G\}.$$

We must show that $\overline{G}$ is a group under composition of functions and find an isomorphism between G and $\overline{G}$. We have closure under composition of functions since

$$(\lambda_g \circ \lambda_h)(a) = \lambda_g(ha) = gha = \lambda_{gh}(a).$$

Also,

$$\lambda_e(a) = ea = a$$

and

$$(\lambda_{g^{-1}} \circ \lambda_g)(a) = \lambda_{g^{-1}}(ga) = g^{-1}ga = a = \lambda_e(a).$$

We can define an isomorphism from G to $\overline{G}$ by $\phi : g \mapsto \lambda_g$. The group operation is preserved since

$$\phi(gh) = \lambda_{gh} = \lambda_g \lambda_h = \phi(g)\phi(h).$$

It is also one-to-one, because if $\phi(g)(a) = \phi(h)(a)$, then

$$ga = \lambda_g a = \lambda_h a = ha.$$

Hence, $g = h$. That ϕ is onto follows from the fact that $\phi(g) = \lambda_g$ for any $\lambda_g \in \overline{G}$. ∎

The isomorphism $g \mapsto \lambda_g$ is known as the ***left regular representation*** of G.

↪   Historical Note   ↩

Arthur Cayley was born in England in 1821, though he spent much of the first part of his life in Russia, where his father was a merchant. Cayley was educated at Cambridge, where he took the first Smith's Prize in mathematics. A lawyer for much of his adult life, he wrote several papers in his early twenties before entering the legal profession at the age of 25. While practicing law he continued his mathematical research, writing more than 300 papers during this period of his life. These included some of his best work. In 1863 he left law to become a professor at Cambridge. Cayley wrote more than 900 papers in fields such as group theory, geometry, and linear algebra. His legal knowledge was very valuable to Cambridge; he participated in the writing of many of the university's statutes. Cayley was also one of the people responsible for the admission of women to Cambridge.

9.2 Direct Products

Given two groups G and H, it is possible to construct a new group from the Cartesian product of G and H, $G \times H$. Conversely, given a large group, it is sometimes possible to decompose the group; that is, a group is sometimes isomorphic to the direct product of two smaller groups. Rather than studying a large group G, it is often easier to study the component groups of G.

External Direct Products

If $(G, \cdot)$ and $(H, \circ)$ are groups, then we can make the Cartesian product of G and H into a new group. As a set, our group is just the ordered pairs $(g, h) \in G \times H$ where $g \in G$ and $h \in H$. We can define a binary operation on $G \times H$ by

$$(g_1, h_1)(g_2, h_2) = (g_1 \cdot g_2, h_1 \circ h_2);$$

that is, we just multiply elements in the first coordinate as we do in G and elements in the second coordinate as we do in H. We have specified the particular operations $\cdot$ and $\circ$ in each group here for the sake of clarity; we usually just write $(g_1, h_1)(g_2, h_2) = (g_1 g_2, h_1 h_2)$.

Proposition 9.13. Let G and H be groups. The set $G \times H$ is a group under the operation $(g_1, h_1)(g_2, h_2) = (g_1 g_2, h_1 h_2)$ where $g_1, g_2 \in G$ and $h_1, h_2 \in H$.

Proof. Clearly the binary operation defined above is closed. If e_G and e_H are the identities of the groups G and H respectively, then (e_G, e_H) is the identity of $G \times H$. The inverse of $(g, h) \in G \times H$ is (g^{-1}, h^{-1}). The fact that the operation is associative follows directly from the associativity of G and H. ∎

Example 9.14. Let $\mathbb{R}$ be the group of real numbers under addition. The Cartesian product of $\mathbb{R}$ with itself, $\mathbb{R} \times \mathbb{R} = \mathbb{R}^2$, is also a group, in which the group operation is just addition in each coordinate; that is, $(a, b) + (c, d) = (a + c, b + d)$. The identity is $(0, 0)$ and the inverse of (a, b) is $(-a, -b)$. □

Example 9.15. Consider

$$\mathbb{Z}_2 \times \mathbb{Z}_2 = \{(0,0),(0,1),(1,0),(1,1)\}.$$

Although $\mathbb{Z}_2 \times \mathbb{Z}_2$ and $\mathbb{Z}_4$ both contain four elements, they are not isomorphic. Every element (a,b) in $\mathbb{Z}_2 \times \mathbb{Z}_2$ other than the identity has order 2, since $(a,b) + (a,b) = (0,0)$; however, $\mathbb{Z}_4$ is cyclic.   □

The group $G \times H$ is called the ***external direct product*** of G and H. Notice that there is nothing special about the fact that we have used only two groups to build a new group. The direct product

$$\prod_{i=1}^{n} G_i = G_1 \times G_2 \times \cdots \times G_n$$

of the groups $G_1, G_2, \ldots, G_n$ is defined in exactly the same manner. If $G = G_1 = G_2 = \cdots = G_n$, we often write G^n instead of $G_1 \times G_2 \times \cdots \times G_n$.

Example 9.16. The group $\mathbb{Z}_2^n$, considered as a set, is just the set of all binary n-tuples. The group operation is the "exclusive or" of two binary n-tuples. For example,

$$(01011101) + (01001011) = (00010110).$$

This group is important in coding theory, in cryptography, and in many areas of computer science.   □

Theorem 9.17. Let $(g,h) \in G \times H$. If g and h have finite orders r and s respectively, then the order of (g,h) in $G \times H$ is the least common multiple of r and s.

Proof. Suppose that m is the least common multiple of r and s and let $n = |(g,h)|$. Then

$$(g,h)^m = (g^m, h^m) = (e_G, e_H)$$
$$(g^n, h^n) = (g,h)^n = (e_G, e_H).$$

Hence, n must divide m, and $n \leq m$. However, by the second equation, both r and s must divide n; therefore, n is a common multiple of r and s. Since m is the *least common multiple* of r and s, $m \leq n$. Consequently, m must be equal to n. ∎

Corollary 9.18. Let $(g_1, \ldots, g_n) \in \prod G_i$. If g_i has finite order r_i in G_i, then the order of $(g_1, \ldots, g_n)$ in $\prod G_i$ is the least common multiple of $r_1, \ldots, r_n$.

Example 9.19. Let $(8,56) \in \mathbb{Z}_{12} \times \mathbb{Z}_{60}$. Since $\gcd(8,12) = 4$, the order of 8 is $12/4 = 3$ in $\mathbb{Z}_{12}$. Similarly, the order of 56 in $\mathbb{Z}_{60}$ is 15. The least common multiple of 3 and 15 is 15; hence, $(8,56)$ has order 15 in $\mathbb{Z}_{12} \times \mathbb{Z}_{60}$.   □

Example 9.20. The group $\mathbb{Z}_2 \times \mathbb{Z}_3$ consists of the pairs

$$(0,0), \qquad (0,1), \qquad (0,2), \qquad (1,0), \qquad (1,1), \qquad (1,2).$$

In this case, unlike that of $\mathbb{Z}_2 \times \mathbb{Z}_2$ and $\mathbb{Z}_4$, it is true that $\mathbb{Z}_2 \times \mathbb{Z}_3 \cong \mathbb{Z}_6$. We need only show that $\mathbb{Z}_2 \times \mathbb{Z}_3$ is cyclic. It is easy to see that $(1,1)$ is a generator for $\mathbb{Z}_2 \times \mathbb{Z}_3$.

$\square$

The next theorem tells us exactly when the direct product of two cyclic groups is cyclic.

Theorem 9.21. The group $\mathbb{Z}_m \times \mathbb{Z}_n$ is isomorphic to $\mathbb{Z}_{mn}$ if and only if $\gcd(m, n) = 1$.

Proof. We will first show that if $\mathbb{Z}_m \times \mathbb{Z}_n \cong \mathbb{Z}_{mn}$, then $\gcd(m, n) = 1$. We will prove the contrapositive; that is, we will show that if $\gcd(m, n) = d > 1$, then $\mathbb{Z}_m \times \mathbb{Z}_n$ cannot be cyclic. Notice that mn/d is divisible by both m and n; hence, for any element $(a, b) \in \mathbb{Z}_m \times \mathbb{Z}_n$,

$$\underbrace{(a, b) + (a, b) + \cdots + (a, b)}_{mn/d \text{ times}} = (0, 0).$$

Therefore, no (a, b) can generate all of $\mathbb{Z}_m \times \mathbb{Z}_n$.

The converse follows directly from Theorem 9.17 since $\operatorname{lcm}(m, n) = mn$ if and only if $\gcd(m, n) = 1$. $\blacksquare$

Corollary 9.22. Let $n_1, \ldots, n_k$ be positive integers. Then

$$\prod_{i=1}^{k} \mathbb{Z}_{n_i} \cong \mathbb{Z}_{n_1 \cdots n_k}$$

if and only if $\gcd(n_i, n_j) = 1$ for $i \neq j$.

Corollary 9.23. If

$$m = p_1^{e_1} \cdots p_k^{e_k},$$

where the p_is are distinct primes, then

$$\mathbb{Z}_m \cong \mathbb{Z}_{p_1^{e_1}} \times \cdots \times \mathbb{Z}_{p_k^{e_k}}.$$

Proof. Since the greatest common divisor of $p_i^{e_i}$ and $p_j^{e_j}$ is 1 for $i \neq j$, the proof follows from Corollary 9.22. $\blacksquare$

In Chapter 13, we will prove that all finite abelian groups are isomorphic to direct products of the form

$$\mathbb{Z}_{p_1^{e_1}} \times \cdots \times \mathbb{Z}_{p_k^{e_k}}$$

where $p_1, \ldots, p_k$ are (not necessarily distinct) primes.

Internal Direct Products

The external direct product of two groups builds a large group out of two smaller groups. We would like to be able to reverse this process and conveniently break down a group into its direct product components; that is, we would like to be able to say when a group is isomorphic to the direct product of two of its subgroups.

Let G be a group with subgroups H and K satisfying the following conditions.

- $G = HK = \{hk : h \in H, k \in K\}$;

- $H \cap K = \{e\}$;

- $hk = kh$ for all $k \in K$ and $h \in H$.

Then G is the *internal direct product* of H and K.

Example 9.24. The group $U(8)$ is the internal direct product of

$$H = \{1, 3\} \quad \text{and} \quad K = \{1, 5\}.$$

$\square$

Example 9.25. The dihedral group D_6 is an internal direct product of its two subgroups

$$H = \{\text{id}, r^3\} \quad \text{and} \quad K = \{\text{id}, r^2, r^4, s, r^2 s, r^4 s\}.$$

It can easily be shown that $K \cong S_3$; consequently, $D_6 \cong \mathbb{Z}_2 \times S_3$. $\square$

Example 9.26. Not every group can be written as the internal direct product of two of its proper subgroups. If the group S_3 were an internal direct product of its proper subgroups H and K, then one of the subgroups, say H, would have to have order 3. In this case H is the subgroup $\{(1), (123), (132)\}$. The subgroup K must have order 2, but no matter which subgroup we choose for K, the condition that $hk = kh$ will never be satisfied for $h \in H$ and $k \in K$. $\square$

Theorem 9.27. *Let G be the internal direct product of subgroups H and K. Then G is isomorphic to $H \times K$.*

Proof. Since G is an internal direct product, we can write any element $g \in G$ as $g = hk$ for some $h \in H$ and some $k \in K$. Define a map $\phi : G \to H \times K$ by $\phi(g) = (h, k)$.

The first problem that we must face is to show that ϕ is a well-defined map; that is, we must show that h and k are uniquely determined by g. Suppose that $g = hk = h'k'$. Then $h^{-1}h' = k(k')^{-1}$ is in both H and K, so it must be the identity. Therefore, $h = h'$ and $k = k'$, which proves that ϕ is, indeed, well-defined.

To show that ϕ preserves the group operation, let $g_1 = h_1 k_1$ and $g_2 = h_2 k_2$ and observe that

$$
\begin{aligned}
\phi(g_1 g_2) &= \phi(h_1 k_1 h_2 k_2) \\
&= \phi(h_1 h_2 k_1 k_2) \\
&= (h_1 h_2, k_1 k_2) \\
&= (h_1, k_1)(h_2, k_2) \\
&= \phi(g_1)\phi(g_2).
\end{aligned}
$$

We will leave the proof that ϕ is one-to-one and onto as an exercise. ∎

Example 9.28. The group $\mathbb{Z}_6$ is an internal direct product isomorphic to $\{0, 2, 4\} \times \{0, 3\}$. □

We can extend the definition of an internal direct product of G to a collection of subgroups $H_1, H_2, \ldots, H_n$ of G, by requiring that

- $G = H_1 H_2 \cdots H_n = \{h_1 h_2 \cdots h_n : h_i \in H_i\}$;
- $H_i \cap \langle \cup_{j \neq i} H_j \rangle = \{e\}$;
- $h_i h_j = h_j h_i$ for all $h_i \in H_i$ and $h_j \in H_j$.

We will leave the proof of the following theorem as an exercise.

Theorem 9.29. Let G be the internal direct product of subgroups H_i, where $i = 1, 2, \ldots, n$. Then G is isomorphic to $\prod_i H_i$.

Sage. Sage can quickly determine if two permutation groups are isomorphic, even though this should, in theory, be a very difficult computation.

9.3 Reading Questions

1. Determine the order of $(1, 2)$ in $\mathbb{Z}_4 \times \mathbb{Z}_8$.

2. List three properties of a group that are preserved by an isomorphism.

3. Find a group isomorphic to $\mathbb{Z}_{15}$ that is an external direct product of two non-trivial groups.

4. Explain why we can now say "*the* infinite cyclic group"?

5. Compare and contrast external direct products and internal direct products.

9.4 Exercises

1. Prove that $\mathbb{Z} \cong n\mathbb{Z}$ for $n \neq 0$.

2. Prove that $\mathbb{C}^*$ is isomorphic to the subgroup of $GL_2(\mathbb{R})$ consisting of matrices of the form

$$\begin{pmatrix} a & b \\ -b & a \end{pmatrix}.$$

3. Prove or disprove: $U(8) \cong \mathbb{Z}_4$.

4. Prove that $U(8)$ is isomorphic to the group of matrices

$$\begin{pmatrix} 1 & 0 \\ 0 & 1 \end{pmatrix}, \begin{pmatrix} 1 & 0 \\ 0 & -1 \end{pmatrix}, \begin{pmatrix} -1 & 0 \\ 0 & 1 \end{pmatrix}, \begin{pmatrix} -1 & 0 \\ 0 & -1 \end{pmatrix}.$$

5. Show that $U(5)$ is isomorphic to $U(10)$, but $U(12)$ is not.

6. Show that the nth roots of unity are isomorphic to $\mathbb{Z}_n$.

7. Show that any cyclic group of order n is isomorphic to $\mathbb{Z}_n$.

8. Prove that $\mathbb{Q}$ is not isomorphic to $\mathbb{Z}$.

9. Let $G = \mathbb{R} \setminus \{-1\}$ and define a binary operation on G by

$$a * b = a + b + ab.$$

   Prove that G is a group under this operation. Show that $(G, *)$ is isomorphic to the multiplicative group of nonzero real numbers.

10. Show that the matrices

$$\begin{pmatrix} 1 & 0 & 0 \\ 0 & 1 & 0 \\ 0 & 0 & 1 \end{pmatrix} \begin{pmatrix} 1 & 0 & 0 \\ 0 & 0 & 1 \\ 0 & 1 & 0 \end{pmatrix} \begin{pmatrix} 0 & 1 & 0 \\ 1 & 0 & 0 \\ 0 & 0 & 1 \end{pmatrix}$$

$$\begin{pmatrix} 0 & 0 & 1 \\ 1 & 0 & 0 \\ 0 & 1 & 0 \end{pmatrix} \begin{pmatrix} 0 & 0 & 1 \\ 0 & 1 & 0 \\ 1 & 0 & 0 \end{pmatrix} \begin{pmatrix} 0 & 1 & 0 \\ 0 & 0 & 1 \\ 1 & 0 & 0 \end{pmatrix}$$

   form a group. Find an isomorphism of G with a more familiar group of order 6.

11. Find five non-isomorphic groups of order 8.

12. Prove S_4 is not isomorphic to D_{12}.

13. Let $\omega = \text{cis}(2\pi/n)$ be a primitive nth root of unity. Prove that the matrices

$$A = \begin{pmatrix} \omega & 0 \\ 0 & \omega^{-1} \end{pmatrix} \quad \text{and} \quad B = \begin{pmatrix} 0 & 1 \\ 1 & 0 \end{pmatrix}$$

generate a multiplicative group isomorphic to D_n.

14. Show that the set of all matrices of the form

$$\begin{pmatrix} \pm 1 & k \\ 0 & 1 \end{pmatrix},$$

is a group isomorphic to D_n, where all entries in the matrix are in $\mathbb{Z}_n$.

15. List all of the elements of $\mathbb{Z}_4 \times \mathbb{Z}_2$.

16. Find the order of each of the following elements.

(a) $(3,4)$ in $\mathbb{Z}_4 \times \mathbb{Z}_6$

(b) $(6,15,4)$ in $\mathbb{Z}_{30} \times \mathbb{Z}_{45} \times \mathbb{Z}_{24}$

(c) $(5,10,15)$ in $\mathbb{Z}_{25} \times \mathbb{Z}_{25} \times \mathbb{Z}_{25}$

(d) $(8,8,8)$ in $\mathbb{Z}_{10} \times \mathbb{Z}_{24} \times \mathbb{Z}_{80}$

17. Prove that D_4 cannot be the internal direct product of two of its proper subgroups.

18. Prove that the subgroup of $\mathbb{Q}^*$ consisting of elements of the form $2^m 3^n$ for $m, n \in \mathbb{Z}$ is an internal direct product isomorphic to $\mathbb{Z} \times \mathbb{Z}$.

19. Prove that $S_3 \times \mathbb{Z}_2$ is isomorphic to D_6. Can you make a conjecture about D_{2n}? Prove your conjecture.

20. Prove or disprove: Every abelian group of order divisible by 3 contains a subgroup of order 3.

21. Prove or disprove: Every nonabelian group of order divisible by 6 contains a subgroup of order 6.

22. Let G be a group of order 20. If G has subgroups H and K of orders 4 and 5 respectively such that $hk = kh$ for all $h \in H$ and $k \in K$, prove that G is the internal direct product of H and K.

23. Prove or disprove the following assertion. Let G, H, and K be groups. If $G \times K \cong H \times K$, then $G \cong H$.

24. Prove or disprove: There is a noncyclic abelian group of order 51.

25. Prove or disprove: There is a noncyclic abelian group of order 52.

26. Let $\phi : G \to H$ be a group isomorphism. Show that $\phi(x) = e_H$ if and only if $x = e_G$, where e_G and e_H are the identities of G and H, respectively.

27. Let $G \cong H$. Show that if G is cyclic, then so is H.

28. Prove that any group G of order p, p prime, must be isomorphic to $\mathbb{Z}_p$.

29. Show that S_n is isomorphic to a subgroup of A_{n+2}.

30. Prove that D_n is isomorphic to a subgroup of S_n.

31. Let $\phi : G_1 \to G_2$ and $\psi : G_2 \to G_3$ be isomorphisms. Show that ϕ^{-1} and $\psi \circ \phi$ are both isomorphisms. Using these results, show that the isomorphism of groups determines an equivalence relation on the class of all groups.

32. Prove $U(5) \cong \mathbb{Z}_4$. Can you generalize this result for $U(p)$, where p is prime?

33. Write out the permutations associated with each element of S_3 in the proof of Cayley's Theorem.

34. An *automorphism* of a group G is an isomorphism with itself. Prove that complex conjugation is an automorphism of the additive group of complex numbers; that is, show that the map $\phi(a + bi) = a - bi$ is an isomorphism from $\mathbb{C}$ to $\mathbb{C}$.

35. Prove that $a + ib \mapsto a - ib$ is an automorphism of $\mathbb{C}^*$.

36. Prove that $A \mapsto B^{-1}AB$ is an automorphism of $SL_2(\mathbb{R})$ for all B in $GL_2(\mathbb{R})$.

37. We will denote the set of all automorphisms of G by $\operatorname{Aut}(G)$. Prove that $\operatorname{Aut}(G)$ is a subgroup of S_G, the group of permutations of G.

38. Find $\operatorname{Aut}(\mathbb{Z}_6)$.

39. Find $\operatorname{Aut}(\mathbb{Z})$.

40. Find two nonisomorphic groups G and H such that $\operatorname{Aut}(G) \cong \operatorname{Aut}(H)$.

41. Let G be a group and $g \in G$. Define a map $i_g : G \to G$ by $i_g(x) = gxg^{-1}$. Prove that i_g defines an automorphism of G. Such an automorphism is called an *inner automorphism*. The set of all inner automorphisms is denoted by $\operatorname{Inn}(G)$.

42. Prove that $\operatorname{Inn}(G)$ is a subgroup of $\operatorname{Aut}(G)$.

43. What are the inner automorphisms of the quaternion group Q_8? Is $\text{Inn}(G) = \text{Aut}(G)$ in this case?

44. Let G be a group and $g \in G$. Define maps $\lambda_g : G \to G$ and $\rho_g : G \to G$ by $\lambda_g(x) = gx$ and $\rho_g(x) = xg^{-1}$. Show that $i_g = \rho_g \circ \lambda_g$ is an automorphism of G. The isomorphism $g \mapsto \rho_g$ is called the *right regular representation* of G.

45. Let G be the internal direct product of subgroups H and K. Show that the map $\phi : G \to H \times K$ defined by $\phi(g) = (h, k)$ for $g = hk$, where $h \in H$ and $k \in K$, is one-to-one and onto.

46. Let G and H be isomorphic groups. If G has a subgroup of order n, prove that H must also have a subgroup of order n.

47. If $G \cong \overline{G}$ and $H \cong \overline{H}$, show that $G \times H \cong \overline{G} \times \overline{H}$.

48. Prove that $G \times H$ is isomorphic to $H \times G$.

49. Let $n_1, \ldots, n_k$ be positive integers. Show that

$$\prod_{i=1}^{k} \mathbb{Z}_{n_i} \cong \mathbb{Z}_{n_1 \cdots n_k}$$

if and only if $\gcd(n_i, n_j) = 1$ for $i \neq j$.

50. Prove that $A \times B$ is abelian if and only if A and B are abelian.

51. If G is the internal direct product of $H_1, H_2, \ldots, H_n$, prove that G is isomorphic to $\prod_i H_i$.

52. Let H_1 and H_2 be subgroups of G_1 and G_2, respectively. Prove that $H_1 \times H_2$ is a subgroup of $G_1 \times G_2$.

53. Let $m, n \in \mathbb{Z}$. Prove that $\langle m, n \rangle = \langle d \rangle$ if and only if $d = \gcd(m, n)$.

54. Let $m, n \in \mathbb{Z}$. Prove that $\langle m \rangle \cap \langle n \rangle = \langle l \rangle$ if and only if $l = \text{lcm}(m, n)$.

55. Groups of order $2p$. In this series of exercises we will classify all groups of order $2p$, where p is an odd prime.

(a) Assume G is a group of order $2p$, where p is an odd prime. If $a \in G$, show that a must have order 1, 2, p, or $2p$.

(b) Suppose that G has an element of order $2p$. Prove that G is isomorphic to $\mathbb{Z}_{2p}$. Hence, G is cyclic.

(c) Suppose that G does not contain an element of order $2p$. Show that G must contain an element of order p. *Hint*: Assume that G does not contain an element of order p.

(d) Suppose that G does not contain an element of order $2p$. Show that G must contain an element of order 2.

(e) Let P be a subgroup of G with order p and $y \in G$ have order 2. Show that $yP = Py$.

(f) Suppose that G does not contain an element of order $2p$ and $P = \langle z \rangle$ is a subgroup of order p generated by z. If y is an element of order 2, then $yz = z^k y$ for some $2 \leq k < p$.

(g) Suppose that G does not contain an element of order $2p$. Prove that G is not abelian.

(h) Suppose that G does not contain an element of order $2p$ and $P = \langle z \rangle$ is a subgroup of order p generated by z and y is an element of order 2. Show that we can list the elements of G as $\{z^i y^j \mid 0 \leq i < p, 0 \leq j < 2\}$.

(i) Suppose that G does not contain an element of order $2p$ and $P = \langle z \rangle$ is a subgroup of order p generated by z and y is an element of order 2. Prove that the product $(z^i y^j)(z^r y^s)$ can be expressed as a uniquely as $z^m y^n$ for some non negative integers m, n. Thus, conclude that there is only one possibility for a non-abelian group of order $2p$, it must therefore be the one we have seen already, the dihedral group.

10

Normal Subgroups and Factor Groups

*I*f H is a subgroup of a group G, then right cosets are not always the same as left cosets; that is, it is not always the case that $gH = Hg$ for all $g \in G$. The subgroups for which this property holds play a critical role in group theory—they allow for the construction of a new class of groups, called factor or quotient groups. Factor groups may be studied directly or by using homomorphisms, a generalization of isomorphisms. We will study homomorphisms in Chapter 11.

10.1 Factor Groups and Normal Subgroups

Normal Subgroups

A subgroup H of a group G is **normal** in G if $gH = Hg$ for all $g \in G$. That is, a normal subgroup of a group G is one in which the right and left cosets are precisely the same.

Example 10.1. Let G be an abelian group. Every subgroup H of G is a normal subgroup. Since $gh = hg$ for all $g \in G$ and $h \in H$, it will always be the case that $gH = Hg$. □

Example 10.2. Let H be the subgroup of S_3 consisting of elements (1) and (12). Since

$$(123)H = \{(123), (13)\} \quad \text{and} \quad H(123) = \{(123), (23)\},$$

H cannot be a normal subgroup of S_3. However, the subgroup N, consisting of the permutations (1), (123), and (132), is normal since the cosets of N are

$$N = \{(1), (123), (132)\}$$
$$(12)N = N(12) = \{(12), (13), (23)\}.$$

□

The following theorem is fundamental to our understanding of normal subgroups.

Theorem 10.3. Let G be a group and N be a subgroup of G. Then the following statements are equivalent.

1. The subgroup N is normal in G.

2. For all $g \in G$, $gNg^{-1} \subset N$.

3. For all $g \in G$, $gNg^{-1} = N$.

Proof. (1) $\Rightarrow$ (2). Since N is normal in G, $gN = Ng$ for all $g \in G$. Hence, for a given $g \in G$ and $n \in N$, there exists an n' in N such that $gn = n'g$. Therefore, $gng^{-1} = n' \in N$ or $gNg^{-1} \subset N$.

(2) $\Rightarrow$ (3). Let $g \in G$. Since $gNg^{-1} \subset N$, we need only show $N \subset gNg^{-1}$. For $n \in N$, $g^{-1}ng = g^{-1}n(g^{-1})^{-1} \in N$. Hence, $g^{-1}ng = n'$ for some $n' \in N$. Therefore, $n = gn'g^{-1}$ is in gNg^{-1}.

(3) $\Rightarrow$ (1). Suppose that $gNg^{-1} = N$ for all $g \in G$. Then for any $n \in N$ there exists an $n' \in N$ such that $gng^{-1} = n'$. Consequently, $gn = n'g$ or $gN \subset Ng$. Similarly, $Ng \subset gN$.   ∎

Factor Groups

If N is a normal subgroup of a group G, then the cosets of N in G form a group G/N under the operation $(aN)(bN) = abN$. This group is called the *factor* or *quotient group* of G and N. Our first task is to prove that G/N is indeed a group.

Theorem 10.4. Let N be a normal subgroup of a group G. The cosets of N in G form a group G/N of order $[G : N]$.

Proof. The group operation on G/N is $(aN)(bN) = abN$. This operation must be shown to be well-defined; that is, group multiplication must be independent of the choice of coset representative. Let $aN = bN$ and $cN = dN$. We must show that

$$(aN)(cN) = acN = bdN = (bN)(dN).$$

Then $a = bn_1$ and $c = dn_2$ for some n_1 and n_2 in N. Hence,

$$
\begin{aligned}
acN &= bn_1dn_2N \\
&= bn_1dN \\
&= bn_1Nd \\
&= bNd \\
&= bdN.
\end{aligned}
$$

The remainder of the theorem is easy: $eN = N$ is the identity and $g^{-1}N$ is the inverse of gN. The order of G/N is, of course, the number of cosets of N in G.   ∎

It is very important to remember that the elements in a factor group are *sets of elements* in the original group.

Example 10.5. Consider the normal subgroup of S_3, $N = \{(1), (123), (132)\}$. The cosets of N in S_3 are N and $(12)N$. The factor group S_3/N has the following multiplication table.

	N	$(12)N$
N	N	$(12)N$
$(12)N$	$(12)N$	N

This group is isomorphic to $\mathbb{Z}_2$. At first, multiplying cosets seems both complicated and strange; however, notice that S_3/N is a smaller group. The factor group displays a certain amount of information about S_3. Actually, $N = A_3$, the group of even permutations, and $(12)N = \{(12), (13), (23)\}$ is the set of odd permutations. The information captured in G/N is parity; that is, multiplying two even or two odd permutations results in an even permutation, whereas multiplying an odd permutation by an even permutation yields an odd permutation. □

Example 10.6. Consider the normal subgroup $3\mathbb{Z}$ of $\mathbb{Z}$. The cosets of $3\mathbb{Z}$ in $\mathbb{Z}$ are

$$0 + 3\mathbb{Z} = \{\ldots, -3, 0, 3, 6, \ldots\}$$
$$1 + 3\mathbb{Z} = \{\ldots, -2, 1, 4, 7, \ldots\}$$
$$2 + 3\mathbb{Z} = \{\ldots, -1, 2, 5, 8, \ldots\}.$$

The group $\mathbb{Z}/3\mathbb{Z}$ is given by the Cayley table below.

+	$0 + 3\mathbb{Z}$	$1 + 3\mathbb{Z}$	$2 + 3\mathbb{Z}$
$0 + 3\mathbb{Z}$	$0 + 3\mathbb{Z}$	$1 + 3\mathbb{Z}$	$2 + 3\mathbb{Z}$
$1 + 3\mathbb{Z}$	$1 + 3\mathbb{Z}$	$2 + 3\mathbb{Z}$	$0 + 3\mathbb{Z}$
$2 + 3\mathbb{Z}$	$2 + 3\mathbb{Z}$	$0 + 3\mathbb{Z}$	$1 + 3\mathbb{Z}$

In general, the subgroup $n\mathbb{Z}$ of $\mathbb{Z}$ is normal. The cosets of $\mathbb{Z}/n\mathbb{Z}$ are

$$n\mathbb{Z}$$
$$1 + n\mathbb{Z}$$
$$2 + n\mathbb{Z}$$
$$\vdots$$
$$(n-1) + n\mathbb{Z}.$$

The sum of the cosets $k + n\mathbb{Z}$ and $l + n\mathbb{Z}$ is $k + l + n\mathbb{Z}$. Notice that we have written our cosets additively, because the group operation is integer addition. □

Example 10.7. Consider the dihedral group D_n, generated by the two elements r and s, satisfying the relations

$$r^n = \mathrm{id}$$
$$s^2 = \mathrm{id}$$
$$srs = r^{-1}.$$

The element r actually generates the cyclic subgroup of rotations, R_n, of D_n. Since $srs^{-1} = srs = r^{-1} \in R_n$, the group of rotations is a normal subgroup of D_n; therefore, D_n/R_n is a group. Since there are exactly two elements in this group, it must be isomorphic to $\mathbb{Z}_2$. □

10.2 The Simplicity of the Alternating Group

Of special interest are groups with no nontrivial normal subgroups. Such groups are called *simple groups*. Of course, we already have a whole class of examples of simple groups, $\mathbb{Z}_p$, where p is prime. These groups are trivially simple since they have no proper subgroups other than the subgroup consisting solely of the identity. Other examples of simple groups are not so easily found. We can, however, show that the alternating group, A_n, is simple for $n \geq 5$. The proof of this result requires several lemmas.

Lemma 10.8. The alternating group A_n is generated by 3-cycles for $n \geq 3$.

Proof. To show that the 3-cycles generate A_n, we need only show that any pair of transpositions can be written as the product of 3-cycles. Since $(a, b) = (b, a)$, every pair of transpositions must be one of the following:

$$(a, b)(a, b) = \mathrm{id}$$
$$(a, b)(c, d) = (a, c, b)(a, c, d)$$
$$(a, b)(a, c) = (a, c, b).$$

■

Lemma 10.9. Let N be a normal subgroup of A_n, where $n \geq 3$. If N contains a 3-cycle, then $N = A_n$.

Proof. We will first show that A_n is generated by 3-cycles of the specific form (i, j, k), where i and j are fixed in $\{1, 2, \ldots, n\}$ and we let k vary. Every 3-cycle is the product of 3-cycles of this form, since

$$(i, a, j) = (i, j, a)^2$$

$$(i, a, b) = (i, j, b)(i, j, a)^2$$
$$(j, a, b) = (i, j, b)^2(i, j, a)$$
$$(a, b, c) = (i, j, a)^2(i, j, c)(i, j, b)^2(i, j, a).$$

Now suppose that N is a nontrivial normal subgroup of A_n for $n \geq 3$ such that N contains a 3-cycle of the form (i, j, a). Using the normality of N, we see that

$$[(i, j)(a, k)](i, j, a)^2[(i, j)(a, k)]^{-1} = (i, j, k)$$

is in N. Hence, N must contain all of the 3-cycles (i, j, k) for $1 \leq k \leq n$. By Lemma 10.8, these 3-cycles generate A_n; hence, $N = A_n$. ∎

Lemma 10.10. For $n \geq 5$, every nontrivial normal subgroup N of A_n contains a 3-cycle.

Proof. Let σ be an arbitrary element in a normal subgroup N. There are several possible cycle structures for σ.

- σ is a 3-cycle.

- σ is the product of disjoint cycles, $\sigma = \tau(a_1, a_2, \ldots, a_r) \in N$, where $r > 3$.

- σ is the product of disjoint cycles, $\sigma = \tau(a_1, a_2, a_3)(a_4, a_5, a_6)$.

- $\sigma = \tau(a_1, a_2, a_3)$, where τ is the product of disjoint 2-cycles.

- $\sigma = \tau(a_1, a_2)(a_3, a_4)$, where τ is the product of an even number of disjoint 2-cycles.

If σ is a 3-cycle, then we are done. If N contains a product of disjoint cycles, σ, and at least one of these cycles has length greater than 3, say $\sigma = \tau(a_1, a_2, \ldots, a_r)$, then

$$(a_1, a_2, a_3)\sigma(a_1, a_2, a_3)^{-1}$$

is in N since N is normal; hence,

$$\sigma^{-1}(a_1, a_2, a_3)\sigma(a_1, a_2, a_3)^{-1}$$

is also in N. Since

$$\sigma^{-1}(a_1, a_2, a_3)\sigma(a_1, a_2, a_3)^{-1} = \sigma^{-1}(a_1, a_2, a_3)\sigma(a_1, a_3, a_2)$$
$$= (a_1, a_2, \ldots, a_r)^{-1}\tau^{-1}(a_1, a_2, a_3)\tau(a_1, a_2, \ldots, a_r)(a_1, a_3, a_2)$$
$$= (a_1, a_r, a_{r-1}, \ldots, a_2)(a_1, a_2, a_3)(a_1, a_2, \ldots, a_r)(a_1, a_3, a_2)$$
$$= (a_1, a_3, a_r),$$

N must contain a 3-cycle; hence, $N = A_n$.

Now suppose that N contains a disjoint product of the form

$$\sigma = \tau(a_1, a_2, a_3)(a_4, a_5, a_6).$$

Then

$$\sigma^{-1}(a_1, a_2, a_4)\sigma(a_1, a_2, a_4)^{-1} \in N$$

since

$$(a_1, a_2, a_4)\sigma(a_1, a_2, a_4)^{-1} \in N.$$

So

$$\sigma^{-1}(a_1, a_2, a_4)\sigma(a_1, a_2, a_4)^{-1}$$
$$= [\tau(a_1, a_2, a_3)(a_4, a_5, a_6)]^{-1}(a_1, a_2, a_4)\tau(a_1, a_2, a_3)(a_4, a_5, a_6)(a_1, a_2, a_4)^{-1}$$
$$= (a_4, a_6, a_5)(a_1, a_3, a_2)\tau^{-1}(a_1, a_2, a_4)\tau(a_1, a_2, a_3)(a_4, a_5, a_6)(a_1, a_4, a_2)$$
$$= (a_4, a_6, a_5)(a_1, a_3, a_2)(a_1, a_2, a_4)(a_1, a_2, a_3)(a_4, a_5, a_6)(a_1, a_4, a_2)$$
$$= (a_1, a_4, a_2, a_6, a_3).$$

So N contains a disjoint cycle of length greater than 3, and we can apply the previous case.

Suppose N contains a disjoint product of the form $\sigma = \tau(a_1, a_2, a_3)$, where τ is the product of disjoint 2-cycles. Since $\sigma \in N$, $\sigma^2 \in N$, and

$$\sigma^2 = \tau(a_1, a_2, a_3)\tau(a_1, a_2, a_3)$$
$$= (a_1, a_3, a_2).$$

So N contains a 3-cycle.

The only remaining possible case is a disjoint product of the form

$$\sigma = \tau(a_1, a_2)(a_3, a_4),$$

where τ is the product of an even number of disjoint 2-cycles. But

$$\sigma^{-1}(a_1, a_2, a_3)\sigma(a_1, a_2, a_3)^{-1}$$

is in N since $(a_1, a_2, a_3)\sigma(a_1, a_2, a_3)^{-1}$ is in N; and so

$$\sigma^{-1}(a_1, a_2, a_3)\sigma(a_1, a_2, a_3)^{-1}$$
$$= \tau^{-1}(a_1, a_2)(a_3, a_4)(a_1, a_2, a_3)\tau(a_1, a_2)(a_3, a_4)(a_1, a_2, a_3)^{-1}$$
$$= (a_1, a_3)(a_2, a_4).$$

Since $n \geq 5$, we can find $b \in \{1, 2, \ldots, n\}$ such that $b \neq a_1, a_2, a_3, a_4$. Let $\mu = (a_1, a_3, b)$. Then

$$\mu^{-1}(a_1, a_3)(a_2, a_4)\mu(a_1, a_3)(a_2, a_4) \in N$$

and

$$\mu^{-1}(a_1, a_3)(a_2, a_4)\mu(a_1, a_3)(a_2, a_4)$$
$$= (a_1, ba_3)(a_1, a_3)(a_2, a_4)(a_1, a_3, b)(a_1, a_3)(a_2, a_4)$$
$$= (a_1 a_3 b).$$

Therefore, N contains a 3-cycle. This completes the proof of the lemma.   ∎

Theorem 10.11. The alternating group, A_n, is simple for $n \geq 5$.

Proof. Let N be a normal subgroup of A_n. By Lemma 10.10 on page 161, N contains a 3-cycle. By Lemma 10.9 on page 160, $N = A_n$; therefore, A_n contains no proper nontrivial normal subgroups for $n \geq 5$.   ∎

Sage. Sage can easily determine if a subgroup is normal or not. If so, it can create the quotient group. However, the construction creates a new permuation group, isomorphic to the quotient group, so its utility is limited.

↪   Historical Note   ↩

One of the foremost problems of group theory has been to classify all simple finite groups. This problem is over a century old and has been solved only in the last few decades of the twentieth century. In a sense, finite simple groups are the building blocks of all finite groups. The first nonabelian simple groups to be discovered were the alternating groups. Galois was the first to prove that A_5 was simple. Later, mathematicians such as C. Jordan and L. E. Dickson found several infinite families of matrix groups that were simple. Other families of simple groups were discovered in the 1950s. At the turn of the century, William Burnside conjectured that all nonabelian simple groups must have even order. In 1963, W. Feit and J. Thompson proved Burnside's conjecture and published their results in the paper "Solvability of Groups of Odd Order," which appeared in the *Pacific Journal of Mathematics*. Their proof, running over 250 pages, gave impetus to a program in the 1960s and 1970s to classify all finite simple groups. Daniel Gorenstein was the organizer of this remarkable effort. One of the last simple groups was the "Monster," discovered by R. Greiss. The Monster, a 196,833 × 196,833 matrix group, is one of the 26 sporadic, or special, simple groups. These sporadic simple groups are groups that fit into no infinite family of simple groups. Some of the sporadic groups play an important role in physics.

10.3 Reading Questions

1. Let G be the group of symmetries of an equilateral triangle, expressed as permutations of the vertices numbered $1, 2, 3$. Let H be the subgroup $H = \langle(12)\rangle$. Build the left and right cosets of H in G.

2. Based on your answer to the previous question, is H normal in G? Explain why or why not.

3. The subgroup $8\mathbb{Z}$ is normal in $\mathbb{Z}$. In the factor group $\mathbb{Z}/8\mathbb{Z}$ perform the computation $(3 + 8\mathbb{Z}) + (7 + 8\mathbb{Z})$.

4. List two statements about a group G and a subgroup H that are equivalent to "H is normal in G."

5. In your own words, what is a factor group?

10.4 Exercises

1. For each of the following groups G, determine whether H is a normal subgroup of G. If H is a normal subgroup, write out a Cayley table for the factor group G/H.

    (a) $G = S_4$ and $H = A_4$

    (b) $G = A_5$ and $H = \{(1), (123), (132)\}$

    (c) $G = S_4$ and $H = D_4$

    (d) $G = Q_8$ and $H = \{1, -1, I, -I\}$

    (e) $G = \mathbb{Z}$ and $H = 5\mathbb{Z}$

2. Find all the subgroups of D_4. Which subgroups are normal? What are all the factor groups of D_4 up to isomorphism?

3. Find all the subgroups of the quaternion group, Q_8. Which subgroups are normal? What are all the factor groups of Q_8 up to isomorphism?

4. Let T be the group of nonsingular upper triangular 2×2 matrices with entries in $\mathbb{R}$; that is, matrices of the form

$$\begin{pmatrix} a & b \\ 0 & c \end{pmatrix},$$

where $a, b, c \in \mathbb{R}$ and $ac \neq 0$. Let U consist of matrices of the form

$$\begin{pmatrix} 1 & x \\ 0 & 1 \end{pmatrix},$$

where $x \in \mathbb{R}$.

 (a) Show that U is a subgroup of T.

 (b) Prove that U is abelian.

 (c) Prove that U is normal in T.

 (d) Show that T/U is abelian.

 (e) Is T normal in $GL_2(\mathbb{R})$?

5. Show that the intersection of two normal subgroups is a normal subgroup.

6. If G is abelian, prove that G/H must also be abelian.

7. Prove or disprove: If H is a normal subgroup of G such that H and G/H are abelian, then G is abelian.

8. If G is cyclic, prove that G/H must also be cyclic.

9. Prove or disprove: If H and G/H are cyclic, then G is cyclic.

10. Let H be a subgroup of index 2 of a group G. Prove that H must be a normal subgroup of G. Conclude that S_n is not simple for $n \geq 3$.

11. If a group G has exactly one subgroup H of order k, prove that H is normal in G.

12. Define the *centralizer* of an element g in a group G to be the set

$$C(g) = \{x \in G : xg = gx\}.$$

Show that $C(g)$ is a subgroup of G. If g generates a normal subgroup of G, prove that $C(g)$ is normal in G.

13. Recall that the *center* of a group G is the set

$$Z(G) = \{x \in G : xg = gx \text{ for all } g \in G\}.$$

(a) Calculate the center of S_3.

(b) Calculate the center of $GL_2(\mathbb{R})$.

(c) Show that the center of any group G is a normal subgroup of G.

(d) If $G/Z(G)$ is cyclic, show that G is abelian.

14. Let G be a group and let $G' = \langle aba^{-1}b^{-1} \rangle$; that is, G' is the subgroup of all finite products of elements in G of the form $aba^{-1}b^{-1}$. The subgroup G' is called the *commutator subgroup* of G.

(a) Show that G' is a normal subgroup of G.

(b) Let N be a normal subgroup of G. Prove that G/N is abelian if and only if N contains the commutator subgroup of G.

Homomorphisms

One of the basic ideas of algebra is the concept of a homomorphism, a natural generalization of an isomorphism. If we relax the requirement that an isomorphism of groups be bijective, we have a homomorphism.

11.1 Group Homomorphisms

A *homomorphism* between groups $(G, \cdot)$ and $(H, \circ)$ is a map $\phi : G \to H$ such that

$$\phi(g_1 \cdot g_2) = \phi(g_1) \circ \phi(g_2)$$

for $g_1, g_2 \in G$. The range of ϕ in H is called the *homomorphic image* of ϕ.

Two groups are related in the strongest possible way if they are isomorphic; however, a weaker relationship may exist between two groups. For example, the symmetric group S_n and the group $\mathbb{Z}_2$ are related by the fact that S_n can be divided into even and odd permutations that exhibit a group structure like that $\mathbb{Z}_2$, as shown in the following multiplication table.

	even	odd
even	even	odd
odd	odd	even

We use homomorphisms to study relationships such as the one we have just described.

Example 11.1. Let G be a group and $g \in G$. Define a map $\phi : \mathbb{Z} \to G$ by $\phi(n) = g^n$. Then ϕ is a group homomorphism, since

$$\phi(m + n) = g^{m+n} = g^m g^n = \phi(m)\phi(n).$$

This homomorphism maps $\mathbb{Z}$ onto the cyclic subgroup of G generated by g. □

Example 11.2. Let $G = GL_2(\mathbb{R})$. If

$$A = \begin{pmatrix} a & b \\ c & d \end{pmatrix}$$

is in G, then the determinant is nonzero; that is, $\det(A) = ad - bc \neq 0$. Also, for any two elements A and B in G, $\det(AB) = \det(A)\det(B)$. Using the determinant, we can define a homomorphism $\phi : GL_2(\mathbb{R}) \to \mathbb{R}^*$ by $A \mapsto \det(A)$.   □

Example 11.3. Recall that the circle group $\mathbb{T}$ consists of all complex numbers z such that $|z| = 1$. We can define a homomorphism ϕ from the additive group of real numbers $\mathbb{R}$ to $\mathbb{T}$ by $\phi : \theta \mapsto \cos\theta + i\sin\theta$. Indeed,

$$\phi(\alpha + \beta) = \cos(\alpha + \beta) + i\sin(\alpha + \beta)$$
$$= (\cos\alpha\cos\beta - \sin\alpha\sin\beta) + i(\sin\alpha\cos\beta + \cos\alpha\sin\beta)$$
$$= (\cos\alpha + i\sin\alpha)(\cos\beta + i\sin\beta)$$
$$= \phi(\alpha)\phi(\beta).$$

Geometrically, we are simply wrapping the real line around the circle in a group-theoretic fashion.   □

The following proposition lists some basic properties of group homomorphisms.

Proposition 11.4. Let $\phi : G_1 \to G_2$ be a homomorphism of groups. Then

1. If e is the identity of G_1, then $\phi(e)$ is the identity of G_2;

2. For any element $g \in G_1$, $\phi(g^{-1}) = [\phi(g)]^{-1}$;

3. If H_1 is a subgroup of G_1, then $\phi(H_1)$ is a subgroup of G_2;

4. If H_2 is a subgroup of G_2, then $\phi^{-1}(H_2) = \{g \in G_1 : \phi(g) \in H_2\}$ is a subgroup of G_1. Furthermore, if H_2 is normal in G_2, then $\phi^{-1}(H_2)$ is normal in G_1.

Proof. (1) Suppose that e and e' are the identities of G_1 and G_2, respectively; then

$$e'\phi(e) = \phi(e) = \phi(ee) = \phi(e)\phi(e).$$

By cancellation, $\phi(e) = e'$.

(2) This statement follows from the fact that

$$\phi(g^{-1})\phi(g) = \phi(g^{-1}g) = \phi(e) = e'.$$

(3) The set $\phi(H_1)$ is nonempty since the identity of G_2 is in $\phi(H_1)$. Suppose that H_1 is a subgroup of G_1 and let x and y be in $\phi(H_1)$. There exist elements $a, b \in H_1$ such that $\phi(a) = x$ and $\phi(b) = y$. Since

$$xy^{-1} = \phi(a)[\phi(b)]^{-1} = \phi(ab^{-1}) \in \phi(H_1),$$

$\phi(H_1)$ is a subgroup of G_2 by Proposition 3.31 on page 49.

(4) Let H_2 be a subgroup of G_2 and define H_1 to be $\phi^{-1}(H_2)$; that is, H_1 is the set of all $g \in G_1$ such that $\phi(g) \in H_2$. The identity is in H_1 since $\phi(e) = e'$. If a and b are in H_1, then $\phi(ab^{-1}) = \phi(a)[\phi(b)]^{-1}$ is in H_2 since H_2 is a subgroup of G_2. Therefore, $ab^{-1} \in H_1$ and H_1 is a subgroup of G_1. If H_2 is normal in G_2, we must show that $g^{-1}hg \in H_1$ for $h \in H_1$ and $g \in G_1$. But

$$\phi(g^{-1}hg) = [\phi(g)]^{-1}\phi(h)\phi(g) \in H_2,$$

since H_2 is a normal subgroup of G_2. Therefore, $g^{-1}hg \in H_1$. ∎

Let $\phi : G \to H$ be a group homomorphism and suppose that e is the identity of H. By Proposition 11.4, $\phi^{-1}(\{e\})$ is a subgroup of G. This subgroup is called the *kernel* of ϕ and will be denoted by ker ϕ. In fact, this subgroup is a normal subgroup of G since the trivial subgroup is normal in H. We state this result in the following theorem, which says that with every homomorphism of groups we can naturally associate a normal subgroup.

Theorem 11.5. Let $\phi : G \to H$ be a group homomorphism. Then the kernel of ϕ is a normal subgroup of G.

Example 11.6. Let us examine the homomorphism $\phi : GL_2(\mathbb{R}) \to \mathbb{R}^*$ defined by $A \mapsto \det(A)$. Since 1 is the identity of $\mathbb{R}^*$, the kernel of this homomorphism is all 2×2 matrices having determinant one. That is, ker $\phi = SL_2(\mathbb{R})$. □

Example 11.7. The kernel of the group homomorphism $\phi : \mathbb{R} \to \mathbb{C}^*$ defined by $\phi(\theta) = \cos\theta + i\sin\theta$ is $\{2\pi n : n \in \mathbb{Z}\}$. Notice that ker $\phi \cong \mathbb{Z}$. □

Example 11.8. Suppose that we wish to determine all possible homomorphisms ϕ from $\mathbb{Z}_7$ to $\mathbb{Z}_{12}$. Since the kernel of ϕ must be a subgroup of $\mathbb{Z}_7$, there are only two possible kernels, $\{0\}$ and all of $\mathbb{Z}_7$. The image of a subgroup of $\mathbb{Z}_7$ must be a subgroup of $\mathbb{Z}_{12}$. Hence, there is no injective homomorphism; otherwise, $\mathbb{Z}_{12}$ would have a subgroup of order 7, which is impossible. Consequently, the only possible homomorphism from $\mathbb{Z}_7$ to $\mathbb{Z}_{12}$ is the one mapping all elements to zero. □

Example 11.9. Let G be a group. Suppose that $g \in G$ and ϕ is the homomorphism from $\mathbb{Z}$ to G given by $\phi(n) = g^n$. If the order of g is infinite, then the kernel of this homomorphism is $\{0\}$ since ϕ maps $\mathbb{Z}$ onto the cyclic subgroup of G generated by g. However, if the order of g is finite, say n, then the kernel of ϕ is $n\mathbb{Z}$. □

11.2 The Isomorphism Theorems

Although it is not evident at first, factor groups correspond exactly to homomorphic images, and we can use factor groups to study homomorphisms. We already know that with every group homomorphism $\phi : G \to H$ we can associate a normal subgroup of G, ker ϕ. The converse is also true; that is, every normal subgroup of a group G gives rise to homomorphism of groups.

Let H be a normal subgroup of G. Define the *natural* or *canonical homomor-phism*

$$\phi : G \to G/H$$

by

$$\phi(g) = gH.$$

This is indeed a homomorphism, since

$$\phi(g_1 g_2) = g_1 g_2 H = g_1 H g_2 H = \phi(g_1)\phi(g_2).$$

The kernel of this homomorphism is H. The following theorems describe the relationships between group homomorphisms, normal subgroups, and factor groups.

Theorem 11.10. First Isomorphism Theorem. If $\psi : G \to H$ is a group homomor-phism with $K = \ker \psi$, then K is normal in G. Let $\phi : G \to G/K$ be the canonical homomorphism. Then there exists a unique isomorphism $\eta : G/K \to \psi(G)$ such that $\psi = \eta\phi$.

Proof. We already know that K is normal in G. Define $\eta : G/K \to \psi(G)$ by $\eta(gK) = \psi(g)$. We first show that η is a well-defined map. If $g_1 K = g_2 K$, then for some $k \in K$, $g_1 k = g_2$; consequently,

$$\eta(g_1 K) = \psi(g_1) = \psi(g_1)\psi(k) = \psi(g_1 k) = \psi(g_2) = \eta(g_2 K).$$

Thus, η does not depend on the choice of coset representatives and the map $\eta : G/K \to \psi(G)$ is uniquely defined since $\psi = \eta\phi$. We must also show that η is a homomorphism. Indeed,

$$\begin{aligned}
\eta(g_1 K g_2 K) &= \eta(g_1 g_2 K) \\
&= \psi(g_1 g_2) \\
&= \psi(g_1)\psi(g_2) \\
&= \eta(g_1 K)\eta(g_2 K).
\end{aligned}$$

Clearly, η is onto $\psi(G)$. To show that η is one-to-one, suppose that $\eta(g_1 K) = \eta(g_2 K)$. Then $\psi(g_1) = \psi(g_2)$. This implies that $\psi(g_1^{-1} g_2) = e$, or $g_1^{-1} g_2$ is in the kernel of ψ; hence, $g_1^{-1} g_2 K = K$; that is, $g_1 K = g_2 K$. ∎

Mathematicians often use diagrams called *commutative diagrams* to describe such theorems. The following diagram "commutes" since $\psi = \eta\phi$.

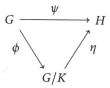

Example 11.11. Let G be a cyclic group with generator g. Define a map $\phi : \mathbb{Z} \to G$ by $n \mapsto g^n$. This map is a surjective homomorphism since

$$\phi(m + n) = g^{m+n} = g^m g^n = \phi(m)\phi(n).$$

Clearly ϕ is onto. If $|g| = m$, then $g^m = e$. Hence, $\ker \phi = m\mathbb{Z}$ and $\mathbb{Z}/\ker \phi = \mathbb{Z}/m\mathbb{Z} \cong G$. On the other hand, if the order of g is infinite, then $\ker \phi = 0$ and ϕ is an isomorphism of G and $\mathbb{Z}$. Hence, two cyclic groups are isomorphic exactly when they have the same order. Up to isomorphism, the only cyclic groups are $\mathbb{Z}$ and $\mathbb{Z}_n$. □

Theorem 11.12. Second Isomorphism Theorem. Let H be a subgroup of a group G (not necessarily normal in G) and N a normal subgroup of G. Then HN is a subgroup of G, $H \cap N$ is a normal subgroup of H, and

$$H/H \cap N \cong HN/N.$$

Proof. We will first show that $HN = \{hn : h \in H, n \in N\}$ is a subgroup of G. Suppose that $h_1 n_1, h_2 n_2 \in HN$. Since N is normal, $(h_2)^{-1} n_1 h_2 \in N$. So

$$(h_1 n_1)(h_2 n_2) = h_1 h_2 ((h_2)^{-1} n_1 h_2) n_2$$

is in HN. The inverse of $hn \in HN$ is in HN since

$$(hn)^{-1} = n^{-1} h^{-1} = h^{-1}(hn^{-1}h^{-1}).$$

Next, we prove that $H \cap N$ is normal in H. Let $h \in H$ and $n \in H \cap N$. Then $h^{-1}nh \in H$ since each element is in H. Also, $h^{-1}nh \in N$ since N is normal in G; therefore, $h^{-1}nh \in H \cap N$.

Now define a map ϕ from H to HN/N by $h \mapsto hN$. The map ϕ is onto, since any coset $hnN = hN$ is the image of h in H. We also know that ϕ is a homomorphism because

$$\phi(hh') = hh'N = hNh'N = \phi(h)\phi(h').$$

By the First Isomorphism Theorem, the image of ϕ is isomorphic to $H/\ker \phi$; that is,

$$HN/N = \phi(H) \cong H/\ker \phi.$$

Since
$$\ker \phi = \{h \in H : h \in N\} = H \cap N,$$

$HN/N = \phi(H) \cong H/H \cap N.$ ∎

Theorem 11.13. Correspondence Theorem. Let N be a normal subgroup of a group G. Then $H \mapsto H/N$ is a one-to-one correspondence between the set of subgroups H of G containing N and the set of subgroups of G/N. Furthermore, the normal subgroups of G containing N correspond to normal subgroups of G/N.

Proof. Let H be a subgroup of G containing N. Since N is normal in H, H/N is a factor group. Let aN and bN be elements of H/N. Then $(aN)(b^{-1}N) = ab^{-1}N \in H/N$; hence, H/N is a subgroup of G/N.

Let S be a subgroup of G/N. This subgroup is a set of cosets of N. If $H = \{g \in G : gN \in S\}$, then for $h_1, h_2 \in H$, we have that $(h_1N)(h_2N) = h_1h_2N \in S$ and $h_1^{-1}N \in S$. Therefore, H must be a subgroup of G. Clearly, H contains N. Therefore, $S = H/N$. Consequently, the map $H \mapsto H/N$ is onto.

Suppose that H_1 and H_2 are subgroups of G containing N such that $H_1/N = H_2/N$. If $h_1 \in H_1$, then $h_1N \in H_1/N$. Hence, $h_1N = h_2N \subset H_2$ for some h_2 in H_2. However, since N is contained in H_2, we know that $h_1 \in H_2$ or $H_1 \subset H_2$. Similarly, $H_2 \subset H_1$. Since $H_1 = H_2$, the map $H \mapsto H/N$ is one-to-one.

Suppose that H is normal in G and N is a subgroup of H. Then it is easy to verify that the map $G/N \to G/H$ defined by $gN \mapsto gH$ is a homomorphism. The kernel of this homomorphism is H/N, which proves that H/N is normal in G/N.

Conversely, suppose that H/N is normal in G/N. The homomorphism given by

$$G \to G/N \to \frac{G/N}{H/N}$$

has kernel H. Hence, H must be normal in G. ∎

Notice that in the course of the proof of Theorem 11.13, we have also proved the following theorem.

Theorem 11.14. Third Isomorphism Theorem. Let G be a group and N and H be normal subgroups of G with $N \subset H$. Then

$$G/H \cong \frac{G/N}{H/N}.$$

Example 11.15. By the Third Isomorphism Theorem,

$$\mathbb{Z}/m\mathbb{Z} \cong (\mathbb{Z}/mn\mathbb{Z})/(m\mathbb{Z}/mn\mathbb{Z}).$$

Since $|\mathbb{Z}/mn\mathbb{Z}| = mn$ and $|\mathbb{Z}/m\mathbb{Z}| = m$, we have $|m\mathbb{Z}/mn\mathbb{Z}| = n$. $\qquad\square$

Sage. Sage can create homomorphisms between groups, which can be used directly as functions, and then queried for their kernels and images. So there is great potential for exploring the many fundamental relationships between groups, normal subgroups, quotient groups and properties of homomorphisms.

11.3 Reading Questions

1. Consider the function $\phi : \mathbb{Z}_{10} \to \mathbb{Z}_{10}$ defined by $\phi(x) = x + x$. Prove that ϕ is a group homomorphism.

2. For ϕ defined in the previous question, explain why ϕ is not a group isomorphism.

3. Compare and contrast isomorphisms and homomorphisms.

4. Paraphrase the First Isomorphism Theorem using *only words*. No symbols allowed *at all*.

5. "For every normal subgroup there is a homomorphism, and for every homomorphism there is a normal subgroup." Explain the (precise) basis for this (vague) statement.

11.4 Exercises

1. Prove that $\det(AB) = \det(A)\det(B)$ for $A, B \in GL_2(\mathbb{R})$. This shows that the determinant is a homomorphism from $GL_2(\mathbb{R})$ to $\mathbb{R}^*$.

2. Which of the following maps are homomorphisms? If the map is a homomorphism, what is the kernel?

(a) $\phi : \mathbb{R}^* \to GL_2(\mathbb{R})$ defined by

$$\phi(a) = \begin{pmatrix} 1 & 0 \\ 0 & a \end{pmatrix}$$

(b) $\phi : \mathbb{R} \to GL_2(\mathbb{R})$ defined by

$$\phi(a) = \begin{pmatrix} 1 & 0 \\ a & 1 \end{pmatrix}$$

(c) $\phi : GL_2(\mathbb{R}) \to \mathbb{R}$ defined by

$$\phi\left(\begin{pmatrix} a & b \\ c & d \end{pmatrix}\right) = a + d$$

(d) $\phi : GL_2(\mathbb{R}) \to \mathbb{R}^*$ defined by

$$\phi\left(\begin{pmatrix} a & b \\ c & d \end{pmatrix}\right) = ad - bc$$

(e) $\phi : \mathbb{M}_2(\mathbb{R}) \to \mathbb{R}$ defined by

$$\phi\left(\begin{pmatrix} a & b \\ c & d \end{pmatrix}\right) = b,$$

where $\mathbb{M}_2(\mathbb{R})$ is the additive group of 2×2 matrices with entries in $\mathbb{R}$.

3. Let A be an $m \times n$ matrix. Show that matrix multiplication, $x \mapsto Ax$, defines a homomorphism $\phi : \mathbb{R}^n \to \mathbb{R}^m$.

4. Let $\phi : \mathbb{Z} \to \mathbb{Z}$ be given by $\phi(n) = 7n$. Prove that ϕ is a group homomorphism. Find the kernel and the image of ϕ.

5. Describe all of the homomorphisms from $\mathbb{Z}_{24}$ to $\mathbb{Z}_{18}$.

6. Describe all of the homomorphisms from $\mathbb{Z}$ to $\mathbb{Z}_{12}$.

7. In the group $\mathbb{Z}_{24}$, let $H = \langle 4 \rangle$ and $N = \langle 6 \rangle$.

   (a) List the elements in HN (we usually write $H + N$ for these additive groups) and $H \cap N$.

   (b) List the cosets in HN/N, showing the elements in each coset.

   (c) List the cosets in $H/(H \cap N)$, showing the elements in each coset.

   (d) Give the correspondence between HN/N and $H/(H \cap N)$ described in the proof of the Second Isomorphism Theorem.

8. If G is an abelian group and $n \in \mathbb{N}$, show that $\phi : G \to G$ defined by $g \mapsto g^n$ is a group homomorphism.

9. If $\phi : G \to H$ is a group homomorphism and G is abelian, prove that $\phi(G)$ is also abelian.

10. If $\phi : G \to H$ is a group homomorphism and G is cyclic, prove that $\phi(G)$ is also cyclic.

11. Show that a homomorphism defined on a cyclic group is completely determined by its action on the generator of the group.

12. If a group G has exactly one subgroup H of order k, prove that H is normal in G.

13. Prove or disprove: $\mathbb{Q}/\mathbb{Z} \cong \mathbb{Q}$.

14. Let G be a finite group and N a normal subgroup of G. If H is a subgroup of G/N, prove that $\phi^{-1}(H)$ is a subgroup in G of order $|H| \cdot |N|$, where $\phi : G \to G/N$ is the canonical homomorphism.

15. Let G_1 and G_2 be groups, and let H_1 and H_2 be normal subgroups of G_1 and G_2 respectively. Let $\phi : G_1 \to G_2$ be a homomorphism. Show that ϕ induces a homomorphism $\overline{\phi} : (G_1/H_1) \to (G_2/H_2)$ if $\phi(H_1) \subset H_2$.

16. If H and K are normal subgroups of G and $H \cap K = \{e\}$, prove that G is isomorphic to a subgroup of $G/H \times G/K$.

17. Let $\phi : G_1 \to G_2$ be a surjective group homomorphism. Let H_1 be a normal subgroup of G_1 and suppose that $\phi(H_1) = H_2$. Prove or disprove that $G_1/H_1 \cong G_2/H_2$.

18. Let $\phi : G \to H$ be a group homomorphism. Show that ϕ is one-to-one if and only if $\phi^{-1}(e) = \{e\}$.

19. Given a homomorphism $\phi : G \to H$ define a relation $\sim$ on G by $a \sim b$ if $\phi(a) = \phi(b)$ for $a, b \in G$. Show this relation is an equivalence relation and describe the equivalence classes.

11.5 Additional Exercises: Automorphisms

1. Let $\mathrm{Aut}(G)$ be the set of all automorphisms of G; that is, isomorphisms from G to itself. Prove this set forms a group and is a subgroup of the group of permutations of G; that is, $\mathrm{Aut}(G) \leq S_G$.

2. An *inner automorphism* of G,

    $$i_g : G \to G,$$

    is defined by the map

    $$i_g(x) = gxg^{-1},$$

    for $g \in G$. Show that $i_g \in \mathrm{Aut}(G)$.

3. The set of all inner automorphisms is denoted by $\mathrm{Inn}(G)$. Show that $\mathrm{Inn}(G)$ is a subgroup of $\mathrm{Aut}(G)$.

4. Find an automorphism of a group G that is not an inner automorphism.

5. Let G be a group and i_g be an inner automorphism of G, and define a map

    $$G \to \mathrm{Aut}(G)$$

    by

    $$g \mapsto i_g.$$

    Prove that this map is a homomorphism with image $\mathrm{Inn}(G)$ and kernel $Z(G)$. Use this result to conclude that

    $$G/Z(G) \cong \mathrm{Inn}(G).$$

6. Compute $\mathrm{Aut}(S_3)$ and $\mathrm{Inn}(S_3)$. Do the same thing for D_4.

7. Find all of the homomorphisms $\phi : \mathbb{Z} \to \mathbb{Z}$. What is $\mathrm{Aut}(\mathbb{Z})$?

8. Find all of the automorphisms of $\mathbb{Z}_8$. Prove that $\mathrm{Aut}(\mathbb{Z}_8) \cong U(8)$.

9. For $k \in \mathbb{Z}_n$, define a map $\phi_k : \mathbb{Z}_n \to \mathbb{Z}_n$ by $a \mapsto ka$. Prove that ϕ_k is a homomorphism.

10. Prove that ϕ_k is an isomorphism if and only if k is a generator of $\mathbb{Z}_n$.

11. Show that every automorphism of $\mathbb{Z}_n$ is of the form ϕ_k, where k is a generator of $\mathbb{Z}_n$.

12. Prove that $\psi : U(n) \to \mathrm{Aut}(\mathbb{Z}_n)$ is an isomorphism, where $\psi : k \mapsto \phi_k$.

Matrix Groups and Symmetry

When Felix Klein (1849–1925) accepted a chair at the University of Erlangen, he outlined in his inaugural address a program to classify different geometries. Central to Klein's program was the theory of groups: he considered geometry to be the study of properties that are left invariant under transformation groups. Groups, especially matrix groups, have now become important in the study of symmetry and have found applications in such disciplines as chemistry and physics. In the first part of this chapter, we will examine some of the classical matrix groups, such as the general linear group, the special linear group, and the orthogonal group. We will then use these matrix groups to investigate some of the ideas behind geometric symmetry.

12.1 Matrix Groups

Some Facts from Linear Algebra

Before we study matrix groups, we must recall some basic facts from linear algebra. One of the most fundamental ideas of linear algebra is that of a linear transformation. A *linear transformation* or *linear map* $T : \mathbb{R}^n \to \mathbb{R}^m$ is a map that preserves vector addition and scalar multiplication; that is, for vectors $\mathbf{x}$ and $\mathbf{y}$ in $\mathbb{R}^n$ and a scalar $\alpha \in \mathbb{R}$,

$$T(\mathbf{x} + \mathbf{y}) = T(\mathbf{x}) + T(\mathbf{y})$$
$$T(\alpha \mathbf{y}) = \alpha T(\mathbf{y}).$$

An $m \times n$ matrix with entries in $\mathbb{R}$ represents a linear transformation from $\mathbb{R}^n$ to $\mathbb{R}^m$. If we write vectors $\mathbf{x} = (x_1, \ldots, x_n)^t$ and $\mathbf{y} = (y_1, \ldots, y_n)^t$ in $\mathbb{R}^n$ as column matrices, then an $m \times n$ matrix

$$A = \begin{pmatrix} a_{11} & a_{12} & \cdots & a_{1n} \\ a_{21} & a_{22} & \cdots & a_{2n} \\ \vdots & \vdots & \ddots & \vdots \\ a_{m1} & a_{m2} & \cdots & a_{mn} \end{pmatrix}$$

maps the vectors to $\mathbb{R}^m$ linearly by matrix multiplication. Observe that if α is a real number,

$$A(\mathbf{x} + \mathbf{y}) = A\mathbf{x} + A\mathbf{y} \qquad \text{and} \qquad \alpha A\mathbf{x} = A(\alpha \mathbf{x}),$$

where

$$\mathbf{x} = \begin{pmatrix} x_1 \\ x_2 \\ \vdots \\ x_n \end{pmatrix}.$$

We will often abbreviate the matrix A by writing (a_{ij}).

Conversely, if $T : \mathbb{R}^n \to \mathbb{R}^m$ is a linear map, we can associate a matrix A with T by considering what T does to the vectors

$$\mathbf{e}_1 = (1, 0, \dots, 0)^{\mathrm{t}}$$
$$\mathbf{e}_2 = (0, 1, \dots, 0)^{\mathrm{t}}$$
$$\vdots$$
$$\mathbf{e}_n = (0, 0, \dots, 1)^{\mathrm{t}}.$$

We can write any vector $\mathbf{x} = (x_1, \dots, x_n)^{\mathrm{t}}$ as

$$x_1 \mathbf{e}_1 + x_2 \mathbf{e}_2 + \cdots + x_n \mathbf{e}_n.$$

Consequently, if

$$T(\mathbf{e}_1) = (a_{11}, a_{21}, \dots, a_{m1})^{\mathrm{t}},$$
$$T(\mathbf{e}_2) = (a_{12}, a_{22}, \dots, a_{m2})^{\mathrm{t}},$$
$$\vdots$$
$$T(\mathbf{e}_n) = (a_{1n}, a_{2n}, \dots, a_{mn})^{\mathrm{t}},$$

then

$$\begin{aligned} T(\mathbf{x}) &= T(x_1 \mathbf{e}_1 + x_2 \mathbf{e}_2 + \cdots + x_n \mathbf{e}_n) \\ &= x_1 T(\mathbf{e}_1) + x_2 T(\mathbf{e}_2) + \cdots + x_n T(\mathbf{e}_n) \\ &= \left(\sum_{k=1}^{n} a_{1k} x_k, \dots, \sum_{k=1}^{n} a_{mk} x_k \right)^{\mathrm{t}} \\ &= A\mathbf{x}. \end{aligned}$$

Example 12.1. If we let $T : \mathbb{R}^2 \to \mathbb{R}^2$ be the map given by

$$T(x_1, x_2) = (2x_1 + 5x_2, -4x_1 + 3x_2),$$

the axioms that T must satisfy to be a linear transformation are easily verified. The column vectors $T\mathbf{e}_1 = (2, -4)^t$ and $T\mathbf{e}_2 = (5, 3)^t$ tell us that T is given by the matrix

$$A = \begin{pmatrix} 2 & 5 \\ -4 & 3 \end{pmatrix}.$$

$\square$

Since we are interested in groups of matrices, we need to know which matrices have multiplicative inverses. Recall that an $n \times n$ matrix A is *invertible* exactly when there exists another matrix A^{-1} such that $AA^{-1} = A^{-1}A = I$, where

$$I = \begin{pmatrix} 1 & 0 & \cdots & 0 \\ 0 & 1 & \cdots & 0 \\ \vdots & \vdots & \ddots & \vdots \\ 0 & 0 & \cdots & 1 \end{pmatrix}$$

is the $n \times n$ identity matrix. From linear algebra we know that A is invertible if and only if the determinant of A is nonzero. Sometimes an invertible matrix is said to be *nonsingular*.

Example 12.2. If A is the matrix

$$\begin{pmatrix} 2 & 1 \\ 5 & 3 \end{pmatrix},$$

then the inverse of A is

$$A^{-1} = \begin{pmatrix} 3 & -1 \\ -5 & 2 \end{pmatrix}.$$

We are guaranteed that A^{-1} exists, since $\det(A) = 2 \cdot 3 - 5 \cdot 1 = 1$ is nonzero. $\square$

Some other facts about determinants will also prove useful in the course of this chapter. Let A and B be $n \times n$ matrices. From linear algebra we have the following properties of determinants.

- The determinant is a homomorphism into the multiplicative group of real numbers; that is, $\det(AB) = (\det A)(\det B)$.

- If A is an invertible matrix, then $\det(A^{-1}) = 1/\det A$.

- If we define the transpose of a matrix $A = (a_{ij})$ to be $A^t = (a_{ji})$, then $\det(A^t) = \det A$.

- Let T be the linear transformation associated with an $n \times n$ matrix A. Then T multiplies volumes by a factor of $|\det A|$. In the case of $\mathbb{R}^2$, this means that T multiplies areas by $|\det A|$.

Linear maps, matrices, and determinants are covered in any elementary linear algebra text; however, if you have not had a course in linear algebra, it is a straightforward process to verify these properties directly for 2×2 matrices, the case with which we are most concerned.

The General and Special Linear Groups

The set of all $n \times n$ invertible matrices forms a group called the **general linear group**. We will denote this group by $GL_n(\mathbb{R})$. The general linear group has several important subgroups. The multiplicative properties of the determinant imply that the set of matrices with determinant one is a subgroup of the general linear group. Stated another way, suppose that $\det(A) = 1$ and $\det(B) = 1$. Then $\det(AB) = \det(A)\det(B) = 1$ and $\det(A^{-1}) = 1/\det A = 1$. This subgroup is called the **special linear group** and is denoted by $SL_n(\mathbb{R})$.

Example 12.3. Given a 2×2 matrix

$$A = \begin{pmatrix} a & b \\ c & d \end{pmatrix},$$

the determinant of A is $ad - bc$. The group $GL_2(\mathbb{R})$ consists of those matrices in which $ad - bc \neq 0$. The inverse of A is

$$A^{-1} = \frac{1}{ad - bc} \begin{pmatrix} d & -b \\ -c & a \end{pmatrix}.$$

If A is in $SL_2(\mathbb{R})$, then

$$A^{-1} = \begin{pmatrix} d & -b \\ -c & a \end{pmatrix}.$$

Geometrically, $SL_2(\mathbb{R})$ is the group that preserves the areas of parallelograms. Let

$$A = \begin{pmatrix} 1 & 1 \\ 0 & 1 \end{pmatrix}$$

be in $SL_2(\mathbb{R})$. In Figure 12.4, the unit square corresponding to the vectors $\mathbf{x} = (1,0)^t$ and $\mathbf{y} = (0,1)^t$ is taken by A to the parallelogram with sides $(1,0)^t$ and

$(1,1)^t$; that is, $A\mathbf{x} = (1,0)^t$ and $A\mathbf{y} = (1,1)^t$. Notice that these two parallelograms have the same area.   □

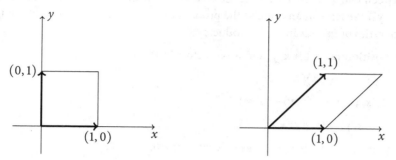

Figure 12.4. $SL_2(\mathbb{R})$ acting on the unit square

The Orthogonal Group $O(n)$

Another subgroup of $GL_n(\mathbb{R})$ is the orthogonal group. A matrix A is **orthogonal** if $A^{-1} = A^t$. The **orthogonal group** consists of the set of all orthogonal matrices. We write $O(n)$ for the $n \times n$ orthogonal group. We leave as an exercise the proof that $O(n)$ is a subgroup of $GL_n(\mathbb{R})$.

Example 12.5. The following matrices are orthogonal:

$$\begin{pmatrix} 3/5 & -4/5 \\ 4/5 & 3/5 \end{pmatrix}, \quad \begin{pmatrix} 1/2 & -\sqrt{3}/2 \\ \sqrt{3}/2 & 1/2 \end{pmatrix}, \quad \begin{pmatrix} -1/\sqrt{2} & 0 & 1/\sqrt{2} \\ 1/\sqrt{6} & -2/\sqrt{6} & 1/\sqrt{6} \\ 1/\sqrt{3} & 1/\sqrt{3} & 1/\sqrt{3} \end{pmatrix}.$$

□

There is a more geometric way of viewing the group $O(n)$. The orthogonal matrices are exactly those matrices that preserve the length of vectors. We can define the length of a vector using the **Euclidean inner product**, or **dot product**, of two vectors. The Euclidean inner product of two vectors $\mathbf{x} = (x_1, \ldots, x_n)^t$ and $\mathbf{y} = (y_1, \ldots, y_n)^t$ is

$$\langle \mathbf{x}, \mathbf{y} \rangle = \mathbf{x}^t \mathbf{y} = (x_1, x_2, \ldots, x_n) \begin{pmatrix} y_1 \\ y_2 \\ \vdots \\ y_n \end{pmatrix} = x_1 y_1 + \cdots + x_n y_n.$$

We define the length of a vector $\mathbf{x} = (x_1, \ldots, x_n)^t$ to be

$$\|\mathbf{x}\| = \sqrt{\langle \mathbf{x}, \mathbf{x} \rangle} = \sqrt{x_1^2 + \cdots + x_n^2}.$$

Associated with the notion of the length of a vector is the idea of the distance between two vectors. We define the *distance* between two vectors $\mathbf{x}$ and $\mathbf{y}$ to be $\|\mathbf{x} - \mathbf{y}\|$. We leave as an exercise the proof of the following proposition about the properties of Euclidean inner products.

Proposition 12.6. Let $\mathbf{x}$, $\mathbf{y}$, and $\mathbf{w}$ be vectors in $\mathbb{R}^n$ and $\alpha \in \mathbb{R}$. Then

1. $\langle \mathbf{x}, \mathbf{y} \rangle = \langle \mathbf{y}, \mathbf{x} \rangle$.

2. $\langle \mathbf{x}, \mathbf{y} + \mathbf{w} \rangle = \langle \mathbf{x}, \mathbf{y} \rangle + \langle \mathbf{x}, \mathbf{w} \rangle$.

3. $\langle \alpha\mathbf{x}, \mathbf{y} \rangle = \langle \mathbf{x}, \alpha\mathbf{y} \rangle = \alpha\langle \mathbf{x}, \mathbf{y} \rangle$.

4. $\langle \mathbf{x}, \mathbf{x} \rangle \geq 0$ with equality exactly when $\mathbf{x} = 0$.

5. If $\langle \mathbf{x}, \mathbf{y} \rangle = 0$ for all $\mathbf{x}$ in $\mathbb{R}^n$, then $\mathbf{y} = 0$.

Example 12.7. The vector $\mathbf{x} = (3, 4)^t$ has length $\sqrt{3^2 + 4^2} = 5$. We can also see that the orthogonal matrix

$$A = \begin{pmatrix} 3/5 & -4/5 \\ 4/5 & 3/5 \end{pmatrix}$$

preserves the length of this vector. The vector $A\mathbf{x} = (-7/5, 24/5)^t$ also has length 5. □

Since $\det(AA^t) = \det(I) = 1$ and $\det(A) = \det(A^t)$, the determinant of any orthogonal matrix is either 1 or -1. Consider the column vectors

$$\mathbf{a}_j = \begin{pmatrix} a_{1j} \\ a_{2j} \\ \vdots \\ a_{nj} \end{pmatrix}$$

of the orthogonal matrix $A = (a_{ij})$. Since $AA^t = I$, $\langle \mathbf{a}_r, \mathbf{a}_s \rangle = \delta_{rs}$, where

$$\delta_{rs} = \begin{cases} 1 & r = s \\ 0 & r \neq s \end{cases}.$$

is the Kronecker delta. Accordingly, column vectors of an orthogonal matrix all have length 1; and the Euclidean inner product of distinct column vectors is zero. Any set of vectors satisfying these properties is called an *orthonormal set*. Conversely, given an $n \times n$ matrix A whose columns form an orthonormal set, it follows that $A^{-1} = A^t$.

We say that a matrix A is *distance-preserving, length-preserving,* or *inner product-preserving* when $\|A\mathbf{x} - A\mathbf{y}\| = \|\mathbf{x} - \mathbf{y}\|$, $\|A\mathbf{x}\| = \|\mathbf{x}\|$, or $\langle A\mathbf{x}, A\mathbf{y} \rangle = \langle \mathbf{x}, \mathbf{y} \rangle$, respectively. The following theorem, which characterizes the orthogonal group, says that these notions are the same.

Theorem 12.8. Let A be an $n \times n$ matrix. The following statements are equivalent.

1. The columns of the matrix A form an orthonormal set.

2. $A^{-1} = A^t$.

3. For vectors $\mathbf{x}$ and $\mathbf{y}$, $\langle A\mathbf{x}, A\mathbf{y} \rangle = \langle \mathbf{x}, \mathbf{y} \rangle$.

4. For vectors $\mathbf{x}$ and $\mathbf{y}$, $\|A\mathbf{x} - A\mathbf{y}\| = \|\mathbf{x} - \mathbf{y}\|$.

5. For any vector $\mathbf{x}$, $\|A\mathbf{x}\| = \|\mathbf{x}\|$.

Proof. We have already shown (1) and (2) to be equivalent.

$(2) \Rightarrow (3)$.

$$
\begin{aligned}
\langle A\mathbf{x}, A\mathbf{y} \rangle &= (A\mathbf{x})^t A\mathbf{y} \\
&= \mathbf{x}^t A^t A\mathbf{y} \\
&= \mathbf{x}^t \mathbf{y} \\
&= \langle \mathbf{x}, \mathbf{y} \rangle.
\end{aligned}
$$

$(3) \Rightarrow (2)$. Since

$$
\begin{aligned}
\langle \mathbf{x}, \mathbf{x} \rangle &= \langle A\mathbf{x}, A\mathbf{x} \rangle \\
&= \mathbf{x}^t A^t A\mathbf{x} \\
&= \langle \mathbf{x}, A^t A\mathbf{x} \rangle,
\end{aligned}
$$

we know that $\langle \mathbf{x}, (A^t A - I)\mathbf{x} \rangle = 0$ for all $\mathbf{x}$. Therefore, $A^t A - I = 0$ or $A^{-1} = A^t$.

$(3) \Rightarrow (4)$. If A is inner product-preserving, then A is distance-preserving, since

$$
\begin{aligned}
\|A\mathbf{x} - A\mathbf{y}\|^2 &= \|A(\mathbf{x} - \mathbf{y})\|^2 \\
&= \langle A(\mathbf{x} - \mathbf{y}), A(\mathbf{x} - \mathbf{y}) \rangle \\
&= \langle \mathbf{x} - \mathbf{y}, \mathbf{x} - \mathbf{y} \rangle \\
&= \|\mathbf{x} - \mathbf{y}\|^2.
\end{aligned}
$$

$(4) \Rightarrow (5)$. If A is distance-preserving, then A is length-preserving. Letting $\mathbf{y} = 0$, we have

$$
\|A\mathbf{x}\| = \|A\mathbf{x} - A\mathbf{y}\| = \|\mathbf{x} - \mathbf{y}\| = \|\mathbf{x}\|.
$$

(5) $\Rightarrow$ (3). We use the following identity to show that length-preserving implies inner product-preserving:

$$\langle \mathbf{x}, \mathbf{y} \rangle = \frac{1}{2} \left[\|\mathbf{x} + \mathbf{y}\|^2 - \|\mathbf{x}\|^2 - \|\mathbf{y}\|^2 \right].$$

Observe that

$$\begin{aligned}
\langle A\mathbf{x}, A\mathbf{y} \rangle &= \frac{1}{2} \left[\|A\mathbf{x} + A\mathbf{y}\|^2 - \|A\mathbf{x}\|^2 - \|A\mathbf{y}\|^2 \right] \\
&= \frac{1}{2} \left[\|A(\mathbf{x} + \mathbf{y})\|^2 - \|A\mathbf{x}\|^2 - \|A\mathbf{y}\|^2 \right] \\
&= \frac{1}{2} \left[\|\mathbf{x} + \mathbf{y}\|^2 - \|\mathbf{x}\|^2 - \|\mathbf{y}\|^2 \right] \\
&= \langle \mathbf{x}, \mathbf{y} \rangle.
\end{aligned}$$

■

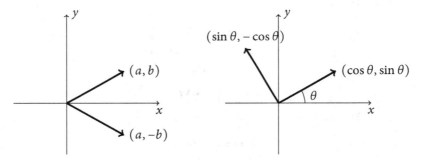

Figure 12.9. $O(2)$ acting on $\mathbb{R}^2$

Example 12.10. Let us examine the orthogonal group on $\mathbb{R}^2$ a bit more closely. An element $A \in O(2)$ is determined by its action on $\mathbf{e}_1 = (1, 0)^t$ and $\mathbf{e}_2 = (0, 1)^t$. If $A\mathbf{e}_1 = (a, b)^t$, then $a^2 + b^2 = 1$, since the length of a vector must be preserved when it is multiplied by A. Since multiplication of an element of $O(2)$ preserves length and orthogonality, $A\mathbf{e}_2 = \pm(-b, a)^t$. If we choose $A\mathbf{e}_2 = (-b, a)^t$, then

$$A = \begin{pmatrix} a & -b \\ b & a \end{pmatrix} = \begin{pmatrix} \cos\theta & -\sin\theta \\ \sin\theta & \cos\theta \end{pmatrix},$$

where $0 \le \theta < 2\pi$. The matrix A rotates a vector in $\mathbb{R}^2$ counterclockwise about the origin by an angle of θ (Figure 12.9).

If we choose $A\mathbf{e}_2 = (b, -a)^t$, then we obtain the matrix

$$B = \begin{pmatrix} a & b \\ b & -a \end{pmatrix} = \begin{pmatrix} \cos\theta & \sin\theta \\ \sin\theta & -\cos\theta \end{pmatrix}.$$

Here, $\det B = -1$ and

$$B^2 = \begin{pmatrix} 1 & 0 \\ 0 & 1 \end{pmatrix}.$$

A reflection about the horizontal axis is given by the matrix

$$C = \begin{pmatrix} 1 & 0 \\ 0 & -1 \end{pmatrix},$$

and $B = AC$ (see Figure 12.9). Thus, a reflection about a line ℓ is simply a reflection about the horizontal axis followed by a rotation. $\qquad\qquad\square$

Two of the other matrix or matrix-related groups that we will consider are the special orthogonal group and the group of Euclidean motions. The *special orthogonal group*, $SO(n)$, is just the intersection of $O(n)$ and $SL_n(\mathbb{R})$; that is, those elements in $O(n)$ with determinant one. The *Euclidean group*, $E(n)$, can be written as ordered pairs $(A, \mathbf{x})$, where A is in $O(n)$ and $\mathbf{x}$ is in $\mathbb{R}^n$. We define multiplication by

$$(A, \mathbf{x})(B, \mathbf{y}) = (AB, A\mathbf{y} + \mathbf{x}).$$

The identity of the group is $(I, \mathbf{0})$; the inverse of $(A, \mathbf{x})$ is $(A^{-1}, -A^{-1}\mathbf{x})$. In Exercise 12.4.6 on page 195, you are asked to check that $E(n)$ is indeed a group under this operation.

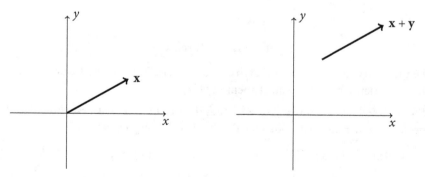

Figure 12.11. Translations in $\mathbb{R}^2$

12.2 Symmetry

An *isometry* or *rigid motion* in $\mathbb{R}^n$ is a distance-preserving function f from $\mathbb{R}^n$ to $\mathbb{R}^n$. This means that f must satisfy

$$\|f(\mathbf{x}) - f(\mathbf{y})\| = \|\mathbf{x} - \mathbf{y}\|$$

for all $\mathbf{x}, \mathbf{y} \in \mathbb{R}^n$. It is not difficult to show that f must be a one-to-one map. By Theorem 12.8 on page 183, any element in $O(n)$ is an isometry on $\mathbb{R}^n$; however, $O(n)$ does not include all possible isometries on $\mathbb{R}^n$. Translation by a vector $\mathbf{x}$, $T_\mathbf{y}(\mathbf{x}) = \mathbf{x} + \mathbf{y}$ is also an isometry (Figure 12.11 on the preceding page); however, T cannot be in $O(n)$ since it is not a linear map.

We are mostly interested in isometries in $\mathbb{R}^2$. In fact, the only isometries in $\mathbb{R}^2$ are rotations and reflections about the origin, translations, and combinations of the two. For example, a *glide reflection* is a translation followed by a reflection (Figure 12.12). In $\mathbb{R}^n$ all isometries are given in the same manner. The proof is very easy to generalize.

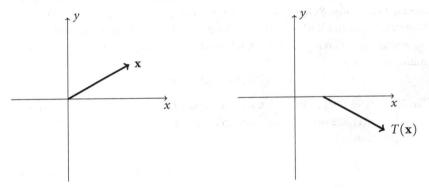

Figure 12.12. Glide reflections

Lemma 12.13. An isometry f that fixes the origin in $\mathbb{R}^2$ is a linear transformation. In particular, f is given by an element in $O(2)$.

Proof. Let f be an isometry in $\mathbb{R}^2$ fixing the origin. We will first show that f preserves inner products. Since $f(0) = 0$, $\|f(\mathbf{x})\| = \|\mathbf{x}\|$; therefore,

$$\begin{aligned}
\|\mathbf{x}\|^2 - 2\langle f(\mathbf{x}), f(\mathbf{y})\rangle + \|\mathbf{y}\|^2 &= \|f(\mathbf{x})\|^2 - 2\langle f(\mathbf{x}), f(\mathbf{y})\rangle + \|f(\mathbf{y})\|^2 \\
&= \langle f(\mathbf{x}) - f(\mathbf{y}), f(\mathbf{x}) - f(\mathbf{y})\rangle \\
&= \|f(\mathbf{x}) - f(\mathbf{y})\|^2 \\
&= \|\mathbf{x} - \mathbf{y}\|^2
\end{aligned}$$

$$= \langle \mathbf{x} - \mathbf{y}, \mathbf{x} - \mathbf{y} \rangle$$
$$= \|\mathbf{x}\|^2 - 2\langle \mathbf{x}, \mathbf{y} \rangle + \|\mathbf{y}\|^2.$$

Consequently,

$$\langle f(\mathbf{x}), f(\mathbf{y}) \rangle = \langle \mathbf{x}, \mathbf{y} \rangle.$$

Now let $\mathbf{e}_1$ and $\mathbf{e}_2$ be $(1, 0)^t$ and $(0, 1)^t$, respectively. If

$$\mathbf{x} = (x_1, x_2) = x_1 \mathbf{e}_1 + x_2 \mathbf{e}_2,$$

then

$$f(\mathbf{x}) = \langle f(\mathbf{x}), f(\mathbf{e}_1) \rangle f(\mathbf{e}_1) + \langle f(\mathbf{x}), f(\mathbf{e}_2) \rangle f(\mathbf{e}_2) = x_1 f(\mathbf{e}_1) + x_2 f(\mathbf{e}_2).$$

The linearity of f easily follows. ∎

For any arbitrary isometry, f, $T_{\mathbf{x}}f$ will fix the origin for some vector $\mathbf{x}$ in $\mathbb{R}^2$; hence, $T_{\mathbf{x}}f(\mathbf{y}) = A\mathbf{y}$ for some matrix $A \in O(2)$. Consequently, $f(\mathbf{y}) = A\mathbf{y} + \mathbf{x}$. Given the isometries

$$f(\mathbf{y}) = A\mathbf{y} + \mathbf{x}_1$$
$$g(\mathbf{y}) = B\mathbf{y} + \mathbf{x}_2,$$

their composition is

$$f(g(\mathbf{y})) = f(B\mathbf{y} + \mathbf{x}_2) = AB\mathbf{y} + A\mathbf{x}_2 + \mathbf{x}_1.$$

This last computation allows us to identify the group of isometries on $\mathbb{R}^2$ with $E(2)$.

Theorem 12.14. The group of isometries on $\mathbb{R}^2$ is the Euclidean group, $E(2)$.

A *symmetry group* in $\mathbb{R}^n$ is a subgroup of the group of isometries on $\mathbb{R}^n$ that fixes a set of points $X \subset \mathbb{R}^n$. It is important to realize that the symmetry group of X depends *both* on $\mathbb{R}^n$ and on X. For example, the symmetry group of the origin in $\mathbb{R}^1$ is $\mathbb{Z}_2$, but the symmetry group of the origin in $\mathbb{R}^2$ is $O(2)$.

Theorem 12.15. The only finite symmetry groups in $\mathbb{R}^2$ are $\mathbb{Z}_n$ and D_n.

Proof. We simply need to find all of the finite subgroups G of $E(2)$. Any finite symmetry group G in $\mathbb{R}^2$ must fix the origin and must be a finite subgroup of $O(2)$, since translations and glide reflections have infinite order. By Example 12.10 on page 184, elements in $O(2)$ are either rotations of the form

$$R_\theta = \begin{pmatrix} \cos \theta & -\sin \theta \\ \sin \theta & \cos \theta \end{pmatrix}$$

or reflections of the form

$$T_\phi = \begin{pmatrix} \cos\phi & -\sin\phi \\ \sin\phi & \cos\phi \end{pmatrix} \begin{pmatrix} 1 & 0 \\ 0 & -1 \end{pmatrix} = \begin{pmatrix} \cos\phi & \sin\phi \\ \sin\phi & -\cos\phi \end{pmatrix}.$$

Notice that $\det(R_\theta) = 1$, $\det(T_\phi) = -1$, and $T_\phi^2 = I$. We can divide the proof up into two cases. In the first case, all of the elements in G have determinant one. In the second case, there exists at least one element in G with determinant -1.

Case 1. The determinant of every element in G is one. In this case every element in G must be a rotation. Since G is finite, there is a smallest angle, say θ_0, such that the corresponding element R_{θ_0} is the smallest rotation in the positive direction. We claim that R_{θ_0} generates G. If not, then for some positive integer n there is an angle θ_1 between $n\theta_0$ and $(n+1)\theta_0$. If so, then $(n+1)\theta_0 - \theta_1$ corresponds to a rotation smaller than θ_0, which contradicts the minimality of θ_0.

Case 2. The group G contains a reflection T. The kernel of the homomorphism $\phi : G \to \{-1, 1\}$ given by $A \mapsto \det(A)$ consists of elements whose determinant is 1. Therefore, $|G/\ker\phi| = 2$. We know that the kernel is cyclic by the first case and is a subgroup of G of, say, order n. Hence, $|G| = 2n$. The elements of G are

$$R_\theta, \ldots, R_\theta^{n-1}, TR_\theta, \ldots, TR_\theta^{n-1}.$$

These elements satisfy the relation

$$TR_\theta T = R_\theta^{-1}.$$

Consequently, G must be isomorphic to D_n in this case.   ■

The Wallpaper Groups

Suppose that we wish to study wallpaper patterns in the plane or crystals in three dimensions. Wallpaper patterns are simply repeating patterns in the plane (Figure 12.16). The analogs of wallpaper patterns in $\mathbb{R}^3$ are crystals, which we can think of as repeating patterns of molecules in three dimensions (Figure 12.17). The mathematical equivalent of a wallpaper or crystal pattern is called a lattice.

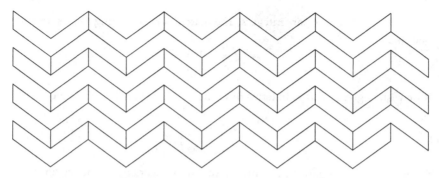

Figure 12.16. A wallpaper pattern in $\mathbb{R}^2$

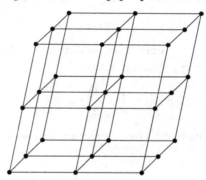

Figure 12.17. A crystal structure in $\mathbb{R}^3$

Let us examine wallpaper patterns in the plane a little more closely. Suppose that **x** and **y** are linearly independent vectors in $\mathbb{R}^2$; that is, one vector cannot be a scalar multiple of the other. A *lattice* of **x** and **y** is the set of all linear combinations $m\mathbf{x} + n\mathbf{y}$, where m and n are integers. The vectors **x** and **y** are said to be a *basis* for the lattice.

Notice that a lattice can have several bases. For example, the vectors $(1,1)^t$ and $(2,0)^t$ have the same lattice as the vectors $(-1,1)^t$ and $(-1,-1)^t$ (Figure 12.18 on the next page). However, any lattice is completely determined by a basis. Given two bases for the same lattice, say $\{\mathbf{x}_1, \mathbf{x}_2\}$ and $\{\mathbf{y}_1, \mathbf{y}_2\}$, we can write

$$\mathbf{y}_1 = \alpha_1\mathbf{x}_1 + \alpha_2\mathbf{x}_2$$
$$\mathbf{y}_2 = \beta_1\mathbf{x}_1 + \beta_2\mathbf{x}_2,$$

where α_1, α_2, β_1, and β_2 are integers. The matrix corresponding to this transformation is

$$U = \begin{pmatrix} \alpha_1 & \alpha_2 \\ \beta_1 & \beta_2 \end{pmatrix}.$$

If we wish to give $\mathbf{x}_1$ and $\mathbf{x}_2$ in terms of $\mathbf{y}_1$ and $\mathbf{y}_2$, we need only calculate U^{-1}; that is,

$$U^{-1} \begin{pmatrix} \mathbf{y}_1 \\ \mathbf{y}_2 \end{pmatrix} = \begin{pmatrix} \mathbf{x}_1 \\ \mathbf{x}_2 \end{pmatrix}.$$

Since U has integer entries, U^{-1} must also have integer entries; hence the determinants of both U and U^{-1} must be integers. Because $UU^{-1} = I$,

$$\det(UU^{-1}) = \det(U)\det(U^{-1}) = 1;$$

consequently, $\det(U) = \pm 1$. A matrix with determinant ± 1 and integer entries is called **unimodular**. For example, the matrix

$$\begin{pmatrix} 3 & 1 \\ 5 & 2 \end{pmatrix}$$

is unimodular. It should be clear that there is a minimum length for vectors in a lattice.

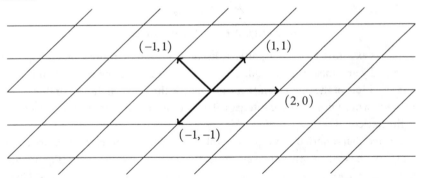

Figure 12.18. A lattice in $\mathbb{R}^2$

We can classify lattices by studying their symmetry groups. The symmetry group of a lattice is the subgroup of $E(2)$ that maps the lattice to itself. We consider two lattices in $\mathbb{R}^2$ to be equivalent if they have the same symmetry group. Similarly, classification of crystals in $\mathbb{R}^3$ is accomplished by associating a symmetry group, called a **space group**, with each type of crystal. Two lattices are considered different if their space groups are not the same. The natural question that now arises is how many space groups exist.

A space group is composed of two parts: a *translation subgroup* and a *point*. The translation subgroup is an infinite abelian subgroup of the space group made up of the translational symmetries of the crystal; the point group is a finite group consisting of rotations and reflections of the crystal about a point. More specifically, a space group is a subgroup of $G \subset E(2)$ whose translations are a set of the form $\{(I, t) : t \in L\}$, where L is a lattice. Space groups are, of course, infinite. Using geometric arguments, we can prove the following theorem (see [5] or [6]).

Theorem 12.19. Every translation group in $\mathbb{R}^2$ is isomorphic to $\mathbb{Z} \times \mathbb{Z}$.

The point group of G is $G_0 = \{A : (A, b) \in G \text{ for some } b\}$. In particular, G_0 must be a subgroup of $O(2)$. Suppose that $\mathbf{x}$ is a vector in a lattice L with space group G, translation group H, and point group G_0. For any element $(A, \mathbf{y})$ in G,

$$(A, \mathbf{y})(I, \mathbf{x})(A, \mathbf{y})^{-1} = (A, A\mathbf{x} + \mathbf{y})(A^{-1}, -A^{-1}\mathbf{y})$$
$$= (AA^{-1}, -AA^{-1}\mathbf{y} + A\mathbf{x} + \mathbf{y})$$
$$= (I, A\mathbf{x});$$

hence, $(I, A\mathbf{x})$ is in the translation group of G. More specifically, $A\mathbf{x}$ must be in the lattice L. It is important to note that G_0 is not usually a subgroup of the space group G; however, if T is the translation subgroup of G, then $G/T \cong G_0$. The proof of the following theorem can be found in [2], [5], or [6].

Theorem 12.20. The point group in the wallpaper groups is isomorphic to $\mathbb{Z}_n$ or D_n, where $n = 1, 2, 3, 4, 6$.

To answer the question of how the point groups and the translation groups can be combined, we must look at the different types of lattices. Lattices can be classified by the structure of a single lattice cell. The possible cell shapes are parallelogram, rectangular, square, rhombic, and hexagonal (Figure 12.21 on the following page). The wallpaper groups can now be classified according to the types of reflections that occur in each group: these are ordinarily reflections, glide reflections, both, or none.

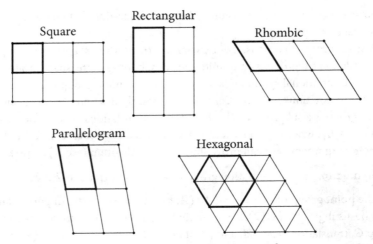

Figure 12.21. Types of lattices in $\mathbb{R}^2$

Table 12.22. The 17 wallpaper groups

Notation and Space Groups	Point Group	Lattice Type	Reflections or Glide Reflections?
p1	$\mathbb{Z}_1$	parallelogram	none
p2	$\mathbb{Z}_2$	parallelogram	none
p3	$\mathbb{Z}_3$	hexagonal	none
p4	$\mathbb{Z}_4$	square	none
p6	$\mathbb{Z}_6$	hexagonal	none
pm	D_1	rectangular	reflections
pg	D_1	rectangular	glide reflections
cm	D_1	rhombic	both
pmm	D_2	rectangular	reflections
pmg	D_2	rectangular	glide reflections
pgg	D_2	rectangular	both
c2mm	D_2	rhombic	both
p3m1, p31m	D_3	hexagonal	both
p4m, p4g	D_4	square	both
p6m	D_6	hexagonal	both

Theorem 12.23. There are exactly 17 wallpaper groups.

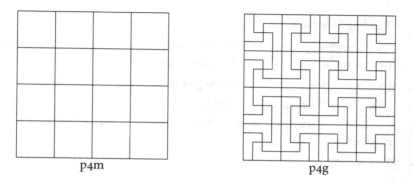

Figure 12.24. The wallpaper groups p4m and p4g

The 17 wallpaper groups are listed in Table 12.22. The groups p3m1 and p31m can be distinguished by whether or not all of their threefold centers lie on the reflection axes: those of p3m1 must, whereas those of p31m may not. Similarly, the fourfold centers of p4m must lie on the reflection axes whereas those of p4g need not (Figure 12.24). The complete proof of this theorem can be found in several of the references at the end of this chapter, including [5], [6], [10], and [11].

Sage. We have not seen how to use Sage profitably with the material in this chapter.

↬   Historical Note   ↫

Symmetry groups have intrigued mathematicians for a long time. Leonardo da Vinci was probably the first person to know all of the point groups. At the International Congress of Mathematicians in 1900, David Hilbert gave a now-famous address outlining 23 problems to guide mathematics in the twentieth century. Hilbert's eighteenth problem asked whether or not crystallographic groups in n dimensions were always finite. In 1910, L. Bieberbach proved that crystallographic groups are finite in every dimension. Finding out how many of these groups there are in each dimension is another matter. In $\mathbb{R}^3$ there are 230 different space groups; in $\mathbb{R}^4$ there are 4783. No one has been able to compute the number of space groups for $\mathbb{R}^5$ and beyond. It is interesting to note that the crystallographic groups were found mathematically for $\mathbb{R}^3$ before the 230 different types of crystals were actually discovered in nature.

12.3 Reading Questions

1. What is a nonsingular matrix? Give an example of a 2 × 2 nonsingular matrix. How do you know your example is nonsingular?

2. What is an isometry in $\mathbb{R}^n$? Can you give an example of an isometry in $\mathbb{R}^2$?

3. What is an orthonormal set of vectors?

4. What is the difference between the orthogonal group and the special orthogonal group?

5. What is a lattice?

12.4 Exercises

1. Prove the identity

$$\langle \mathbf{x}, \mathbf{y} \rangle = \frac{1}{2}\left[\|\mathbf{x}+\mathbf{y}\|^2 - \|\mathbf{x}\|^2 - \|\mathbf{y}\|^2 \right].$$

2. Show that $O(n)$ is a group.

3. Prove that the following matrices are orthogonal. Are any of these matrices in $SO(n)$?

   (a)
   $$\begin{pmatrix} 1/\sqrt{2} & -1/\sqrt{2} \\ 1/\sqrt{2} & 1/\sqrt{2} \end{pmatrix}$$

   (b)
   $$\begin{pmatrix} 1/\sqrt{5} & 2/\sqrt{5} \\ -2/\sqrt{5} & 1/\sqrt{5} \end{pmatrix}$$

   (c)
   $$\begin{pmatrix} 4/5 & 0 & 3/5 \\ -3/5 & 0 & 4/5 \\ 0 & -1 & 0 \end{pmatrix}$$

   (d)
   $$\begin{pmatrix} 1/3 & 2/3 & -2/3 \\ -2/3 & 2/3 & 1/3 \\ 2/3 & 1/3 & 2/3 \end{pmatrix}$$

4. Determine the symmetry group of each of the figures in Figure 12.25.

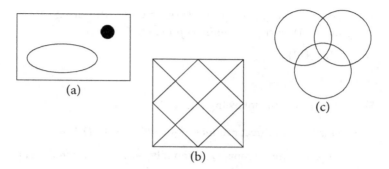

(a)

(b)

(c)

Figure 12.25.

5. Let **x**, **y**, and **w** be vectors in $\mathbb{R}^n$ and $\alpha \in \mathbb{R}$. Prove each of the following properties of inner products.

    (a) $\langle \mathbf{x}, \mathbf{y} \rangle = \langle \mathbf{y}, \mathbf{x} \rangle$.

    (b) $\langle \mathbf{x}, \mathbf{y} + \mathbf{w} \rangle = \langle \mathbf{x}, \mathbf{y} \rangle + \langle \mathbf{x}, \mathbf{w} \rangle$.

    (c) $\langle \alpha\mathbf{x}, \mathbf{y} \rangle = \langle \mathbf{x}, \alpha\mathbf{y} \rangle = \alpha\langle \mathbf{x}, \mathbf{y} \rangle$.

    (d) $\langle \mathbf{x}, \mathbf{x} \rangle \geq 0$ with equality exactly when $\mathbf{x} = 0$.

    (e) If $\langle \mathbf{x}, \mathbf{y} \rangle = 0$ for all **x** in $\mathbb{R}^n$, then $\mathbf{y} = 0$.

6. Verify that
$$E(n) = \{(A, \mathbf{x}) : A \in O(n) \text{ and } \mathbf{x} \in \mathbb{R}^n\}$$
    is a group.

7. Prove that $\{(2,1), (1,1)\}$ and $\{(12,5), (7,3)\}$ are bases for the same lattice.

8. Let G be a subgroup of $E(2)$ and suppose that T is the translation subgroup of G. Prove that the point group of G is isomorphic to G/T.

9. Let $A \in SL_2(\mathbb{R})$ and suppose that the vectors **x** and **y** form two sides of a parallelogram in $\mathbb{R}^2$. Prove that the area of this parallelogram is the same as the area of the parallelogram with sides $A\mathbf{x}$ and $A\mathbf{y}$.

10. Prove that $SO(n)$ is a normal subgroup of $O(n)$.

11. Show that any isometry f in $\mathbb{R}^n$ is a one-to-one map.

12. Prove or disprove: an element in $E(2)$ of the form $(A, \mathbf{x})$, where $\mathbf{x} \neq 0$, has infinite order.

13. Prove or disprove: There exists an infinite abelian subgroup of $O(n)$.

14. Let $\mathbf{x} = (x_1, x_2)$ be a point on the unit circle in $\mathbb{R}^2$; that is, $x_1^2 + x_2^2 = 1$. If $A \in O(2)$, show that $A\mathbf{x}$ is also a point on the unit circle.

15. Let G be a group with a subgroup H (not necessarily normal) and a normal subgroup N. Then G is a *semidirect product* of N by H if

   - $H \cap N = \{\mathrm{id}\}$;
   - $HN = G$.

   Show that each of the following is true.

   (a) S_3 is the semidirect product of A_3 by $H = \{(1), (12)\}$.

   (b) The quaternion group, Q_8, cannot be written as a semidirect product.

   (c) $E(2)$ is the semidirect product of $O(2)$ by H, where H consists of all translations in $\mathbb{R}^2$.

16. Determine which of the 17 wallpaper groups preserves the symmetry of the pattern in Figure 12.16 on page 189.

17. Determine which of the 17 wallpaper groups preserves the symmetry of the pattern in Figure 12.26.

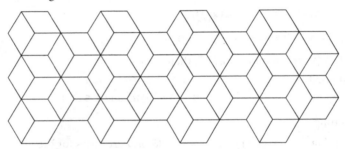

Figure 12.26.

18. Find the rotation group of a dodecahedron.

19. For each of the 17 wallpaper groups, draw a wallpaper pattern having that group as a symmetry group.

12.5 References and Suggested Readings

[1]   Coxeter, H. M. and Moser, W. O. J. *Generators and Relations for Discrete Groups*, 3rd ed. Springer-Verlag, New York, 1972.

[2]   Grove, L. C. and Benson, C. T. *Finite Reflection Groups*. 2nd ed. Springer-Verlag, New York, 1985.

[3]   Hiller, H. "Crystallography and Cohomology of Groups," *American Mathematical Monthly* **93** (1986), 765–79.

[4]   Lockwood, E. H. and Macmillan, R. H. *Geometric Symmetry.* Cambridge University Press, Cambridge, 1978.

[5]   Mackiw, G. *Applications of Abstract Algebra.* Wiley, New York, 1985.

[6]   Martin, G. *Transformation Groups: An Introduction to Symmetry.* Springer-Verlag, New York, 1982.

[7]   Milnor, J. "Hilbert's Problem 18: On Crystallographic Groups, Fundamental Domains, and Sphere Packing," t *Proceedings of Symposia in Pure Mathematics* 18, American Mathematical Society, 1976.

[8]   Phillips, F. C. *An Introduction to Crystallography.* 4th ed. Wiley, New York, 1971.

[9]   Rose, B. I. and Stafford, R. D. "An Elementary Course in Mathematical Symmetry," *American Mathematical Monthly* 88 (1980), 54–64.

[10] Schattschneider, D. "The Plane Symmetry Groups: Their Recognition and Their Notation," *American Mathematical Monthly* 85 (1978), 439–50.

[11] Schwarzenberger, R. L. "The 17 Plane Symmetry Groups," *Mathematical Gazette* 58 (1974), 123–31.

[12] Weyl, H. *Symmetry.* Princeton University Press, Princeton, NJ, 1952.

13

The Structure of Groups

*T*he ultimate goal of group theory is to classify all groups up to isomorphism; that is, given a particular group, we should be able to match it up with a known group via an isomorphism. For example, we have already proved that any finite cyclic group of order n is isomorphic to $\mathbb{Z}_n$; hence, we "know" all finite cyclic groups. It is probably not reasonable to expect that we will ever know all groups; however, we can often classify certain types of groups or distinguish between groups in special cases.

In this chapter we will characterize all finite abelian groups. We shall also investigate groups with sequences of subgroups. If a group has a sequence of subgroups, say

$$G = H_n \supset H_{n-1} \supset \cdots \supset H_1 \supset H_0 = \{e\},$$

where each subgroup H_i is normal in H_{i+1} and each of the factor groups H_{i+1}/H_i is abelian, then G is a solvable group. In addition to allowing us to distinguish between certain classes of groups, solvable groups turn out to be central to the study of solutions to polynomial equations.

13.1 Finite Abelian Groups

In our investigation of cyclic groups we found that every group of prime order was isomorphic to $\mathbb{Z}_p$, where p was a prime number. We also determined that $\mathbb{Z}_{mn} \cong \mathbb{Z}_m \times \mathbb{Z}_n$ when $\gcd(m, n) = 1$. In fact, much more is true. Every finite abelian group is isomorphic to a direct product of cyclic groups of prime power order; that is, every finite abelian group is isomorphic to a group of the type

$$\mathbb{Z}_{p_1^{\alpha_1}} \times \cdots \times \mathbb{Z}_{p_n^{\alpha_n}},$$

where each p_k is prime (not necessarily distinct).

First, let us examine a slight generalization of finite abelian groups. Suppose that G is a group and let $\{g_i\}$ be a set of elements in G, where i is in some index set I (not necessarily finite). The smallest subgroup of G containing all of the g_i's is the subgroup of G **generated** by the g_i's. If this subgroup of G is in fact all of G, then G is generated by the set $\{g_i : i \in I\}$. In this case the g_i's are said to be the *generators* of G. If there is a finite set $\{g_i : i \in I\}$ that generates G, then G is *finitely generated*.

Example 13.1. Obviously, all finite groups are finitely generated. For example, the group S_3 is generated by the permutations $(1\,2)$ and $(1\,2\,3)$. The group $\mathbb{Z} \times \mathbb{Z}_n$ is an infinite group but is finitely generated by $\{(1, 0), (0, 1)\}$. □

Example 13.2. Not all groups are finitely generated. Consider the rational numbers $\mathbb{Q}$ under the operation of addition. Suppose that $\mathbb{Q}$ is finitely generated with generators $p_1/q_1, \ldots, p_n/q_n$, where each p_i/q_i is a fraction expressed in its lowest terms. Let p be some prime that does not divide any of the denominators $q_1, \ldots, q_n$. We claim that $1/p$ cannot be in the subgroup of $\mathbb{Q}$ that is generated by $p_1/q_1, \ldots, p_n/q_n$, since p does not divide the denominator of any element in this subgroup. This fact is easy to see since the sum of any two generators is

$$p_i/q_i + p_j/q_j = (p_iq_j + p_jq_i)/(q_iq_j).$$

□

Proposition 13.3. Let H be the subgroup of a group G that is generated by $\{g_i \in G : i \in I\}$. Then $h \in H$ exactly when it is a product of the form

$$h = g_{i_1}^{\alpha_1} \cdots g_{i_n}^{\alpha_n},$$

where the g_{i_k}s are not necessarily distinct.

Proof. Let K be the set of all products of the form $g_{i_1}^{\alpha_1} \cdots g_{i_n}^{\alpha_n}$, where the g_{i_k}s are not necessarily distinct. Certainly K is a subset of H. We need only show that K is a subgroup of G. If this is the case, then $K = H$, since H is the smallest subgroup containing all the g_is.

Clearly, the set K is closed under the group operation. Since $g_i^0 = 1$, the identity is in K. It remains to show that the inverse of an element $g = g_{i_1}^{k_1} \cdots g_{i_n}^{k_n}$ in K must also be in K. However,

$$g^{-1} = (g_{i_1}^{k_1} \cdots g_{i_n}^{k_n})^{-1} = (g_{i_n}^{-k_n} \cdots g_{i_1}^{-k_1}).$$

∎

The reason that powers of a fixed g_i may occur several times in the product is that we may have a nonabelian group. However, if the group is abelian, then the g_is need occur only once. For example, a product such as $a^{-3}b^5a^7$ in an abelian group could always be simplified (in this case, to a^4b^5).

Now let us restrict our attention to finite abelian groups. We can express any finite abelian group as a finite direct product of cyclic groups. More specifically, letting p be prime, we define a group G to be a *p-group* if every element in G

has as its order a power of p. For example, both $\mathbb{Z}_2 \times \mathbb{Z}_2$ and $\mathbb{Z}_4$ are 2-groups, whereas $\mathbb{Z}_{27}$ is a 3-group. We shall prove the Fundamental Theorem of Finite Abelian Groups which tells us that every finite abelian group is isomorphic to a direct product of cyclic p-groups.

Theorem 13.4. Fundamental Theorem of Finite Abelian Groups. Every finite abelian group G is isomorphic to a direct product of cyclic groups of the form

$$\mathbb{Z}_{p_1^{\alpha_1}} \times \mathbb{Z}_{p_2^{\alpha_2}} \times \cdots \times \mathbb{Z}_{p_n^{\alpha_n}}$$

here the p_i's are primes (not necessarily distinct).

Example 13.5. Suppose that we wish to classify all abelian groups of order $540 = 2^2 \cdot 3^3 \cdot 5$. The Fundamental Theorem of Finite Abelian Groups tells us that we have the following six possibilities.

- $\mathbb{Z}_2 \times \mathbb{Z}_2 \times \mathbb{Z}_3 \times \mathbb{Z}_3 \times \mathbb{Z}_3 \times \mathbb{Z}_5$;

- $\mathbb{Z}_2 \times \mathbb{Z}_2 \times \mathbb{Z}_3 \times \mathbb{Z}_9 \times \mathbb{Z}_5$;

- $\mathbb{Z}_2 \times \mathbb{Z}_2 \times \mathbb{Z}_{27} \times \mathbb{Z}_5$;

- $\mathbb{Z}_4 \times \mathbb{Z}_3 \times \mathbb{Z}_3 \times \mathbb{Z}_3 \times \mathbb{Z}_5$;

- $\mathbb{Z}_4 \times \mathbb{Z}_3 \times \mathbb{Z}_9 \times \mathbb{Z}_5$;

- $\mathbb{Z}_4 \times \mathbb{Z}_{27} \times \mathbb{Z}_5$.

$\square$

The proof of the Fundamental Theorem of Finite Abelian Groups depends on several lemmas.

Lemma 13.6. Let G be a finite abelian group of order n. If p is a prime that divides n, then G contains an element of order p.

Proof. We will prove this lemma by induction. If $n = 1$, then there is nothing to show. Now suppose that the lemma is true for all groups of order k, where $k < n$. Furthermore, let p be a prime that divides n.

If G has no proper nontrivial subgroups, then $G = \langle a \rangle$, where a is any element other than the identity. By Exercise 4.5.39 on page 73, the order of G must be prime. Since p divides n, we know that $p = n$, and G contains $p - 1$ elements of order p.

Now suppose that G contains a nontrivial proper subgroup H. Then $1 < |H| < n$. If $p \mid |H|$, then H contains an element of order p by induction and the lemma is true. Suppose that p does not divide the order of H. Since G is abelian, it must be the case that H is a normal subgroup of G, and $|G| = |H| \cdot |G/H|$. Consequently, p

must divide $|G/H|$. Since $|G/H| < |G| = n$, we know that G/H contains an element aH of order p by the induction hypothesis. Thus,

$$H = (aH)^p = a^p H,$$

and $a^p \in H$ but $a \notin H$. If $|H| = r$, then p and r are relatively prime, and there exist integers s and t such that $sp + tr = 1$. Furthermore, the order of a^p must divide r, and $(a^p)^r = (a^r)^p = 1$.

We claim that a^r has order p. We must show that $a^r \neq 1$. Suppose $a^r = 1$. Then

$$\begin{aligned}
a &= a^{sp+tr} \\
&= a^{sp} a^{tr} \\
&= (a^p)^s (a^r)^t \\
&= (a^p)^s 1 \\
&= (a^p)^s.
\end{aligned}$$

Since $a^p \in H$, it must be the case that $a = (a^p)^s \in H$, which is a contradiction. Therefore, $a^r \neq 1$ is an element of order p in G. ∎

Lemma 13.6 is a special case of Cauchy's Theorem (Theorem 15.1 on page 228), which states that if G is a finite group and p a prime such that p divides the order of G, then G contains a subgroup of order p. We will prove Cauchy's Theorem in Chapter 15.

Lemma 13.7. A finite abelian group is a p-group if and only if its order is a power of p.

Proof. If $|G| = p^n$ then by Lagrange's theorem, then the order of any $g \in G$ must divide p^n, and therefore must be a power of p. Conversely, if $|G|$ is not a power of p, then it has some other prime divisor q, so by Lemma 13.6, G has an element of order q and thus is not a p-group. ∎

Lemma 13.8. Let G be a finite abelian group of order $n = p_1^{\alpha_1} \cdots p_k^{\alpha_k}$, where where $p_1, \ldots, p_k$ are distinct primes and $\alpha_1, \alpha_2, \ldots, \alpha_k$ are positive integers. Then G is the internal direct product of subgroups $G_1, G_2, \ldots, G_k$, where G_i is the subgroup of G consisting of all elements of order p_i^r for some integer r.

Proof. Since G is an abelian group, we are guaranteed that G_i is a subgroup of G for $i = 1, \ldots, k$. Since the identity has order $p_i^0 = 1$, we know that $1 \in G_i$. If $g \in G_i$ has order p_i^r, then g^{-1} must also have order p_i^r. Finally, if $h \in G_i$ has order p_i^s, then

$$(gh)^{p_i^t} = g^{p_i^t} h^{p_i^t} = 1 \cdot 1 = 1,$$

where t is the maximum of r and s.

We must show that

$$G = G_1 G_2 \cdots G_k$$

and $G_i \cap G_j = \{1\}$ for $i \neq j$. Suppose that $g_1 \in G_1$ is in the subgroup generated by $G_2, G_3, \ldots, G_k$. Then $g_1 = g_2 g_3 \cdots g_k$ for $g_i \in G_i$. Since g_i has order p^{α_i}, we know that $g_i^{p^{\alpha_i}} = 1$ for $i = 2, 3, \ldots, k$, and $g_1^{p_2^{\alpha_2} \cdots p_k^{\alpha_k}} = 1$. Since the order of g_1 is a power of p_1 and $\gcd(p_1, p_2^{\alpha_2} \cdots p_k^{\alpha_k}) = 1$, it must be the case that $g_1 = 1$ and the intersection of G_1 with any of the subgroups $G_2, G_3, \ldots, G_k$ is the identity. A similar argument shows that $G_i \cap G_j = \{1\}$ for $i \neq j$.

Next, we must show that it possible to write every $g \in G$ as a product $g_1 \cdots g_k$, where $g_i \in G_i$. Since the order of g divides the order of G, we know that

$$|g| = p_1^{\beta_1} p_2^{\beta_2} \cdots p_k^{\beta_k}$$

for some integers $\beta_1, \ldots, \beta_k$. Letting $a_i = |g|/p_i^{\beta_i}$, the a_i's are relatively prime; hence, there exist integers $b_1, \ldots, b_k$ such that $a_1 b_1 + \cdots + a_k b_k = 1$. Consequently,

$$g = g^{a_1 b_1 + \cdots + a_k b_k} = g^{a_1 b_1} \cdots g^{a_k b_k}.$$

Since

$$g^{(a_i b_i) p_i^{\beta_i}} = g^{b_i |g|} = e,$$

it follows that $g^{a_i b_i}$ must be in G_i. Let $g_i = g^{a_i b_i}$. Then $g = g_1 \cdots g_k \in G_1 G_2 \cdots G_k$. Therefore, $G = G_1 G_2 \cdots G_k$ is an internal direct product of subgroups. ∎

If remains for us to determine the possible structure of each p_i-group G_i in Lemma 13.8 on the preceding page.

Lemma 13.9.   Let G be a finite abelian p-group and suppose that $g \in G$ has maximal order. Then G is isomorphic to $\langle g \rangle \times H$ for some subgroup H of G.

Proof. By Lemma 13.7 on the previous page, we may assume that the order of G is p^n. We shall induct on n. If $n = 1$, then G is cyclic of order p and must be generated by g. Suppose now that the statement of the lemma holds for all integers k with $1 \leq k < n$ and let g be of maximal order in G, say $|g| = p^m$. Then $a^{p^m} = e$ for all $a \in G$. Now choose h in G such that $h \notin \langle g \rangle$, where h has the smallest possible order. Certainly such an h exists; otherwise, $G = \langle g \rangle$ and we are done. Let $H = \langle h \rangle$.

We claim that $\langle g \rangle \cap H = \{e\}$. It suffices to show that $|H| = p$. Since $|h^p| = |h|/p$, the order of h^p is smaller than the order of h and must be in $\langle g \rangle$ by the minimality of h; that is, $h^p = g^r$ for some number r. Hence,

$$(g^r)^{p^{m-1}} = (h^p)^{p^{m-1}} = h^{p^m} = e,$$

and the order of g^r must be less than or equal to p^{m-1}. Therefore, g^r cannot generate $\langle g \rangle$. Notice that p must occur as a factor of r, say $r = ps$, and $h^p = g^r = g^{ps}$. Define a to be $g^{-s}h$. Then a cannot be in $\langle g \rangle$; otherwise, h would also have to be in $\langle g \rangle$. Also,

$$a^p = g^{-sp}h^p = g^{-r}h^p = h^{-p}h^p = e.$$

We have now formed an element a with order p such that $a \notin \langle g \rangle$. Since h was chosen to have the smallest order of all of the elements that are not in $\langle g \rangle$, $|H| = p$.

Now we will show that the order of gH in the factor group G/H must be the same as the order of g in G. If $|gH| < |g| = p^m$, then

$$H = (gH)^{p^{m-1}} = g^{p^{m-1}}H;$$

hence, $g^{p^{m-1}}$ must be in $\langle g \rangle \cap H = \{e\}$, which contradicts the fact that the order of g is p^m. Therefore, gH must have maximal order in G/H. By the Correspondence Theorem and our induction hypothesis,

$$G/H \cong \langle gH \rangle \times K/H$$

for some subgroup K of G containing H. We claim that $\langle g \rangle \cap K = \{e\}$. If $b \in \langle g \rangle \cap K$, then $bH \in \langle gH \rangle \cap K/H = \{H\}$ and $b \in \langle g \rangle \cap H = \{e\}$. It follows that $G = \langle g \rangle K$ implies that $G \cong \langle g \rangle \times K$. $\blacksquare$

The proof of the Fundamental Theorem of Finite Abelian Groups follows very quickly from Lemma 13.9. Suppose that G is a finite abelian group and let g be an element of maximal order in G. If $\langle g \rangle = G$, then we are done; otherwise, $G \cong \mathbb{Z}_{|g|} \times H$ for some subgroup H contained in G by the lemma. Since $|H| < |G|$, we can apply mathematical induction.

We now state the more general theorem for all finitely generated abelian groups. The proof of this theorem can be found in any of the references at the end of this chapter.

Theorem 13.10. Fundamental Theorem of Finitely Generated Abelian Groups.
Every finitely generated abelian group G is isomorphic to a direct product of cyclic groups of the form

$$\mathbb{Z}_{p_1^{\alpha_1}} \times \mathbb{Z}_{p_2^{\alpha_2}} \times \cdots \times \mathbb{Z}_{p_n^{\alpha_n}} \times \mathbb{Z} \times \cdots \times \mathbb{Z},$$

where the p_i's are primes (not necessarily distinct).

13.2 Solvable Groups

A *subnormal series* of a group G is a finite sequence of subgroups

$$G = H_n \supset H_{n-1} \supset \cdots \supset H_1 \supset H_0 = \{e\},$$

where H_i is a normal subgroup of H_{i+1}. If each subgroup H_i is normal in G, then the series is called a *normal series*. The *length* of a subnormal or normal series is the number of proper inclusions.

Example 13.11. Any series of subgroups of an abelian group is a normal series. Consider the following series of groups:

$$\mathbb{Z} \supset 9\mathbb{Z} \supset 45\mathbb{Z} \supset 180\mathbb{Z} \supset \{0\},$$
$$\mathbb{Z}_{24} \supset \langle 2 \rangle \supset \langle 6 \rangle \supset \langle 12 \rangle \supset \{0\}.$$

□

Example 13.12. A subnormal series need not be a normal series. Consider the following subnormal series of the group D_4:

$$D_4 \supset \{(1), (12)(34), (13)(24), (14)(23)\} \supset \{(1), (12)(34)\} \supset \{(1)\}.$$

The subgroup $\{(1), (12)(34)\}$ is not normal in D_4; consequently, this series is not a normal series.   □

A subnormal (normal) series $\{K_j\}$ is a *refinement of a subnormal (normal) series* $\{H_i\}$ if $\{H_i\} \subset \{K_j\}$. That is, each H_i is one of the K_j.

Example 13.13. The series

$$\mathbb{Z} \supset 3\mathbb{Z} \supset 9\mathbb{Z} \supset 45\mathbb{Z} \supset 90\mathbb{Z} \supset 180\mathbb{Z} \supset \{0\}$$

is a refinement of the series

$$\mathbb{Z} \supset 9\mathbb{Z} \supset 45\mathbb{Z} \supset 180\mathbb{Z} \supset \{0\}.$$

□

The best way to study a subnormal or normal series of subgroups, $\{H_i\}$ of G, is actually to study the factor groups H_{i+1}/H_i. We say that two subnormal (normal) series $\{H_i\}$ and $\{K_j\}$ of a group G are *isomorphic* if there is a one-to-one correspondence between the collections of factor groups $\{H_{i+1}/H_i\}$ and $\{K_{j+1}/K_j\}$.

Example 13.14. The two normal series

$$\mathbb{Z}_{60} \supset \langle 3 \rangle \supset \langle 15 \rangle \supset \{0\}$$

$$\mathbb{Z}_{60} \supset \langle 4 \rangle \supset \langle 20 \rangle \supset \{0\}$$

of the group $\mathbb{Z}_{60}$ are isomorphic since

$$\mathbb{Z}_{60}/\langle 3 \rangle \cong \langle 20 \rangle/\{0\} \cong \mathbb{Z}_3$$
$$\langle 3 \rangle/\langle 15 \rangle \cong \langle 4 \rangle/\langle 20 \rangle \cong \mathbb{Z}_5$$
$$\langle 15 \rangle/\{0\} \cong \mathbb{Z}_{60}/\langle 4 \rangle \cong \mathbb{Z}_4.$$

$\square$

A subnormal series $\{H_i\}$ of a group G is a *composition series* if all the factor groups are simple; that is, if none of the factor groups of the series contains a normal subgroup. A normal series $\{H_i\}$ of G is a *principal series* if all the factor groups are simple.

Example 13.15. The group $\mathbb{Z}_{60}$ has a composition series

$$\mathbb{Z}_{60} \supset \langle 3 \rangle \supset \langle 15 \rangle \supset \langle 30 \rangle \supset \{0\}$$

with factor groups

$$\mathbb{Z}_{60}/\langle 3 \rangle \cong \mathbb{Z}_3$$
$$\langle 3 \rangle/\langle 15 \rangle \cong \mathbb{Z}_5$$
$$\langle 15 \rangle/\langle 30 \rangle \cong \mathbb{Z}_2$$
$$\langle 30 \rangle/\{0\} \cong \mathbb{Z}_2.$$

Since $\mathbb{Z}_{60}$ is an abelian group, this series is automatically a principal series. Notice that a composition series need not be unique. The series

$$\mathbb{Z}_{60} \supset \langle 2 \rangle \supset \langle 4 \rangle \supset \langle 20 \rangle \supset \{0\}$$

is also a composition series. $\square$

Example 13.16. For $n \geq 5$, the series

$$S_n \supset A_n \supset \{(1)\}$$

is a composition series for S_n since $S_n/A_n \cong \mathbb{Z}_2$ and A_n is simple. $\square$

Example 13.17. Not every group has a composition series or a principal series. Suppose that

$$\{0\} = H_0 \subset H_1 \subset \cdots \subset H_{n-1} \subset H_n = \mathbb{Z}$$

is a subnormal series for the integers under addition. Then H_1 must be of the form $k\mathbb{Z}$ for some $k \in \mathbb{N}$. In this case $H_1/H_0 \cong k\mathbb{Z}$ is an infinite cyclic group with many nontrivial proper normal subgroups. □

Although composition series need not be unique as in the case of $\mathbb{Z}_{60}$, it turns out that any two composition series are related. The factor groups of the two composition series for $\mathbb{Z}_{60}$ are $\mathbb{Z}_2$, $\mathbb{Z}_2$, $\mathbb{Z}_3$, and $\mathbb{Z}_5$; that is, the two composition series are isomorphic. The Jordan-Hölder Theorem says that this is always the case.

Theorem 13.18. Jordan-Hölder. Any two composition series of G are isomorphic.

Proof. We shall employ mathematical induction on the length of the composition series. If the length of a composition series is 1, then G must be a simple group. In this case any two composition series are isomorphic.

Suppose now that the theorem is true for all groups having a composition series of length k, where $1 \le k < n$. Let

$$G = H_n \supset H_{n-1} \supset \cdots \supset H_1 \supset H_0 = \{e\}$$
$$G = K_m \supset K_{m-1} \supset \cdots \supset K_1 \supset K_0 = \{e\}$$

be two composition series for G. We can form two new subnormal series for G since $H_i \cap K_{m-1}$ is normal in $H_{i+1} \cap K_{m-1}$ and $K_j \cap H_{n-1}$ is normal in $K_{j+1} \cap H_{n-1}$:

$$G = H_n \supset H_{n-1} \supset H_{n-1} \cap K_{m-1} \supset \cdots \supset H_0 \cap K_{m-1} = \{e\}$$
$$G = K_m \supset K_{m-1} \supset K_{m-1} \cap H_{n-1} \supset \cdots \supset K_0 \cap H_{n-1} = \{e\}.$$

Since $H_i \cap K_{m-1}$ is normal in $H_{i+1} \cap K_{m-1}$, the Second Isomorphism Theorem (Theorem 11.12 on page 171) implies that

$$(H_{i+1} \cap K_{m-1})/(H_i \cap K_{m-1}) = (H_{i+1} \cap K_{m-1})/(H_i \cap (H_{i+1} \cap K_{m-1}))$$
$$\cong H_i(H_{i+1} \cap K_{m-1})/H_i,$$

where H_i is normal in $H_i(H_{i+1} \cap K_{m-1})$. Since $\{H_i\}$ is a composition series, H_{i+1}/H_i must be simple; consequently, $H_i(H_{i+1} \cap K_{m-1})/H_i$ is either H_{i+1}/H_i or H_i/H_i. That is, $H_i(H_{i+1} \cap K_{m-1})$ must be either H_i or H_{i+1}. Removing any nonproper inclusions from the series

$$H_{n-1} \supset H_{n-1} \cap K_{m-1} \supset \cdots \supset H_0 \cap K_{m-1} = \{e\},$$

we have a composition series for H_{n-1}. Our induction hypothesis says that this series must be equivalent to the composition series

$$H_{n-1} \supset \cdots \supset H_1 \supset H_0 = \{e\}.$$

Hence, the composition series

$$G = H_n \supset H_{n-1} \supset \cdots \supset H_1 \supset H_0 = \{e\}$$

and

$$G = H_n \supset H_{n-1} \supset H_{n-1} \cap K_{m-1} \supset \cdots \supset H_0 \cap K_{m-1} = \{e\}$$

are equivalent. If $H_{n-1} = K_{m-1}$, then the composition series $\{H_i\}$ and $\{K_j\}$ are equivalent and we are done; otherwise, $H_{n-1}K_{m-1}$ is a normal subgroup of G properly containing H_{n-1}. In this case $H_{n-1}K_{m-1} = G$ and we can apply the Second Isomorphism Theorem once again; that is,

$$K_{m-1}/(K_{m-1} \cap H_{n-1}) \cong (H_{n-1}K_{m-1})/H_{n-1} = G/H_{n-1}.$$

Therefore,

$$G = H_n \supset H_{n-1} \supset H_{n-1} \cap K_{m-1} \supset \cdots \supset H_0 \cap K_{m-1} = \{e\}$$

and

$$G = K_m \supset K_{m-1} \supset K_{m-1} \cap H_{n-1} \supset \cdots \supset K_0 \cap H_{n-1} = \{e\}$$

are equivalent and the proof of the theorem is complete.   ∎

A group G is *solvable* if it has a subnormal series $\{H_i\}$ such that all of the factor groups H_{i+1}/H_i are abelian. Solvable groups will play a fundamental role when we study Galois theory and the solution of polynomial equations.

Example 13.19. The group S_4 is solvable since

$$S_4 \supset A_4 \supset \{(1), (12)(34), (13)(24), (14)(23)\} \supset \{(1)\}$$

has abelian factor groups; however, for $n \geq 5$ the series

$$S_n \supset A_n \supset \{(1)\}$$

is a composition series for S_n with a nonabelian factor group. Therefore, S_n is not a solvable group for $n \geq 5$.   □

Sage. Sage is able to create direct products of cyclic groups, though they are realized as permutation groups. This is a situation that should improve. However, with a classification of finite abelian groups, we can describe how to construct in Sage every group of order less than 16.

13.3 Reading Questions

1. How many abelian groups are there of order $200 = 2^3 5^2$?

2. How many abelian groups are there of order $729 = 3^6$?

3. Find a subgroup of order 6 in $\mathbb{Z}_8 \times \mathbb{Z}_3 \times \mathbb{Z}_3$.

4. It can be shown that an abelian group of order 72 contains a subgroup of order 8. What are the possibilities for this subgroup?

5. What is a principal series of the group G? Your answer should not use new terms defined in this chapter.

13.4 Exercises

1. Find all of the abelian groups of order less than or equal to 40 up to isomorphism.

2. Find all of the abelian groups of order 200 up to isomorphism.

3. Find all of the abelian groups of order 720 up to isomorphism.

4. Find all of the composition series for each of the following groups.

(a) $\mathbb{Z}_{12}$ (e) $S_3 \times \mathbb{Z}_4$

(b) $\mathbb{Z}_{48}$ (f) S_4

(c) The quaternions, Q_8 (g) S_n, $n \geq 5$

(d) D_4 (h) $\mathbb{Q}$

5. Show that the infinite direct product $G = \mathbb{Z}_2 \times \mathbb{Z}_2 \times \cdots$ is not finitely generated.

6. Let G be an abelian group of order m. If n divides m, prove that G has a subgroup of order n.

7. A group G is a *torsion group* if every element of G has finite order. Prove that a finitely generated abelian torsion group must be finite.

8. Let G, H, and K be finitely generated abelian groups. Show that if $G \times H \cong G \times K$, then $H \cong K$. Give a counterexample to show that this cannot be true in general.

9. Let G and H be solvable groups. Show that $G \times H$ is also solvable.

10. If G has a composition (principal) series and if N is a proper normal subgroup of G, show there exists a composition (principal) series containing N.

11. Prove or disprove: Let N be a normal subgroup of G. If N and G/N have composition series, then G must also have a composition series.

12. Let N be a normal subgroup of G. If N and G/N are solvable groups, show that G is also a solvable group.

13. Prove that G is a solvable group if and only if G has a series of subgroups

$$G = P_n \supset P_{n-1} \supset \cdots \supset P_1 \supset P_0 = \{e\}$$

where P_i is normal in P_{i+1} and the order of P_{i+1}/P_i is prime.

14. Let G be a solvable group. Prove that any subgroup of G is also solvable.

15. Let G be a solvable group and N a normal subgroup of G. Prove that G/N is solvable.

16. Prove that D_n is solvable for all integers n.

17. Suppose that G has a composition series. If N is a normal subgroup of G, show that N and G/N also have composition series.

18. Let G be a cyclic p-group with subgroups H and K. Prove that either H is contained in K or K is contained in H.

19. Suppose that G is a solvable group with order $n \geq 2$. Show that G contains a normal nontrivial abelian subgroup.

20. Recall that the *commutator subgroup* G' of a group G is defined as the subgroup of G generated by elements of the form $a^{-1}b^{-1}ab$ for $a, b \in G$. We can define a series of subgroups of G by $G^{(0)} = G$, $G^{(1)} = G'$, and $G^{(i+1)} = (G^{(i)})'$.

    (a) Prove that $G^{(i+1)}$ is normal in $(G^{(i)})'$. The series of subgroups

$$G^{(0)} = G \supset G^{(1)} \supset G^{(2)} \supset \cdots$$

    is called the *derived series* of G.

    (b) Show that G is solvable if and only if $G^{(n)} = \{e\}$ for some integer n.

21. Suppose that G is a solvable group with order $n \geq 2$. Show that G contains a normal nontrivial abelian factor group.

22. **Zassenhaus Lemma.** Let H and K be subgroups of a group G. Suppose also that H^* and K^* are normal subgroups of H and K respectively. Then

    (a) $H^*(H \cap K^*)$ is a normal subgroup of $H^*(H \cap K)$.

    (b) $K^*(H^* \cap K)$ is a normal subgroup of $K^*(H \cap K)$.

    (c) $H^*(H \cap K)/H^*(H \cap K^*) \cong K^*(H \cap K)/K^*(H^* \cap K) \cong (H \cap K)/(H^* \cap K)(H \cap K^*)$.

23. **Schreier's Theorem.** Use the Zassenhaus Lemma to prove that two subnormal (normal) series of a group G have isomorphic refinements.

24. Use Schreier's Theorem to prove the Jordan-Hölder Theorem.

13.5 Programming Exercises

1. Write a program that will compute all possible abelian groups of order n. What is the largest n for which your program will work?

13.6 References and Suggested Readings

[1]   Hungerford, T. W. *Algebra*. Springer, New York, 1974.

[2]   Lang, S. *Algebra*. 3rd ed. Springer, New York, 2002.

[3]   Rotman, J. J. *An Introduction to the Theory of Groups*. 4th ed. Springer, New York, 1995.

14

Group Actions

$\mathcal{G}$roup actions generalize group multiplication. If G is a group and X is an arbitrary set, a group action of an element $g \in G$ and $x \in X$ is a product, gx, living in X. Many problems in algebra are best be attacked via group actions. For example, the proofs of the Sylow theorems and of Burnside's Counting Theorem are most easily understood when they are formulated in terms of group actions.

14.1 Groups Acting on Sets

Let X be a set and G be a group. A *(left) action* of G on X is a map $G \times X \to X$ given by $(g, x) \mapsto gx$, where

1. $ex = x$ for all $x \in X$;

2. $(g_1 g_2)x = g_1(g_2 x)$ for all $x \in X$ and all $g_1, g_2 \in G$.

Under these considerations X is called a *G-set*. Notice that we are not requiring X to be related to G in any way. It is true that every group G acts on every set X by the trivial action $(g, x) \mapsto x$; however, group actions are more interesting if the set X is somehow related to the group G.

Example 14.1. Let $G = GL_2(\mathbb{R})$ and $X = \mathbb{R}^2$. Then G acts on X by left multiplication. If $v \in \mathbb{R}^2$ and I is the identity matrix, then $Iv = v$. If A and B are 2×2 invertible matrices, then $(AB)v = A(Bv)$ since matrix multiplication is associative. $\square$

Example 14.2. Let $G = D_4$ be the symmetry group of a square. If $X = \{1, 2, 3, 4\}$ is the set of vertices of the square, then we can consider D_4 to consist of the following permutations:

$$\{(1), (13), (24), (1432), (1234), (12)(34), (14)(23), (13)(24)\}.$$

The elements of D_4 act on X as functions. The permutation $(13)(24)$ acts on vertex 1 by sending it to vertex 3, on vertex 2 by sending it to vertex 4, and so on. It is easy to see that the axioms of a group action are satisfied. $\square$

In general, if X is any set and G is a subgroup of S_X, the group of all permutations acting on X, then X is a G-set under the group action

$$(\sigma, x) \mapsto \sigma(x)$$

for $\sigma \in G$ and $x \in X$.

Example 14.3. If we let $X = G$, then every group G acts on itself by the left regular representation; that is, $(g, x) \mapsto \lambda_g(x) = gx$, where λ_g is left multiplication:

$$e \cdot x = \lambda_e x = ex = x$$
$$(gh) \cdot x = \lambda_{gh} x = \lambda_g \lambda_h x = \lambda_g(hx) = g \cdot (h \cdot x).$$

If H is a subgroup of G, then G is an H-set under left multiplication by elements of H.   □

Example 14.4. Let G be a group and suppose that $X = G$. If H is a subgroup of G, then G is an H-set under *conjugation*; that is, we can define an action of H on G,

$$H \times G \to G,$$

via

$$(h, g) \mapsto hgh^{-1}$$

for $h \in H$ and $g \in G$. Clearly, the first axiom for a group action holds. Observing that

$$\begin{aligned}
(h_1 h_2, g) &= h_1 h_2 g (h_1 h_2)^{-1} \\
&= h_1(h_2 g h_2^{-1}) h_1^{-1} \\
&= (h_1, (h_2, g)),
\end{aligned}$$

we see that the second condition is also satisfied.   □

Example 14.5. Let H be a subgroup of G and $\mathcal{L}_H$ the set of left cosets of H. The set $\mathcal{L}_H$ is a G-set under the action

$$(g, xH) \mapsto gxH.$$

Again, it is easy to see that the first axiom is true. Since $(gg')xH = g(g'xH)$, the second axiom is also true.   □

If G acts on a set X and $x, y \in X$, then x is said to be *G-equivalent* to y if there exists a $g \in G$ such that $gx = y$. We write $x \sim_G y$ or $x \sim y$ if two elements are G-equivalent.

Proposition 14.6. Let X be a G-set. Then G-equivalence is an equivalence relation on X.

Proof. The relation ~ is reflexive since $ex = x$. Suppose that $x \sim y$ for $x, y \in X$. Then there exists a g such that $gx = y$. In this case $g^{-1}y = x$; hence, $y \sim x$. To show that the relation is transitive, suppose that $x \sim y$ and $y \sim z$. Then there must exist group elements g and h such that $gx = y$ and $hy = z$. So $z = hy = (hg)x$, and x is equivalent to z.   ∎

If X is a G-set, then each partition of X associated with G-equivalence is called an *orbit* of X under G. We will denote the orbit that contains an element x of X by $\mathcal{O}_x$.

Example 14.7. Let G be the permutation group defined by

$$G = \{(1), (123), (132), (45), (123)(45), (132)(45)\}$$

and $X = \{1, 2, 3, 4, 5\}$. Then X is a G-set. The orbits are $\mathcal{O}_1 = \mathcal{O}_2 = \mathcal{O}_3 = \{1, 2, 3\}$ and $\mathcal{O}_4 = \mathcal{O}_5 = \{4, 5\}$.   □

Now suppose that G is a group acting on a set X and let g be an element of G. The *fixed point set* of g in X, denoted by X_g, is the set of all $x \in X$ such that $gx = x$. We can also study the group elements g that fix a given $x \in X$. This set is more than a subset of G, it is a subgroup. This subgroup is called the *stabilizer subgroup* or *isotropy subgroup* of x. We will denote the stabilizer subgroup of x by G_x.

Remark 14.8. It is important to remember that $X_g \subset X$ and $G_x \subset G$.

Example 14.9. Let $X = \{1, 2, 3, 4, 5, 6\}$ and suppose that G is the permutation group given by the permutations

$$\{(1), (12)(3456), (35)(46), (12)(3654)\}.$$

Then the fixed point sets of X under the action of G are

$$X_{(1)} = X,$$
$$X_{(35)(46)} = \{1, 2\},$$
$$X_{(12)(3456)} = X_{(12)(3654)} = \varnothing,$$

and the stabilizer subgroups are

$$G_1 = G_2 = \{(1), (35)(46)\},$$
$$G_3 = G_4 = G_5 = G_6 = \{(1)\}.$$

It is easily seen that G_x is a subgroup of G for each $x \in X$.   □

Proposition 14.10. Let G be a group acting on a set X and $x \in X$. The stabilizer group of x, G_x, is a subgroup of G.

Proof. Clearly, $e \in G_x$ since the identity fixes every element in the set X. Let $g, h \in G_x$. Then $gx = x$ and $hx = x$. So $(gh)x = g(hx) = gx = x$; hence, the product of two elements in G_x is also in G_x. Finally, if $g \in G_x$, then $x = ex = (g^{-1}g)x = (g^{-1})gx = g^{-1}x$. So g^{-1} is in G_x. ∎

We will denote the number of elements in the fixed point set of an element $g \in G$ by $|X_g|$ and denote the number of elements in the orbit of $x \in X$ by $|\mathcal{O}_x|$. The next theorem demonstrates the relationship between orbits of an element $x \in X$ and the left cosets of G_x in G.

Theorem 14.11. Let G be a finite group and X a finite G-set. If $x \in X$, then $|\mathcal{O}_x| = [G : G_x]$.

Proof. We know that $|G|/|G_x|$ is the number of left cosets of G_x in G by Lagrange's Theorem (Theorem 6.10 on page 93). We will define a bijective map ϕ between the orbit $\mathcal{O}_x$ of X and the set of left cosets $\mathcal{L}_{G_x}$ of G_x in G. Let $y \in \mathcal{O}_x$. Then there exists a g in G such that $gx = y$. Define ϕ by $\phi(y) = gG_x$. To show that ϕ is one-to-one, assume that $\phi(y_1) = \phi(y_2)$. Then

$$\phi(y_1) = g_1 G_x = g_2 G_x = \phi(y_2),$$

where $g_1 x = y_1$ and $g_2 x = y_2$. Since $g_1 G_x = g_2 G_x$, there exists a $g \in G_x$ such that $g_2 = g_1 g$,

$$y_2 = g_2 x = g_1 g x = g_1 x = y_1;$$

consequently, the map ϕ is one-to-one. Finally, we must show that the map ϕ is onto. Let gG_x be a left coset. If $gx = y$, then $\phi(y) = gG_x$. ∎

14.2 The Class Equation

Let X be a finite G-set and X_G be the set of fixed points in X; that is,

$$X_G = \{x \in X : gx = x \text{ for all } g \in G\}.$$

Since the orbits of the action partition X,

$$|X| = |X_G| + \sum_{i=k}^{n} |\mathcal{O}_{x_i}|,$$

where $x_k, \ldots, x_n$ are representatives from the distinct nontrivial orbits of X.

Now consider the special case in which G acts on itself by conjugation, where $(g, x) \mapsto gxg^{-1}$. The *center* of G,

$$Z(G) = \{x : xg = gx \text{ for all } g \in G\},$$

is the set of points that are fixed by conjugation. The nontrivial orbits of the action are called the *conjugacy classes* of G. If $x_1, \ldots, x_k$ are representatives from each of the nontrivial conjugacy classes of G and $|\mathcal{O}_{x_1}| = n_1, \ldots, |\mathcal{O}_{x_k}| = n_k$, then

$$|G| = |Z(G)| + n_1 + \cdots + n_k.$$

The stabilizer subgroups of each of the x_i's, $C(x_i) = \{g \in G : gx_i = x_ig\}$, are called the *centralizer subgroups* of the x_i's. From Theorem 14.11, we obtain the *class equation*:

$$|G| = |Z(G)| + [G : C(x_1)] + \cdots + [G : C(x_k)].$$

One of the consequences of the class equation is that the order of each conjugacy class must divide the order of G.

Example 14.12. It is easy to check that the conjugacy classes in S_3 are the following:

$$\{(1)\}, \quad \{(1\,2\,3),(1\,3\,2)\}, \quad \{(1\,2),(1\,3),(2\,3)\}.$$

The class equation is $6 = 1 + 2 + 3$. □

Example 14.13. The center of D_4 is $\{(1),(1\,3)(2\,4)\}$, and the conjugacy classes are

$$\{(1\,3),(2\,4)\}, \quad \{(1\,4\,3\,2),(1\,2\,3\,4)\}, \quad \{(1\,2)(3\,4),(1\,4)(2\,3)\}.$$

Thus, the class equation for D_4 is $8 = 2 + 2 + 2 + 2$. □

Example 14.14. For S_n it takes a bit of work to find the conjugacy classes. We begin with cycles. Suppose that $\sigma = (a_1, \ldots, a_k)$ is a cycle and let $\tau \in S_n$. By Theorem 6.16 on page 95,

$$\tau\sigma\tau^{-1} = (\tau(a_1), \ldots, \tau(a_k)).$$

Consequently, any two cycles of the same length are conjugate. Now let $\sigma = \sigma_1\sigma_2\cdots\sigma_r$ be a cycle decomposition, where the length of each cycle σ_i is r_i. Then σ is conjugate to every other $\tau \in S_n$ whose cycle decomposition has the same lengths.

The number of conjugate classes in S_n is the number of ways in which n can be partitioned into sums of positive integers. In the case of S_3 for example, we can partition the integer 3 into the following three sums:

$$3 = 1 + 1 + 1$$
$$3 = 1 + 2$$
$$3 = 3;$$

therefore, there are three conjugacy classes. There are variations to problem of finding the number of such partitions for any positive integer n that are what computer scientists call **NP-complete**. This effectively means that the problem cannot be solved for a large n because the computations would be too time-consuming for even the largest computer.   □

Theorem 14.15. Let G be a group of order p^n where p is prime. Then G has a nontrivial center.

Proof. We apply the class equation

$$|G| = |Z(G)| + n_1 + \cdots + n_k.$$

Since each $n_i > 1$ and $n_i \mid |G|$, it follows that p must divide each n_i. Also, $p \mid |G|$; hence, p must divide $|Z(G)|$. Since the identity is always in the center of G, $|Z(G)| \geq 1$. Therefore, $|Z(G)| \geq p$, and there exists some $g \in Z(G)$ such that $g \neq 1$.
∎

Corollary 14.16. Let G be a group of order p^2 where p is prime. Then G is abelian.

Proof. By Theorem 14.15, $|Z(G)| = p$ or p^2. Suppose that $|Z(G)| = p$. Then $Z(G)$ and $G/Z(G)$ both have order p and must both be cyclic groups. Choosing a generator $aZ(G)$ for $G/Z(G)$, we can write any element $gZ(G)$ in the quotient group as $a^m Z(G)$ for some integer m; hence, $g = a^m x$ for some x in the center of G. Similarly, if $hZ(G) \in G/Z(G)$, there exists a y in $Z(G)$ such that $h = a^n y$ for some integer n. Since x and y are in the center of G, they commute with all other elements of G; therefore,

$$gh = a^m x a^n y = a^{m+n} xy = a^n y a^m x = hg,$$

and G must be abelian. Hence, $|Z(G)| = p^2$.
∎

14.3 Burnside's Counting Theorem

Suppose that we wish to color the vertices of a square with two different colors, say black and white. We might suspect that there would be $2^4 = 16$ different colorings. However, some of these colorings are equivalent. If we color the first vertex black and the remaining vertices white, it is the same as coloring the second vertex black and the remaining ones white since we could obtain the second coloring simply by rotating the square $90°$ (Figure 14.17).

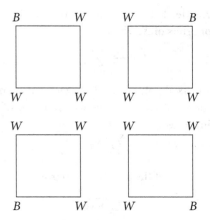

Figure 14.17. Equivalent colorings of square

Burnside's Counting Theorem offers a method of computing the number of distinguishable ways in which something can be done. In addition to its geometric applications, the theorem has interesting applications to areas in switching theory and chemistry. The proof of Burnside's Counting Theorem depends on the following lemma.

Lemma 14.18. Let X be a G-set and suppose that $x \sim y$. Then G_x is isomorphic to G_y. In particular, $|G_x| = |G_y|$.

Proof. Let G act on X by $(g, x) \mapsto g \cdot x$. Since $x \sim y$, there exists a $g \in G$ such that $g \cdot x = y$. Let $a \in G_x$. Since

$$gag^{-1} \cdot y = ga \cdot g^{-1}y = ga \cdot x = g \cdot x = y,$$

we can define a map $\phi : G_x \to G_y$ by $\phi(a) = gag^{-1}$. The map ϕ is a homomorphism since

$$\phi(ab) = gabg^{-1} = gag^{-1}gbg^{-1} = \phi(a)\phi(b).$$

Suppose that $\phi(a) = \phi(b)$. Then $gag^{-1} = gbg^{-1}$ or $a = b$; hence, the map is injective. To show that ϕ is onto, let b be in G_y; then $g^{-1}bg$ is in G_x since

$$g^{-1}bg \cdot x = g^{-1}b \cdot gx = g^{-1}b \cdot y = g^{-1} \cdot y = x;$$

and $\phi(g^{-1}bg) = b$. ∎

Theorem 14.19. Burnside. Let G be a finite group acting on a set X and let k denote the number of orbits of X. Then

$$k = \frac{1}{|G|} \sum_{g \in G} |X_g|.$$

Proof. We look at all the fixed points x of all the elements in $g \in G$; that is, we look at all g's and all x's such that $gx = x$. If viewed in terms of fixed point sets, the number of all g's fixing x's is

$$\sum_{g \in G} |X_g|.$$

However, if viewed in terms of the stabilizer subgroups, this number is

$$\sum_{x \in X} |G_x|;$$

hence, $\sum_{g \in G} |X_g| = \sum_{x \in X} |G_x|$. By Lemma 14.18 on the previous page,

$$\sum_{y \in \mathcal{O}_x} |G_y| = |\mathcal{O}_x| \cdot |G_x|.$$

By Theorem 14.11 on page 214 and Lagrange's Theorem, this expression is equal to $|G|$. Summing over all of the k distinct orbits, we conclude that

$$\sum_{g \in G} |X_g| = \sum_{x \in X} |G_x| = k \cdot |G|.$$

∎

Example 14.20. Let $X = \{1, 2, 3, 4, 5\}$ and suppose that G is the permutation group $G = \{(1), (13), (13)(25), (25)\}$. The orbits of X are $\{1, 3\}$, $\{2, 5\}$, and $\{4\}$. The fixed point sets are

$$X_{(1)} = X$$
$$X_{(13)} = \{2, 4, 5\}$$
$$X_{(13)(25)} = \{4\}$$
$$X_{(25)} = \{1, 3, 4\}.$$

Burnside's Theorem says that

$$k = \frac{1}{|G|} \sum_{g \in G} |X_g| = \frac{1}{4}(5 + 3 + 1 + 3) = 3.$$

□

A Geometric Example

Before we apply Burnside's Theorem to switching-theory problems, let us examine the number of ways in which the vertices of a square can be colored black or white. Notice that we can sometimes obtain equivalent colorings by simply applying a rigid motion to the square. For instance, as we have pointed out, if we color one of the vertices black and the remaining three white, it does not matter which vertex was colored black since a rotation will give an equivalent coloring.

The symmetry group of a square, D_4, is given by the following permutations:

$$(1) \qquad (13) \qquad (24) \qquad (1432)$$
$$(1234) \qquad (12)(34) \qquad (14)(23) \qquad (13)(24)$$

The group G acts on the set of vertices $\{1, 2, 3, 4\}$ in the usual manner. We can describe the different colorings by mappings from X into $Y = \{B, W\}$ where B and W represent the colors black and white, respectively. Each map $f : X \to Y$ describes a way to color the corners of the square. Every $\sigma \in D_4$ induces a permutation $\widetilde{\sigma}$ of the possible colorings given by $\widetilde{\sigma}(f) = f \circ \sigma$ for $f : X \to Y$. For example, suppose that f is defined by

$$f(1) = B$$
$$f(2) = W$$
$$f(3) = W$$
$$f(4) = W$$

and $\sigma = (12)(34)$. Then $\widetilde{\sigma}(f) = f \circ \sigma$ sends vertex 2 to B and the remaining vertices to W. The set of all such $\widetilde{\sigma}$ is a permutation group $\widetilde{G}$ on the set of possible colorings. Let $\widetilde{X}$ denote the set of all possible colorings; that is, $\widetilde{X}$ is the set of all possible maps from X to Y. Now we must compute the number of $\widetilde{G}$-equivalence classes.

1. $\widetilde{X}_{(1)} = \widetilde{X}$ since the identity fixes every possible coloring. $|\widetilde{X}| = 2^4 = 16$.

2. $\widetilde{X}_{(1234)}$ consists of all $f \in \widetilde{X}$ such that f is unchanged by the permutation (1234). In this case $f(1) = f(2) = f(3) = f(4)$, so that all values of f must be the same; that is, either $f(x) = B$ or $f(x) = W$ for every vertex x of the square. So $|\widetilde{X}_{(1234)}| = 2$.

3. $|\widetilde{X}_{(1432)}| = 2$.

4. For $\widetilde{X}_{(13)(24)}$, $f(1) = f(3)$ and $f(2) = f(4)$. Thus, $|\widetilde{X}_{(13)(24)}| = 2^2 = 4$.

5. $|\widetilde{X}_{(12)(34)}| = 4$.

6. $|\widetilde{X}_{(14)(23)}| = 4.$

7. For $\widetilde{X}_{(13)}$, $f(1) = f(3)$ and the other corners can be of any color; hence, $|\widetilde{X}_{(13)}| = 2^3 = 8$.

8. $|\widetilde{X}_{(24)}| = 8$.

By Burnside's Theorem, we can conclude that there are exactly

$$\frac{1}{8}(2^4 + 2^1 + 2^2 + 2^1 + 2^2 + 2^2 + 2^3 + 2^3) = 6$$

ways to color the vertices of the square.

Proposition 14.21. Let G be a permutation group of X and $\widetilde{X}$ the set of functions from X to Y. Then there exists a permutation group $\widetilde{G}$ acting on $\widetilde{X}$, where $\widetilde{\sigma} \in \widetilde{G}$ is defined by $\widetilde{\sigma}(f) = f \circ \sigma$ for $\sigma \in G$ and $f \in \widetilde{X}$. Furthermore, if n is the number of cycles in the cycle decomposition of σ, then $|\widetilde{X}_\sigma| = |Y|^n$.

Proof. Let $\sigma \in G$ and $f \in \widetilde{X}$. Clearly, $f \circ \sigma$ is also in $\widetilde{X}$. Suppose that g is another function from X to Y such that $\widetilde{\sigma}(f) = \widetilde{\sigma}(g)$. Then for each $x \in X$,

$$f(\sigma(x)) = \widetilde{\sigma}(f)(x) = \widetilde{\sigma}(g)(x) = g(\sigma(x)).$$

Since σ is a permutation of X, every element x' in X is the image of some x in X under σ; hence, f and g agree on all elements of X. Therefore, $f = g$ and $\widetilde{\sigma}$ is injective. The map $\sigma \mapsto \widetilde{\sigma}$ is onto, since the two sets are the same size.

Suppose that σ is a permutation of X with cycle decomposition $\sigma = \sigma_1 \sigma_2 \cdots \sigma_n$. Any f in $\widetilde{X}_\sigma$ must have the same value on each cycle of σ. Since there are n cycles and $|Y|$ possible values for each cycle, $|\widetilde{X}_\sigma| = |Y|^n$. ∎

Example 14.22. Let $X = \{1, 2, \ldots, 7\}$ and suppose that $Y = \{A, B, C\}$. If g is the permutation of X given by $(13)(245) = (13)(245)(6)(7)$, then $n = 4$. Any $f \in \widetilde{X}_g$ must have the same value on each cycle in g. There are $|Y| = 3$ such choices for any value, so $|\widetilde{X}_g| = 3^4 = 81$. □

Example 14.23. Suppose that we wish to color the vertices of a square using four different colors. By Proposition 14.21, we can immediately decide that there are

$$\frac{1}{8}(4^4 + 4^1 + 4^2 + 4^1 + 4^2 + 4^2 + 4^3 + 4^3) = 55$$

possible ways. □

Switching Functions

In switching theory we are concerned with the design of electronic circuits with binary inputs and outputs. The simplest of these circuits is a switching function that has n inputs and a single output (Figure 14.24). Large electronic circuits can often be constructed by combining smaller modules of this kind. The inherent problem here is that even for a simple circuit a large number of different switching functions can be constructed. With only four inputs and a single output, we can construct 65,536 different switching functions. However, we can often replace one switching function with another merely by permuting the input leads to the circuit (Figure 14.25).

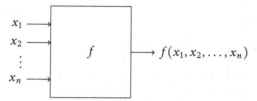

Figure 14.24. A switching function of n variables

We define a *switching* or *Boolean function* of n variables to be a function from $\mathbb{Z}_2^n$ to $\mathbb{Z}_2$. Since any switching function can have two possible values for each binary n-tuple and there are 2^n binary n-tuples, 2^{2^n} switching functions are possible for n variables. In general, allowing permutations of the inputs greatly reduces the number of different kinds of modules that are needed to build a large circuit.

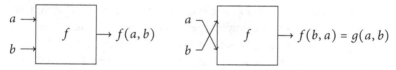

Figure 14.25. A switching function of two variables

The possible switching functions with two input variables a and b are listed in Table 14.26 on the following page. Two switching functions f and g are equivalent if g can be obtained from f by a permutation of the input variables. For example, $g(a, b, c) = f(b, c, a)$. In this case $g \sim f$ via the permutation (a, c, b). In the case of switching functions of two variables, the permutation (a, b) reduces 16 possible switching functions to 12 equivalent functions since

$$f_2 \sim f_4$$
$$f_3 \sim f_5$$
$$f_{10} \sim f_{12}$$
$$f_{11} \sim f_{13}.$$

Table 14.26. Switching functions in two variables

Inputs		Outputs							
		f_0	f_1	f_2	f_3	f_4	f_5	f_6	f_7
0	0	0	0	0	0	0	0	0	0
0	1	0	0	0	0	1	1	1	1
1	0	0	0	1	1	0	0	1	1
1	1	0	1	0	1	0	1	0	1
Inputs		Outputs							
		f_8	f_9	f_{10}	f_{11}	f_{12}	f_{13}	f_{14}	f_{15}
0	0	1	1	1	1	1	1	1	1
0	1	0	0	0	0	1	1	1	1
1	0	0	0	1	1	0	0	1	1
1	1	0	1	0	1	0	1	0	1

For three input variables there are 2^{2^3} = 256 possible switching functions; in the case of four variables there are 2^{2^4} = 65,536. The number of equivalence classes is too large to reasonably calculate directly. It is necessary to employ Burnside's Theorem.

Consider a switching function with three possible inputs, a, b, and c. As we have mentioned, two switching functions f and g are equivalent if a permutation of the input variables of f gives g. It is important to notice that a permutation of the switching functions is not simply a permutation of the input values $\{a, b, c\}$. A switching function is a set of output values for the inputs a, b, and c, so when we consider equivalent switching functions, we are permuting 2^3 possible outputs, not just three input values. For example, each binary triple (a, b, c) has a specific output associated with it. The permutation (acb) changes outputs as follows:

$$(0, 0, 0) \mapsto (0, 0, 0)$$
$$(0, 0, 1) \mapsto (0, 1, 0)$$
$$(0, 1, 0) \mapsto (1, 0, 0)$$

$$\vdots$$

$$(1, 1, 0) \mapsto (1, 0, 1)$$
$$(1, 1, 1) \mapsto (1, 1, 1).$$

Let X be the set of output values for a switching function in n variables. Then $|X| = 2^n$. We can enumerate these values as follows:

$$(0, \ldots, 0, 1) \mapsto 0$$
$$(0, \ldots, 1, 0) \mapsto 1$$
$$(0, \ldots, 1, 1) \mapsto 2$$
$$\vdots$$
$$(1, \ldots, 1, 1) \mapsto 2^n - 1.$$

Now let us consider a circuit with four input variables and a single output. Suppose that we can permute the leads of any circuit according to the following permutation group:

$$(a), \quad (a, c), \quad (b, d), \quad (a, d, c, b),$$
$$(a, b, c, d), \quad (a, b)(c, d), \quad (a, d)(b, c), \quad (a, c)(b, d).$$

The permutations of the four possible input variables induce the permutations of the output values in Table 14.27.

Hence, there are

$$\frac{1}{8}(2^{16} + 2 \cdot 2^{12} + 2 \cdot 2^6 + 3 \cdot 2^{10}) = 9616$$

possible switching functions of four variables under this group of permutations. This number will be even smaller if we consider the full symmetric group on four letters.

Table 14.27. Permutations of switching functions in four variables

Group Permutation	Switching Function Permutation	Number of Cycles
(a)	(0)	16
(a, c)	$(2, 8)(3, 9)(6, 12)(7, 13)$	12
(b, d)	$(1, 4)(3, 6)(9, 12)(11, 14)$	12
(a, d, c, b)	$(1, 2, 4, 8)(3, 6.12, 9)(5, 10)(7, 14, 13, 11)$	6
(a, b, c, d)	$(1, 8, 4, 2)(3, 9, 12, 6)(5, 10)(7, 11, 13, 14)$	6
$(a, b)(c, d)$	$(1, 2)(4, 8)(5, 10)(6, 9)(7, 11)(13, 14)$	10
$(a, d)(b, c)$	$(1, 8)(2, 4)(3, 12)(5, 10)(7, 14)(11, 13)$	10
$(a, c)(b, d)$	$(1, 4)(2, 8)(3, 12)(6, 9)(7, 13)(11, 14)$	10

Sage.   Sage has many commands related to conjugacy, which is a group action. It also has commands for orbits and stabilizers of permutation groups. In the supplement, we illustrate the automorphism group of a (combinatorial) graph as another example of a group action on the vertex set of the graph.

∽   Historical Note   ∾

William Burnside was born in London in 1852. He attended Cambridge University from 1871 to 1875 and won the Smith's Prize in his last year. After his graduation he lectured at Cambridge. He was made a member of the Royal Society in 1893. Burnside wrote approximately 150 papers on topics in applied mathematics, differential geometry, and probability, but his most famous contributions were in group theory. Several of Burnside's conjectures have stimulated research to this day. One such conjecture was that every group of odd order is solvable; that is, for a group G of odd order, there exists a sequence of subgroups

$$G = H_n \supset H_{n-1} \supset \cdots \supset H_1 \supset H_0 = \{e\}$$

such that H_i is normal in H_{i+1} and H_{i+1}/H_i is abelian. This conjecture was finally proven by W. Feit and J. Thompson in 1963. Burnside's *The Theory of Groups of Finite Order*, published in 1897, was one of the first books to treat groups in a modern context as opposed to permutation groups. The second edition, published in 1911, is still a classic.

14.4 Reading Questions

1. Give an informal description of a group action.

2. Describe the class equation.

3. What are the groups of order 49?

4. How many switching functions are there with 5 inputs? (Give both a simple expression and the total number as a single integer.)

5. The "Historical Note" mentions the proof of Burnside's Conjecture. How long was the proof?

14.5 Exercises

1. Examples 14.1 on page 211–14.5 on page 212 in the first section each describe an action of a group G on a set X, which will give rise to the equivalence relation defined by G-equivalence. For each example, compute the equivalence classes of the equivalence relation, the *G-equivalence classes*.

2. Compute all X_g and all G_x for each of the following permutation groups.

    (a) $X = \{1, 2, 3\}$, $G = S_3 = \{(1), (12), (13), (23), (123), (132)\}$

    (b) $X = \{1, 2, 3, 4, 5, 6\}$,
       $G = \{(1), (12), (345), (354), (12)(345), (12)(354)\}$

3. Compute the G-equivalence classes of X for each of the G-sets in Exercise 14.5.2. For each $x \in X$ verify that $|G| = |\mathcal{O}_x| \cdot |G_x|$.

4. Let G be the additive group of real numbers. Let the action of $\theta \in G$ on the real plane $\mathbb{R}^2$ be given by rotating the plane counterclockwise about the origin through θ radians. Let P be a point on the plane other than the origin.

    (a) Show that $\mathbb{R}^2$ is a G-set.

    (b) Describe geometrically the orbit containing P.

    (c) Find the group G_P.

5. Let $G = A_4$ and suppose that G acts on itself by conjugation: $(g, h) \mapsto ghg^{-1}$.

    (a) Determine the conjugacy classes (orbits) of each element of G.

    (b) Determine all of the isotropy subgroups for each element of G.

6. Find the conjugacy classes and the class equation for each of the following groups.

    (a) S_4 (b) D_5 (c) $\mathbb{Z}_9$ (d) Q_8

7. Write the class equation for S_5 and for A_5.

8. If a square remains fixed in the plane, how many different ways can the corners of the square be colored if three colors are used?

9. How many ways can the vertices of an equilateral triangle be colored using three different colors?

10. Find the number of ways a six-sided die can be constructed if each side is marked differently with $1, \ldots, 6$ dots.

11. Up to a rotation, how many ways can the faces of a cube be colored with three different colors?

12. Consider 12 straight wires of equal lengths with their ends soldered together to form the edges of a cube. Either silver or copper wire can be used for each edge. How many different ways can the cube be constructed?

13. Suppose that we color each of the eight corners of a cube. Using three different colors, how many ways can the corners be colored up to a rotation of the cube?

14. Each of the faces of a regular tetrahedron can be painted either red or white. Up to a rotation, how many different ways can the tetrahedron be painted?

15. Suppose that the vertices of a regular hexagon are to be colored either red or white. How many ways can this be done up to a symmetry of the hexagon?

16. A molecule of benzene is made up of six carbon atoms and six hydrogen atoms, linked together in a hexagonal shape as in Figure 14.28.

   (a) How many different compounds can be formed by replacing one or more of the hydrogen atoms with a chlorine atom?

   (b) Find the number of different chemical compounds that can be formed by replacing three of the six hydrogen atoms in a benzene ring with a CH_3 radical.

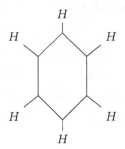

Figure 14.28. A benzene ring

17. How many equivalence classes of switching functions are there if the input variables x_1, x_2, and x_3 can be permuted by any permutation in S_3? What if the input variables x_1, x_2, x_3, and x_4 can be permuted by any permutation in S_4?

18. How many equivalence classes of switching functions are there if the input variables x_1, x_2, x_3, and x_4 can be permuted by any permutation in the subgroup of S_4 generated by the permutation (x_1, x_2, x_3, x_4)?

19. A striped necktie has 12 bands of color. Each band can be colored by one of four possible colors. How many possible different-colored neckties are there?

20. A group acts *faithfully* on a G-set X if the identity is the only element of G that leaves every element of X fixed. Show that G acts faithfully on X if and only if no two distinct elements of G have the same action on each element of X.

21. Let p be prime. Show that the number of different abelian groups of order p^n (up to isomorphism) is the same as the number of conjugacy classes in S_n.

22. Let $a \in G$. Show that for any $g \in G$, $gC(a)g^{-1} = C(gag^{-1})$.

23. Let $|G| = p^n$ be a nonabelian group for p prime. Prove that $|Z(G)| < p^{n-1}$.

24. Let G be a group with order p^n where p is prime and X a finite G-set. If $X_G = \{x \in X : gx = x \text{ for all } g \in G\}$ is the set of elements in X fixed by the group action, then prove that $|X| \equiv |X_G| \pmod{p}$.

25. If G is a group of order p^n, where p is prime and $n \geq 2$, show that G must have a proper subgroup of order p. If $n \geq 3$, is it true that G will have a proper subgroup of order p^2?

14.6 Programming Exercise

1. Write a program to compute the number of conjugacy classes in S_n. What is the largest n for which your program will work?

14.7 References and Suggested Reading

[1] De Bruijin, N. G. "Pólya's Theory of Counting," in *Applied Combinatorial Mathematics*, Beckenbach, E. F., ed. Wiley, New York, 1964.

[2] Eidswick, J. A. "Cubelike Puzzles—What Are They and How Do You Solve Them?" *American Mathematical Monthly* 93 (1986), 157–76.

[3] Harary, F., Palmer, E. M., and Robinson, R. W. "Pólya's Contributions to Chemical Enumeration," in *Chemical Applications of Graph Theory*, Balaban, A. T., ed. Academic Press, London, 1976.

[4] Gårding, L. and Tambour, T. *Algebra for Computer Science*. Springer-Verlag, New York, 1988.

[5] Laufer, H. B. *Discrete Mathematics and Applied Modern Algebra*. PWS-KENT, Boston, 1984.

[6] Pólya, G. and Read, R. C. *Combinatorial Enumeration of Groups, Graphs, and Chemical Compounds*. Springer-Verlag, New York, 1985.

[7] Shapiro, L. W. "Finite Groups Acting on Sets with Applications," *Mathematics Magazine*, May–June 1973, 136–47.

15

The Sylow Theorems

$\mathcal{W}$e already know that the converse of Lagrange's Theorem is false. If G is a group of order m and n divides m, then G does not necessarily possess a subgroup of order n. For example, A_4 has order 12 but does not possess a subgroup of order 6. However, the Sylow Theorems do provide a partial converse for Lagrange's Theorem—in certain cases they guarantee us subgroups of specific orders. These theorems yield a powerful set of tools for the classification of all finite nonabelian groups.

15.1 The Sylow Theorems

We will use what we have learned about group actions to prove the Sylow Theorems. Recall for a moment what it means for G to act on itself by conjugation and how conjugacy classes are distributed in the group according to the class equation, discussed in Chapter 14. A group G acts on itself by conjugation via the map $(g, x) \mapsto gxg^{-1}$. Let $x_1, \ldots, x_k$ be representatives from each of the distinct conjugacy classes of G that consist of more than one element. Then the class equation can be written as

$$|G| = |Z(G)| + [G : C(x_1)] + \cdots + [G : C(x_k)],$$

where $Z(G) = \{g \in G : gx = xg \text{ for all } x \in G\}$ is the center of G and $C(x_i) = \{g \in G : gx_i = x_i g\}$ is the centralizer subgroup of x_i.

We begin our investigation of the Sylow Theorems by examining subgroups of order p, where p is prime. A group G is a *p-group* if every element in G has as its order a power of p, where p is a prime number. A subgroup of a group G is a *p-subgroup* if it is a p-group.

Theorem 15.1. Cauchy. Let G be a finite group and p a prime such that p divides the order of G. Then G contains a subgroup of order p.

Proof. We will use induction on the order of G. If $|G| = p$, then clearly G itself is the required subgroup. We now assume that every group of order k, where $p \leq k < n$ and p divides k, has an element of order p. Assume that $|G| = n$ and

$p \mid n$ and consider the class equation of G:

$$|G| = |Z(G)| + [G : C(x_1)] + \cdots + [G : C(x_k)].$$

We have two cases.

Case 1. Suppose the order of one of the centralizer subgroups, $C(x_i)$, is divisible by p for some i, $i = 1, \ldots, k$. In this case, by our induction hypothesis, we are done. Since $C(x_i)$ is a proper subgroup of G and p divides $|C(x_i)|$, $C(x_i)$ must contain an element of order p. Hence, G must contain an element of order p.

Case 2. Suppose the order of no centralizer subgroup is divisible by p. Then p divides $[G : C(x_i)]$, the order of each conjugacy class in the class equation; hence, p must divide the center of G, $Z(G)$. Since $Z(G)$ is abelian, it must have a subgroup of order p by the Fundamental Theorem of Finite Abelian Groups. Therefore, the center of G contains an element of order p.   ∎

Corollary 15.2. Let G be a finite group. Then G is a p-group if and only if $|G| = p^n$.

Example 15.3. Let us consider the group A_5. We know that $|A_5| = 60 = 2^2 \cdot 3 \cdot 5$. By Cauchy's Theorem, we are guaranteed that A_5 has subgroups of orders 2, 3 and 5. The Sylow Theorems will give us even more information about the possible subgroups of A_5.   □

We are now ready to state and prove the first of the Sylow Theorems. The proof is very similar to the proof of Cauchy's Theorem.

Theorem 15.4. First Sylow Theorem. Let G be a finite group and p a prime such that p^r divides $|G|$. Then G contains a subgroup of order p^r.

Proof. We induct on the order of G once again. If $|G| = p$, then we are done. Now suppose that the order of G is n with $n > p$ and that the theorem is true for all groups of order less than n, where p divides n. We shall apply the class equation once again:

$$|G| = |Z(G)| + [G : C(x_1)] + \cdots + [G : C(x_k)].$$

First suppose that p does not divide $[G : C(x_i)]$ for some i. Then $p^r \mid |C(x_i)|$, since p^r divides $|G| = |C(x_i)| \cdot [G : C(x_i)]$. Now we can apply the induction hypothesis to $C(x_i)$.

Hence, we may assume that p divides $[G : C(x_i)]$ for all i. Since p divides $|G|$, the class equation says that p must divide $|Z(G)|$; hence, by Cauchy's Theorem, $Z(G)$ has an element of order p, say g. Let N be the group generated by g. Clearly, N is a normal subgroup of $Z(G)$ since $Z(G)$ is abelian; therefore, N is normal in G since every element in $Z(G)$ commutes with every element in G. Now consider the factor group G/N of order $|G|/p$. By the induction hypothesis, G/N contains a subgroup H of order p^{r-1}. The inverse image of H under the canonical homomorphism $\phi : G \to G/N$ is a subgroup of order p^r in G.   ∎

A *Sylow p-subgroup* P of a group G is a maximal p-subgroup of G. To prove the other two Sylow Theorems, we need to consider conjugate subgroups as opposed to conjugate elements in a group. For a group G, let $\mathcal{S}$ be the collection of all subgroups of G. For any subgroup H, $\mathcal{S}$ is a H-set, where H acts on $\mathcal{S}$ by conjugation. That is, we have an action

$$H \times \mathcal{S} \to \mathcal{S}$$

defined by

$$h \cdot K \mapsto hKh^{-1}$$

for K in $\mathcal{S}$.

The set

$$N(H) = \{g \in G : gHg^{-1} = H\}$$

is a subgroup of G called the the *normalizer* of H in G. Notice that H is a normal subgroup of $N(H)$. In fact, $N(H)$ is the largest subgroup of G in which H is normal.

Lemma 15.5. Let P be a Sylow p-subgroup of a finite group G and let x have as its order a power of p. If $x^{-1}Px = P$, then $x \in P$.

Proof. Certainly $x \in N(P)$, and the cyclic subgroup, $\langle xP \rangle \subset N(P)/P$, has as its order a power of p. By the Correspondence Theorem there exists a subgroup H of $N(P)$ containing P such that $H/P = \langle xP \rangle$. Since $|H| = |P| \cdot |\langle xP \rangle|$, the order of H must be a power of p. However, P is a Sylow p-subgroup contained in H. Since the order of P is the largest power of p dividing $|G|$, $H = P$. Therefore, H/P is the trivial subgroup and $xP = P$, or $x \in P$.   ∎

Lemma 15.6.   Let H and K be subgroups of G. The number of distinct H-conjugates of K is $[H : N(K) \cap H]$.

Proof. We define a bijection between the conjugacy classes of K and the right cosets of $N(K) \cap H$ by $h^{-1}Kh \mapsto (N(K) \cap H)h$. To show that this map is a bijection, let $h_1, h_2 \in H$ and suppose that $(N(K) \cap H)h_1 = (N(K) \cap H)h_2$. Then $h_2 h_1^{-1} \in N(K)$. Therefore, $K = h_2 h_1^{-1} K h_1 h_2^{-1}$ or $h_1^{-1} K h_1 = h_2^{-1} K h_2$, and the map is an injection. It is easy to see that this map is surjective; hence, we have a one-to-one and onto map between the H-conjugates of K and the right cosets of $N(K) \cap H$ in H.   ∎

Theorem 15.7. Second Sylow Theorem. Let G be a finite group and p a prime dividing $|G|$. Then all Sylow p-subgroups of G are conjugate. That is, if P_1 and P_2 are two Sylow p-subgroups, there exists a $g \in G$ such that $gP_1g^{-1} = P_2$.

Proof. Let P be a Sylow p-subgroup of G and suppose that $|G| = p^r m$ with $|P| = p^r$. Let

$$\mathcal{S} = \{P = P_1, P_2, \ldots, P_k\}$$

consist of the distinct conjugates of P in G. By Lemma 15.6, $k = [G : N(P)]$. Notice that

$$|G| = p^r m = |N(P)| \cdot [G : N(P)] = |N(P)| \cdot k.$$

Since p^r divides $|N(P)|$, p cannot divide k.

Given any other Sylow p-subgroup Q, we must show that $Q \in \mathcal{S}$. Consider the Q-conjugacy classes of each P_i. Clearly, these conjugacy classes partition $\mathcal{S}$. The size of the partition containing P_i is $[Q : N(P_i) \cap Q]$ by Lemma 15.6, and Lagrange's Theorem tells us that $|Q| = [Q : N(P_i) \cap Q]|N(P_i) \cap Q|$. Thus, $[Q : N(P_i) \cap Q]$ must be a divisor of $|Q| = p^r$. Hence, the number of conjugates in every equivalence class of the partition is a power of p. However, since p does not divide k, one of these equivalence classes must contain only a single Sylow p-subgroup, say P_j. In this case, $x^{-1}P_j x = P_j$ for all $x \in Q$. By Lemma 15.5, $P_j = Q$. ∎

Theorem 15.8. Third Sylow Theorem. Let G be a finite group and let p be a prime dividing the order of G. Then the number of Sylow p-subgroups is congruent to 1 (mod p) and divides $|G|$.

Proof. Let P be a Sylow p-subgroup acting on the set of Sylow p-subgroups,

$$\mathcal{S} = \{P = P_1, P_2, \ldots, P_k\},$$

by conjugation. From the proof of the Second Sylow Theorem, the only P-conjugate of P is itself and the order of the other P-conjugacy classes is a power of p. Each P-conjugacy class contributes a positive power of p toward $|\mathcal{S}|$ except the equivalence class $\{P\}$. Since $|\mathcal{S}|$ is the sum of positive powers of p and 1, $|\mathcal{S}| \equiv 1$ (mod p).

Now suppose that G acts on $\mathcal{S}$ by conjugation. Since all Sylow p-subgroups are conjugate, there can be only one orbit under this action. For $P \in \mathcal{S}$,

$$|\mathcal{S}| = |\text{orbit of } P| = [G : N(P)]$$

by Lemma 15.6. But $[G : N(P)]$ is a divisor of $|G|$; consequently, the number of Sylow p-subgroups of a finite group must divide the order of the group. ∎

Peter Ludvig Mejdell Sylow was born in 1832 in Christiania, Norway (now Oslo). After attending Christiania University, Sylow taught high school. In 1862 he obtained a temporary appointment at Christiania University. Even though his appointment was relatively brief, he influenced students such as Sophus Lie (1842–1899). Sylow had a chance at a permanent chair in 1869, but failed to obtain the appointment. In 1872, he published a 10-page paper presenting the theorems that now bear his name. Later Lie and Sylow collaborated on a new edition of Abel's works. In 1898, a chair at Christiania University was finally created for Sylow through the efforts of his student and colleague Lie. Sylow died in 1918.

15.2 Examples and Applications

Example 15.9. Using the Sylow Theorems, we can determine that A_5 has subgroups of orders 2, 3, 4, and 5. The Sylow p-subgroups of A_5 have orders 3, 4, and 5. The Third Sylow Theorem tells us exactly how many Sylow p-subgroups A_5 has. Since the number of Sylow 5-subgroups must divide 60 and also be congruent to 1 (mod 5), there are either one or six Sylow 5-subgroups in A_5. All Sylow 5-subgroups are conjugate. If there were only a single Sylow 5-subgroup, it would be conjugate to itself; that is, it would be a normal subgroup of A_5. Since A_5 has no normal subgroups, this is impossible; hence, we have determined that there are exactly six distinct Sylow 5-subgroups of A_5. □

The Sylow Theorems allow us to prove many useful results about finite groups. By using them, we can often conclude a great deal about groups of a particular order if certain hypotheses are satisfied.

Theorem 15.10. If p and q are distinct primes with $p < q$, then every group G of order pq has a single subgroup of order q and this subgroup is normal in G. Hence, G cannot be simple. Furthermore, if $q \not\equiv 1 \pmod{p}$, then G is cyclic.

Proof. We know that G contains a subgroup H of order q. The number of conjugates of H divides pq and is equal to $1 + kq$ for $k = 0, 1, \ldots$. However, $1 + q$ is already too large to divide the order of the group; hence, H can only be conjugate to itself. That is, H must be normal in G.

The group G also has a Sylow p-subgroup, say K. The number of conjugates of K must divide q and be equal to $1 + kp$ for $k = 0, 1, \ldots$. Since q is prime, either $1 + kp = q$ or $1 + kp = 1$. If $1 + kp = 1$, then K is normal in G. In this case, we can easily show that G satisfies the criteria, given in Chapter 9, for the internal direct product of H and K. Since H is isomorphic to $\mathbb{Z}_q$ and K is isomorphic to $\mathbb{Z}_p$, $G \cong \mathbb{Z}_p \times \mathbb{Z}_q \cong \mathbb{Z}_{pq}$ by Theorem 9.21 on page 149.   ∎

Example 15.11. Every group of order 15 is cyclic. This is true because $15 = 5 \cdot 3$ and $5 \not\equiv 1 \pmod 3$.   □

Example 15.12. Let us classify all of the groups of order $99 = 3^2 \cdot 11$ up to isomorphism. First we will show that every group G of order 99 is abelian. By the Third Sylow Theorem, there are $1 + 3k$ Sylow 3-subgroups, each of order 9, for some $k = 0, 1, 2, \ldots$. Also, $1 + 3k$ must divide 11; hence, there can only be a single normal Sylow 3-subgroup H in G. Similarly, there are $1 + 11k$ Sylow 11-subgroups and $1 + 11k$ must divide 9. Consequently, there is only one Sylow 11-subgroup K in G. By Corollary 14.16 on page 216, any group of order p^2 is abelian for p prime; hence, H is isomorphic either to $\mathbb{Z}_3 \times \mathbb{Z}_3$ or to $\mathbb{Z}_9$. Since K has order 11, it must be isomorphic to $\mathbb{Z}_{11}$. Therefore, the only possible groups of order 99 are $\mathbb{Z}_3 \times \mathbb{Z}_3 \times \mathbb{Z}_{11}$ or $\mathbb{Z}_9 \times \mathbb{Z}_{11}$ up to isomorphism.   □

To determine all of the groups of order $5 \cdot 7 \cdot 47 = 1645$, we need the following theorem.

Theorem 15.13. Let $G' = \langle aba^{-1}b^{-1} : a, b \in G \rangle$ be the subgroup consisting of all finite products of elements of the form $aba^{-1}b^{-1}$ in a group G. Then G' is a normal subgroup of G and G/G' is abelian.

The subgroup G' of G is called the *commutator subgroup* of G. We leave the proof of this theorem as an exercise (Exercise 10.4.14 on page 166).

Example 15.14. We will now show that every group of order $5 \cdot 7 \cdot 47 = 1645$ is abelian, and cyclic by Theorem 9.21 on page 149. By the Third Sylow Theorem, G has only one subgroup H_1 of order 47. So G/H_1 has order 35 and must be abelian by Theorem 15.10. Hence, the commutator subgroup of G is contained in H which tells us that $|G'|$ is either 1 or 47. If $|G'| = 1$, we are done. Suppose that $|G'| = 47$. The Third Sylow Theorem tells us that G has only one subgroup of order 5 and one subgroup of order 7. So there exist normal subgroups H_2 and H_3 in G, where $|H_2| = 5$ and $|H_3| = 7$. In either case the quotient group is abelian; hence, G' must be a subgroup of H_i, $i = 1, 2$. Therefore, the order of G' is 1, 5, or 7. However, we already have determined that $|G'| = 1$ or 47. So the commutator subgroup of G is trivial, and consequently G is abelian.   □

Finite Simple Groups

Given a finite group, one can ask whether or not that group has any normal subgroups. Recall that a simple group is one with no proper nontrivial normal subgroups. As in the case of A_5, proving a group to be simple can be a very difficult task; however, the Sylow Theorems are useful tools for proving that a group is not simple. Usually, some sort of counting argument is involved.

Example 15.15. Let us show that no group G of order 20 can be simple. By the Third Sylow Theorem, G contains one or more Sylow 5-subgroups. The number of such subgroups is congruent to 1 (mod 5) and must also divide 20. The only possible such number is 1. Since there is only a single Sylow 5-subgroup and all Sylow 5-subgroups are conjugate, this subgroup must be normal.   □

Example 15.16. Let G be a finite group of order p^n, $n > 1$ and p prime. By Theorem 14.15 on page 216, G has a nontrivial center. Since the center of any group G is a normal subgroup, G cannot be a simple group. Therefore, groups of orders 4, 8, 9, 16, 25, 27, 32, 49, 64, and 81 are not simple. In fact, the groups of order 4, 9, 25, and 49 are abelian by Corollary 14.16 on page 216.   □

Example 15.17. No group of order $56 = 2^3 \cdot 7$ is simple. We have seen that if we can show that there is only one Sylow p-subgroup for some prime p dividing 56, then this must be a normal subgroup and we are done. By the Third Sylow Theorem, there are either one or eight Sylow 7-subgroups. If there is only a single Sylow 7-subgroup, then it must be normal.

On the other hand, suppose that there are eight Sylow 7-subgroups. Then each of these subgroups must be cyclic; hence, the intersection of any two of these subgroups contains only the identity of the group. This leaves $8 \cdot 6 = 48$ distinct elements in the group, each of order 7. Now let us count Sylow 2-subgroups. There are either one or seven Sylow 2-subgroups. Any element of a Sylow 2-subgroup other than the identity must have as its order a power of 2; and therefore cannot be one of the 48 elements of order 7 in the Sylow 7-subgroups. Since a Sylow 2-subgroup has order 8, there is only enough room for a single Sylow 2-subgroup in a group of order 56. If there is only one Sylow 2-subgroup, it must be normal.
          □

For other groups G, it is more difficult to prove that G is not simple. Suppose G has order 48. In this case the technique that we employed in the last example will not work. We need the following lemma to prove that no group of order 48 is simple.

Lemma 15.18. Let H and K be finite subgroups of a group G. Then

$$|HK| = \frac{|H| \cdot |K|}{|H \cap K|}.$$

Proof. Recall that $HK = \{hk : h \in H, k \in K\}$. Certainly, $|HK| \leq |H| \cdot |K|$ since some element in HK could be written as the product of different elements in H and K. It is quite possible that $h_1 k_1 = h_2 k_2$ for $h_1, h_2 \in H$ and $k_1, k_2 \in K$. If this is the case, let $a = (h_1)^{-1} h_2 = k_1 (k_2)^{-1}$. Notice that $a \in H \cap K$, since $(h_1)^{-1} h_2$ is in H and $k_2 (k_1)^{-1}$ is in K; consequently,

$$h_2 = h_1 a^{-1} \quad \text{and} \quad k_2 = a k_1.$$

Conversely, let $h = h_1 b^{-1}$ and $k = b k_1$ for $b \in H \cap K$. Then $hk = h_1 k_1$, where $h \in H$ and $k \in K$. Hence, any element $hk \in HK$ can be written in the form $h_i k_i$ for $h_i \in H$ and $k_i \in K$, as many times as there are elements in $H \cap K$; that is, $|H \cap K|$ times. Therefore, $|HK| = (|H| \cdot |K|)/|H \cap K|$. ∎

Example 15.19. To demonstrate that a group G of order 48 is not simple, we will show that G contains either a normal subgroup of order 8 or a normal subgroup of order 16. By the Third Sylow Theorem, G has either one or three Sylow 2-subgroups of order 16. If there is only one subgroup, then it must be a normal subgroup.

Suppose that the other case is true, and two of the three Sylow 2-subgroups are H and K. We claim that $|H \cap K| = 8$. If $|H \cap K| \leq 4$, then by Lemma 15.18,

$$|HK| \geq \frac{16 \cdot 16}{4} = 64,$$

which is impossible. Notice that $H \cap K$ has index two in both of H and K, so is normal in both, and thus H and K are each in the normalizer of $H \cap K$. Because H is a subgroup of $N(H \cap K)$ and because $N(H \cap K)$ has strictly more than 16 elements, $|N(H \cap K)|$ must be a multiple of 16 greater than 1, as well as dividing 48. The only possibility is that $|N(H \cap K)| = 48$. Hence, $N(H \cap K) = G$. □

The following famous conjecture of Burnside was proved in a long and difficult paper by Feit and Thompson [2].

Theorem 15.20. Odd Order Theorem. Every finite simple group of nonprime order must be of even order.

The proof of this theorem laid the groundwork for a program in the 1960s and 1970s that classified all finite simple groups. The success of this program is one of the outstanding achievements of modern mathematics.

Sage. Sage will compute a single Sylow p-subgroup for each prime divisor p of the order of the group. Then, with conjugacy, all of the Sylow p-subgroups can be enumerated. It is also possible to compute the normalizer of a subgroup.

15.3 Reading Questions

1. State Sylow's First Theorem.

2. How many groups are there of order 69? Why?

3. Give two descriptions, fundamentally different in character, of the normalizer of a subgroup.

4. Suppose that G is an abelian group. What is the commutator subgroup of G, and how do you know?

5. What's all the fuss about Sylow's Theorems?

15.4 Exercises

1. What are the orders of all Sylow p-subgroups where G has order 18, 24, 54, 72, and 80?

2. Find all the Sylow 3-subgroups of S_4 and show that they are all conjugate.

3. Show that every group of order 45 has a normal subgroup of order 9.

4. Let H be a Sylow p-subgroup of G. Prove that H is the only Sylow p-subgroup of G contained in $N(H)$.

5. Prove that no group of order 96 is simple.

6. Prove that no group of order 160 is simple.

7. If H is a normal subgroup of a finite group G and $|H| = p^k$ for some prime p, show that H is contained in every Sylow p-subgroup of G.

8. Let G be a group of order $p^2 q^2$, where p and q are distinct primes such that $q \nmid p^2 - 1$ and $p \nmid q^2 - 1$. Prove that G must be abelian. Find a pair of primes for which this is true.

9. Show that a group of order 33 has only one Sylow 3-subgroup.

10. Let H be a subgroup of a group G. Prove or disprove that the normalizer of H is normal in G.

11. Let G be a finite group whose order is divisible by a prime p. Prove that if there is only one Sylow p-subgroup in G, it must be a normal subgroup of G.

12. Let G be a group of order p^r, p prime. Prove that G contains a normal subgroup of order p^{r-1}.

13. Suppose that G is a finite group of order $p^n k$, where $k < p$. Show that G must contain a normal subgroup.

14. Let H be a subgroup of a finite group G. Prove that $gN(H)g^{-1} = N(gHg^{-1})$ for any $g \in G$.

15. Prove that a group of order 108 must have a normal subgroup.

16. Classify all the groups of order 175 up to isomorphism.

17. Show that every group of order 255 is cyclic.

18. Let G have order $p_1^{e_1} \cdots p_n^{e_n}$ and suppose that G has n Sylow p-subgroups $P_1, \ldots, P_n$ where $|P_i| = p_i^{e_i}$. Prove that G is isomorphic to $P_1 \times \cdots \times P_n$.

19. Let P be a normal Sylow p-subgroup of G. Prove that every inner automorphism of G fixes P.

20. What is the smallest possible order of a group G such that G is nonabelian and $|G|$ is odd? Can you find such a group?

21. **The Frattini Lemma.** If H is a normal subgroup of a finite group G and P is a Sylow p-subgroup of H, for each $g \in G$ show that there is an h in H such that $gPg^{-1} = hPh^{-1}$. Also, show that if N is the normalizer of P, then $G = HN$.

22. Show that if the order of G is $p^n q$, where p and q are primes and $p > q$, then G contains a normal subgroup.

23. Prove that the number of distinct conjugates of a subgroup H of a finite group G is $[G : N(H)]$.

24. Prove that a Sylow 2-subgroup of S_5 is isomorphic to D_4.

25. **Another Proof of the Sylow Theorems.**

    (a) Suppose p is prime and p does not divide m. Show that

    $$p \nmid \binom{p^k m}{p^k}.$$

    (b) Let $\mathcal{S}$ denote the set of all p^k element subsets of G. Show that p does not divide $|\mathcal{S}|$.

    (c) Define an action of G on $\mathcal{S}$ by left multiplication, $aT = \{at : t \in T\}$ for $a \in G$ and $T \in \mathcal{S}$. Prove that this is a group action.

    (d) Prove $p \nmid |\mathcal{O}_T|$ for some $T \in \mathcal{S}$.

    (e) Let $\{T_1, \ldots, T_u\}$ be an orbit such that $p \nmid u$ and $H = \{g \in G : gT_1 = T_1\}$. Prove that H is a subgroup of G and show that $|G| = u|H|$.

    (f) Show that p^k divides $|H|$ and $p^k \le |H|$.

    (g) Show that $|H| = |\mathcal{O}_T| \le p^k$; conclude that therefore $p^k = |H|$.

26. Let G be a group. Prove that $G' = \langle aba^{-1}b^{-1} : a, b \in G \rangle$ is a normal subgroup of G and G/G' is abelian. Find an example to show that $\{aba^{-1}b^{-1} : a, b \in G\}$ is not necessarily a group.

15.5 A Project

The main objective of finite group theory is to classify all possible finite groups up to isomorphism. This problem is very difficult even if we try to classify the groups of order less than or equal to 60. However, we can break the problem down into several intermediate problems. This is a challenging project that requires a working knowledge of the group theory you have learned up to this point. Even if you do not complete it, it will teach you a great deal about finite groups. You can use Table 15.21 as a guide.

Table 15.21. Numbers of distinct groups G, $|G| \leq 60$

Order	Number	Order	Number	Order	Number	Order	Number
1	?	16	14	31	1	46	2
2	?	17	1	32	51	47	1
3	?	18	?	33	1	48	52
4	?	19	?	34	?	49	?
5	?	20	5	35	1	50	5
6	?	21	?	36	14	51	?
7	?	22	2	37	1	52	?
8	?	23	1	38	?	53	?
9	?	24	?	39	2	54	15
10	?	25	2	40	14	55	2
11	?	26	2	41	1	56	?
12	5	27	5	42	?	57	2
13	?	28	?	43	1	58	?
14	?	29	1	44	4	59	1
15	1	30	4	45	?	60	13

1. Find all simple groups G ($|G| \leq 60$). *Do not use the Odd Order Theorem unless you are prepared to prove it.*

2. Find the number of distinct groups G, where the order of G is n for $n = 1, \ldots, 60$.

3. Find the actual groups (up to isomorphism) for each n.

15.6 References and Suggested Readings

[1] Edwards, H. "A Short History of the Fields Medal," *Mathematical Intelligencer* 1 (1978), 127–29.

[2] Feit, W. and Thompson, J. G. "Solvability of Groups of Odd Order," *Pacific Journal of Mathematics* 13 (1963), 775–1029.

[3] Gallian, J. A. "The Search for Finite Simple Groups," *Mathematics Magazine* 49 (1976), 163–79.

[4] Gorenstein, D. "Classifying the Finite Simple Groups," *Bulletin of the American Mathematical Society* 14 (1986), 1–98.

[5] Gorenstein, D. *Finite Groups.* AMS Chelsea Publishing, Providence RI, 1968.

[6] Gorenstein, D., Lyons, R., and Solomon, R. *The Classification of Finite Simple Groups.* American Mathematical Society, Providence RI, 1994.

16

Rings

$\mathcal{U}$p to this point we have studied sets with a single binary operation satisfying certain axioms, but we are often more interested in working with sets that have two binary operations. For example, one of the most natural algebraic structures to study is the integers with the operations of addition and multiplication. These operations are related to one another by the distributive property. If we consider a set with two such related binary operations satisfying certain axioms, we have an algebraic structure called a ring. In a ring we add and multiply elements such as real numbers, complex numbers, matrices, and functions.

16.1 Rings

A nonempty set R is a *ring* if it has two closed binary operations, addition and multiplication, satisfying the following conditions.

1. $a + b = b + a$ for $a, b \in R$.

2. $(a + b) + c = a + (b + c)$ for $a, b, c \in R$.

3. There is an element 0 in R such that $a + 0 = a$ for all $a \in R$.

4. For every element $a \in R$, there exists an element $-a$ in R such that $a + (-a) = 0$.

5. $(ab)c = a(bc)$ for $a, b, c \in R$.

6. For $a, b, c \in R$,

$$a(b + c) = ab + ac$$
$$(a + b)c = ac + bc.$$

This last condition, the distributive axiom, relates the binary operations of addition and multiplication. Notice that the first four axioms simply require that a ring be an abelian group under addition, so we could also have defined a ring to be an abelian group $(R, +)$ together with a second binary operation satisfying the fifth and sixth conditions given above.

If there is an element $1 \in R$ such that $1 \neq 0$ and $1a = a1 = a$ for each element $a \in R$, we say that R is a ring with **unity** or **identity**. A ring R for which $ab = ba$ for all a, b in R is called a **commutative ring**. A commutative ring R with identity is called an **integral domain** if, for every $a, b \in R$ such that $ab = 0$, either $a = 0$ or $b = 0$. A **division ring** is a ring R, with an identity, in which every nonzero element in R is a **unit**; that is, for each $a \in R$ with $a \neq 0$, there exists a unique element a^{-1} such that $a^{-1}a = aa^{-1} = 1$. A commutative division ring is called a **field**. The relationship among rings, integral domains, division rings, and fields is shown in Figure 16.1.

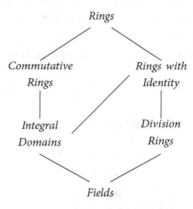

Figure 16.1. Types of rings

Example 16.2. As we have mentioned previously, the integers form a ring. In fact, $\mathbb{Z}$ is an integral domain. Certainly if $ab = 0$ for two integers a and b, either $a = 0$ or $b = 0$. However, $\mathbb{Z}$ is not a field. There is no integer that is the multiplicative inverse of 2, since $1/2$ is not an integer. The only integers with multiplicative inverses are 1 and -1. □

Example 16.3. Under the ordinary operations of addition and multiplication, all of the familiar number systems are rings: the rationals, $\mathbb{Q}$; the real numbers, $\mathbb{R}$; and the complex numbers, $\mathbb{C}$. Each of these rings is a field. □

Example 16.4. We can define the product of two elements a and b in $\mathbb{Z}_n$ by $ab \pmod{n}$. For instance, in $\mathbb{Z}_{12}$, $5 \cdot 7 \equiv 11 \pmod{12}$. This product makes the abelian group $\mathbb{Z}_n$ into a ring. Certainly $\mathbb{Z}_n$ is a commutative ring; however, it may fail to be an integral domain. If we consider $3 \cdot 4 \equiv 0 \pmod{12}$ in $\mathbb{Z}_{12}$, it is easy to see that a product of two nonzero elements in the ring can be equal to zero. □

A nonzero element a in a commutative ring R is called a ***zero divisor*** if there is a nonzero element b in R such that $ab = 0$. In the previous example, 3 and 4 are zero divisors in $\mathbb{Z}_{12}$.

Example 16.5. In calculus the continuous real-valued functions on an interval $[a, b]$ form a commutative ring. We add or multiply two functions by adding or multiplying the values of the functions. If $f(x) = x^2$ and $g(x) = \cos x$, then $(f + g)(x) = f(x) + g(x) = x^2 + \cos x$ and $(fg)(x) = f(x)g(x) = x^2 \cos x$. ☐

Example 16.6. The 2×2 matrices with entries in $\mathbb{R}$ form a ring under the usual operations of matrix addition and multiplication. This ring is noncommutative, since it is usually the case that $AB \neq BA$. Also, notice that we can have $AB = 0$ when neither A nor B is zero. ☐

Example 16.7. For an example of a noncommutative division ring, let

$$1 = \begin{pmatrix} 1 & 0 \\ 0 & 1 \end{pmatrix}, \quad \mathbf{i} = \begin{pmatrix} 0 & 1 \\ -1 & 0 \end{pmatrix}, \quad \mathbf{j} = \begin{pmatrix} 0 & i \\ i & 0 \end{pmatrix}, \quad \mathbf{k} = \begin{pmatrix} i & 0 \\ 0 & -i \end{pmatrix},$$

where $i^2 = -1$. These elements satisfy the following relations:

$$\mathbf{i}^2 = \mathbf{j}^2 = \mathbf{k}^2 = -1$$
$$\mathbf{ij} = \mathbf{k}$$
$$\mathbf{jk} = \mathbf{i}$$
$$\mathbf{ki} = \mathbf{j}$$
$$\mathbf{ji} = -\mathbf{k}$$
$$\mathbf{kj} = -\mathbf{i}$$
$$\mathbf{ik} = -\mathbf{j}.$$

Let $\mathbb{H}$ consist of elements of the form $a + b\mathbf{i} + c\mathbf{j} + d\mathbf{k}$, where a, b, c, d are real numbers. Equivalently, $\mathbb{H}$ can be considered to be the set of all 2×2 matrices of the form

$$\begin{pmatrix} \alpha & \beta \\ -\overline{\beta} & \overline{\alpha} \end{pmatrix},$$

where $\alpha = a + di$ and $\beta = b + ci$ are complex numbers. We can define addition and multiplication on $\mathbb{H}$ either by the usual matrix operations or in terms of the generators 1, $\mathbf{i}$, $\mathbf{j}$, and $\mathbf{k}$:

$$(a_1 + b_1\mathbf{i} + c_1\mathbf{j} + d_1\mathbf{k}) + (a_2 + b_2\mathbf{i} + c_2\mathbf{j} + d_2\mathbf{k})$$
$$= (a_1 + a_2) + (b_1 + b_2)\mathbf{i} + (c_1 + c_2)\mathbf{j} + (d_1 + d_2)\mathbf{k}$$

and

$$(a_1 + b_1\mathbf{i} + c_1\mathbf{j} + d_1\mathbf{k})(a_2 + b_2\mathbf{i} + c_2\mathbf{j} + d_2\mathbf{k}) = \alpha + \beta\mathbf{i} + \gamma\mathbf{j} + \delta\mathbf{k},$$

where

$$\alpha = a_1a_2 - b_1b_2 - c_1c_2 - d_1d_2$$
$$\beta = a_1b_2 + a_2b_1 + c_1d_2 - d_1c_2$$
$$\gamma = a_1c_2 - b_1d_2 + c_1a_2 + d_1b_2$$
$$\delta = a_1d_2 + b_1c_2 - c_1b_2 + d_1a_2.$$

Though multiplication looks complicated, it is actually a straightforward computation if we remember that we just add and multiply elements in $\mathbb{H}$ like polynomials and keep in mind the relationships between the generators $\mathbf{i}$, $\mathbf{j}$, and $\mathbf{k}$. The ring $\mathbb{H}$ is called the ring of *quaternions*.

To show that the quaternions are a division ring, we must be able to find an inverse for each nonzero element. Notice that

$$(a + b\mathbf{i} + c\mathbf{j} + d\mathbf{k})(a - b\mathbf{i} - c\mathbf{j} - d\mathbf{k}) = a^2 + b^2 + c^2 + d^2.$$

This element can be zero only if a, b, c, and d are all zero. So if $a + b\mathbf{i} + c\mathbf{j} + d\mathbf{k} \neq 0$,

$$(a + b\mathbf{i} + c\mathbf{j} + d\mathbf{k}) \left(\frac{a - b\mathbf{i} - c\mathbf{j} - d\mathbf{k}}{a^2 + b^2 + c^2 + d^2} \right) = 1.$$

$\square$

Proposition 16.8. Let R be a ring with $a, b \in R$. Then

1. $a0 = 0a = 0$;
2. $a(-b) = (-a)b = -ab$;
3. $(-a)(-b) = ab$.

Proof. To prove (1), observe that

$$a0 = a(0 + 0) = a0 + a0;$$

hence, $a0 = 0$. Similarly, $0a = 0$. For (2), we have $ab + a(-b) = a(b - b) = a0 = 0$; consequently, $-ab = a(-b)$. Similarly, $-ab = (-a)b$. Part (3) follows directly from (2) since $(-a)(-b) = -(a(-b)) = -(-ab) = ab$. $\blacksquare$

Just as we have subgroups of groups, we have an analogous class of substructures for rings. A *subring* S of a ring R is a subset S of R such that S is also a ring under the inherited operations from R.

Example 16.9. The ring $n\mathbb{Z}$ is a subring of $\mathbb{Z}$. Notice that even though the original ring may have an identity, we do not require that its subring have an identity. We have the following chain of subrings:

$$\mathbb{Z} \subset \mathbb{Q} \subset \mathbb{R} \subset \mathbb{C}.$$

□

The following proposition gives us some easy criteria for determining whether or not a subset of a ring is indeed a subring. (We will leave the proof of this proposition as an exercise.)

Proposition 16.10. Let R be a ring and S a subset of R. Then S is a subring of R if and only if the following conditions are satisfied.

1. $S \neq \varnothing$.

2. $rs \in S$ for all $r, s \in S$.

3. $r - s \in S$ for all $r, s \in S$.

Example 16.11. Let $R = \mathbb{M}_2(\mathbb{R})$ be the ring of 2×2 matrices with entries in $\mathbb{R}$. If T is the set of upper triangular matrices in R; i.e.,

$$T = \left\{ \begin{pmatrix} a & b \\ 0 & c \end{pmatrix} : a, b, c \in \mathbb{R} \right\},$$

then T is a subring of R. If

$$A = \begin{pmatrix} a & b \\ 0 & c \end{pmatrix} \quad \text{and} \quad B = \begin{pmatrix} a' & b' \\ 0 & c' \end{pmatrix}$$

are in T, then clearly $A - B$ is also in T. Also,

$$AB = \begin{pmatrix} aa' & ab' + bc' \\ 0 & cc' \end{pmatrix}$$

is in T.   □

16.2 Integral Domains and Fields

Let us briefly recall some definitions. If R is a commutative ring and r is a nonzero element in R, then r is said to be a *zero divisor* if there is some nonzero element $s \in R$ such that $rs = 0$. A commutative ring with identity is said to be an *integral domain* if it has no zero divisors. If an element a in a ring R with identity has a multiplicative inverse, we say that a is a *unit*. If every nonzero element in a ring R is a unit, then R is called a *division ring*. A commutative division ring is called a *field*.

Example 16.12. If $i^2 = -1$, then the set $\mathbb{Z}[i] = \{m + ni : m, n \in \mathbb{Z}\}$ forms a ring known as the *Gaussian integers*. It is easily seen that the Gaussian integers are a subring of the complex numbers since they are closed under addition and multiplication. Let $\alpha = a + bi$ be a unit in $\mathbb{Z}[i]$. Then $\bar{\alpha} = a - bi$ is also a unit since if $\alpha\beta = 1$, then $\bar{\alpha}\bar{\beta} = 1$. If $\beta = c + di$, then

$$1 = \alpha\beta\bar{\alpha}\bar{\beta} = (a^2 + b^2)(c^2 + d^2).$$

Therefore, $a^2 + b^2$ must either be 1 or -1; or, equivalently, $a + bi = \pm 1$ or $a + bi = \pm i$. Therefore, units of this ring are ± 1 and $\pm i$; hence, the Gaussian integers are not a field. We will leave it as an exercise to prove that the Gaussian integers are an integral domain. $\square$

Example 16.13. The set of matrices

$$F = \left\{ \begin{pmatrix} 1 & 0 \\ 0 & 1 \end{pmatrix}, \begin{pmatrix} 1 & 1 \\ 1 & 0 \end{pmatrix}, \begin{pmatrix} 0 & 1 \\ 1 & 1 \end{pmatrix}, \begin{pmatrix} 0 & 0 \\ 0 & 0 \end{pmatrix} \right\}$$

with entries in $\mathbb{Z}_2$ forms a field. $\square$

Example 16.14. The set $\mathbb{Q}(\sqrt{2}) = \{a + b\sqrt{2} : a, b \in \mathbb{Q}\}$ is a field. The inverse of an element $a + b\sqrt{2}$ in $\mathbb{Q}(\sqrt{2})$ is

$$\frac{a}{a^2 - 2b^2} + \frac{-b}{a^2 - 2b^2}\sqrt{2}.$$

$\square$

We have the following alternative characterization of integral domains.

Proposition 16.15. Cancellation Law. Let D be a commutative ring with identity. Then D is an integral domain if and only if for all nonzero elements $a \in D$ with $ab = ac$, we have $b = c$.

Proof. Let D be an integral domain. Then D has no zero divisors. Let $ab = ac$ with $a \neq 0$. Then $a(b - c) = 0$. Hence, $b - c = 0$ and $b = c$.

Conversely, let us suppose that cancellation is possible in D. That is, suppose that $ab = ac$ implies $b = c$. Let $ab = 0$. If $a \neq 0$, then $ab = a0$ or $b = 0$. Therefore, a cannot be a zero divisor. ∎

The following surprising theorem is due to Wedderburn.

Theorem 16.16. Every finite integral domain is a field.

Proof. Let D be a finite integral domain and D^* be the set of nonzero elements of D. We must show that every element in D^* has an inverse. For each $a \in D^*$ we can define a map $\lambda_a : D^* \to D^*$ by $\lambda_a(d) = ad$. This map makes sense, because if

$a \neq 0$ and $d \neq 0$, then $ad \neq 0$. The map λ_a is one-to-one, since for $d_1, d_2 \in D^*$,

$$ad_1 = \lambda_a(d_1) = \lambda_a(d_2) = ad_2$$

implies $d_1 = d_2$ by left cancellation. Since D^* is a finite set, the map λ_a must also be onto; hence, for some $d \in D^*$, $\lambda_a(d) = ad = 1$. Therefore, a has a left inverse. Since D is commutative, d must also be a right inverse for a. Consequently, D is a field.   ∎

For any nonnegative integer n and any element r in a ring R we write $r + \cdots + r$ (n times) as nr. We define the *characteristic* of a ring R to be the least positive integer n such that $nr = 0$ for all $r \in R$. If no such integer exists, then the characteristic of R is defined to be 0. We will denote the characteristic of R by char R.

Example 16.17. For every prime p, $\mathbb{Z}_p$ is a field of characteristic p. By Proposition 3.4 on page 37, every nonzero element in $\mathbb{Z}_p$ has an inverse; hence, $\mathbb{Z}_p$ is a field. If a is any nonzero element in the field, then $pa = 0$, since the order of any nonzero element in the abelian group $\mathbb{Z}_p$ is p.   □

Lemma 16.18. Let R be a ring with identity. If 1 has order n, then the characteristic of R is n.

Proof. If 1 has order n, then n is the least positive integer such that $n1 = 0$. Thus, for all $r \in R$,

$$nr = n(1r) = (n1)r = 0r = 0.$$

On the other hand, if no positive n exists such that $n1 = 0$, then the characteristic of R is zero.   ∎

Theorem 16.19. The characteristic of an integral domain is either prime or zero.

Proof. Let D be an integral domain and suppose that the characteristic of D is n with $n \neq 0$. If n is not prime, then $n = ab$, where $1 < a < n$ and $1 < b < n$. By Lemma 16.18, we need only consider the case $n1 = 0$. Since $0 = n1 = (ab)1 = (a1)(b1)$ and there are no zero divisors in D, either $a1 = 0$ or $b1 = 0$. Hence, the characteristic of D must be less than n, which is a contradiction. Therefore, n must be prime.   ∎

16.3 Ring Homomorphisms and Ideals

In the study of groups, a homomorphism is a map that preserves the operation of the group. Similarly, a homomorphism between rings preserves the operations of addition and multiplication in the ring. More specifically, if R and S are rings, then a *ring homomorphism* is a map $\phi : R \to S$ satisfying

$$\phi(a + b) = \phi(a) + \phi(b)$$

$$\phi(ab) = \phi(a)\phi(b)$$

for all $a, b \in R$. If $\phi : R \to S$ is a one-to-one and onto homomorphism, then ϕ is called an *isomorphism* of rings.

The set of elements that a ring homomorphism maps to 0 plays a fundamental role in the theory of rings. For any ring homomorphism $\phi : R \to S$, we define the *kernel* of a ring homomorphism to be the set

$$\ker \phi = \{r \in R : \phi(r) = 0\}.$$

Example 16.20. For any integer n we can define a ring homomorphism $\phi : \mathbb{Z} \to \mathbb{Z}_n$ by $a \mapsto a \pmod{n}$. This is indeed a ring homomorphism, since

$$\begin{aligned}
\phi(a + b) &= (a + b) \pmod{n} \\
&= a \pmod{n} + b \pmod{n} \\
&= \phi(a) + \phi(b)
\end{aligned}$$

and

$$\begin{aligned}
\phi(ab) &= ab \pmod{n} \\
&= a \pmod{n} \cdot b \pmod{n} \\
&= \phi(a)\phi(b).
\end{aligned}$$

The kernel of the homomorphism ϕ is $n\mathbb{Z}$. □

Example 16.21. Let $C[a, b]$ be the ring of continuous real-valued functions on an interval $[a, b]$ as in Example 16.5 on page 242. For a fixed $\alpha \in [a, b]$, we can define a ring homomorphism $\phi_\alpha : C[a, b] \to \mathbb{R}$ by $\phi_\alpha(f) = f(\alpha)$. This is a ring homomorphism since

$$\phi_\alpha(f + g) = (f + g)(\alpha) = f(\alpha) + g(\alpha) = \phi_\alpha(f) + \phi_\alpha(g)$$
$$\phi_\alpha(fg) = (fg)(\alpha) = f(\alpha)g(\alpha) = \phi_\alpha(f)\phi_\alpha(g).$$

Ring homomorphisms of the type ϕ_α are called *evaluation homomorphisms*. □

In the next proposition we will examine some fundamental properties of ring homomorphisms. The proof of the proposition is left as an exercise.

Proposition 16.22. Let $\phi : R \to S$ be a ring homomorphism.

1. If R is a commutative ring, then $\phi(R)$ is a commutative ring.

2. $\phi(0) = 0$.

3. Let 1_R and 1_S be the identities for R and S, respectively. If ϕ is onto, then $\phi(1_R) = 1_S$.

4. If R is a field and $\phi(R) \neq \{0\}$, then $\phi(R)$ is a field.

In group theory we found that normal subgroups play a special role. These subgroups have nice characteristics that make them more interesting to study than arbitrary subgroups. In ring theory the objects corresponding to normal subgroups are a special class of subrings called ideals. An *ideal* in a ring R is a subring I of R such that if a is in I and r is in R, then both ar and ra are in I; that is, $rI \subset I$ and $Ir \subset I$ for all $r \in R$.

Example 16.23. Every ring R has at least two ideals, $\{0\}$ and R. These ideals are called the *trivial ideals*.   □

Let R be a ring with identity and suppose that I is an ideal in R such that 1 is in I. Since for any $r \in R$, $r1 = r \in I$ by the definition of an ideal, $I = R$.

Example 16.24. If a is any element in a commutative ring R with identity, then the set

$$\langle a \rangle = \{ar : r \in R\}$$

is an ideal in R. Certainly, $\langle a \rangle$ is nonempty since both $0 = a0$ and $a = a1$ are in $\langle a \rangle$. The sum of two elements in $\langle a \rangle$ is again in $\langle a \rangle$ since $ar + ar' = a(r + r')$. The inverse of ar is $-ar = a(-r) \in \langle a \rangle$. Finally, if we multiply an element $ar \in \langle a \rangle$ by an arbitrary element $s \in R$, we have $s(ar) = a(sr)$. Therefore, $\langle a \rangle$ satisfies the definition of an ideal.   □

If R is a commutative ring with identity, then an ideal of the form $\langle a \rangle = \{ar : r \in R\}$ is called a *principal ideal*.

Theorem 16.25. Every ideal in the ring of integers $\mathbb{Z}$ is a principal ideal.

Proof. The zero ideal $\{0\}$ is a principal ideal since $\langle 0 \rangle = \{0\}$. If I is any nonzero ideal in $\mathbb{Z}$, then I must contain some positive integer m. There exists a least positive integer n in I by the Principle of Well-Ordering. Now let a be any element in I. Using the division algorithm, we know that there exist integers q and r such that

$$a = nq + r$$

where $0 \leq r < n$. This equation tells us that $r = a - nq \in I$, but r must be 0 since n is the least positive element in I. Therefore, $a = nq$ and $I = \langle n \rangle$.   ■

Example 16.26. The set $n\mathbb{Z}$ is ideal in the ring of integers. If na is in $n\mathbb{Z}$ and b is in $\mathbb{Z}$, then nab is in $n\mathbb{Z}$ as required. In fact, by Theorem 16.25, these are the only ideals of $\mathbb{Z}$.   □

Proposition 16.27. The kernel of any ring homomorphism $\phi : R \to S$ is an ideal in R.

Proof. We know from group theory that ker ϕ is an additive subgroup of R. Suppose that $r \in R$ and $a \in$ ker ϕ. Then we must show that ar and ra are in ker ϕ. However,

$$\phi(ar) = \phi(a)\phi(r) = 0\phi(r) = 0$$

and

$$\phi(ra) = \phi(r)\phi(a) = \phi(r)0 = 0.$$

∎

Remark 16.28. In our definition of an ideal we have required that $rI \subset I$ and $Ir \subset I$ for all $r \in R$. Such ideals are sometimes referred to as *two-sided ideals*. We can also consider *one-sided ideals*; that is, we may require only that either $rI \subset I$ or $Ir \subset I$ for $r \in R$ hold but not both. Such ideals are called *left ideals* and *right ideals*, respectively. Of course, in a commutative ring any ideal must be two-sided. In this text we will concentrate on two-sided ideals.

Theorem 16.29. Let I be an ideal of R. The factor group R/I is a ring with multiplication defined by

$$(r + I)(s + I) = rs + I.$$

Proof. We already know that R/I is an abelian group under addition. Let $r + I$ and $s + I$ be in R/I. We must show that the product $(r + I)(s + I) = rs + I$ is independent of the choice of coset; that is, if $r' \in r + I$ and $s' \in s + I$, then $r's'$ must be in $rs + I$. Since $r' \in r + I$, there exists an element a in I such that $r' = r + a$. Similarly, there exists a $b \in I$ such that $s' = s + b$. Notice that

$$r's' = (r + a)(s + b) = rs + as + rb + ab$$

and $as + rb + ab \in I$ since I is an ideal; consequently, $r's' \in rs + I$. We will leave as an exercise the verification of the associative law for multiplication and the distributive laws. ∎

The ring R/I in Theorem 16.29 is called the *factor* or *quotient ring*. Just as with group homomorphisms and normal subgroups, there is a relationship between ring homomorphisms and ideals.

Theorem 16.30. Let I be an ideal of R. The map $\phi : R \to R/I$ defined by $\phi(r) = r+I$ is a ring homomorphism of R onto R/I with kernel I.

Proof. Certainly $\phi : R \to R/I$ is a surjective abelian group homomorphism. It remains to show that ϕ works correctly under ring multiplication. Let r and s be in R. Then

$$\phi(r)\phi(s) = (r + I)(s + I) = rs + I = \phi(rs),$$

which completes the proof of the theorem.   ∎

The map $\phi : R \to R/I$ is often called the **natural** or **canonical homomor-phism**. In ring theory we have isomorphism theorems relating ideals and ring homomorphisms similar to the isomorphism theorems for groups that relate normal subgroups and homomorphisms in Chapter 11. We will prove only the First Isomorphism Theorem for rings in this chapter and leave the proofs of the other two theorems as exercises. All of the proofs are similar to the proofs of the isomorphism theorems for groups.

Theorem 16.31. First Isomorphism Theorem. Let $\psi : R \to S$ be a ring homo-morphism. Then $\ker \psi$ is an ideal of R. If $\phi : R \to R/\ker \psi$ is the canonical homomorphism, then there exists a unique isomorphism $\eta : R/\ker \psi \to \psi(R)$ such that $\psi = \eta \phi$.

Proof. Let $K = \ker \psi$. By the First Isomorphism Theorem for groups, there exists a well-defined group homomorphism $\eta : R/K \to \psi(R)$ defined by $\eta(r + K) = \psi(r)$ for the additive abelian groups R and R/K. To show that this is a ring homomorphism, we need only show that $\eta((r + K)(s + K)) = \eta(r + K)\eta(s + K)$; but

$$\begin{aligned}
\eta((r + K)(s + K)) &= \eta(rs + K) \\
&= \psi(rs) \\
&= \psi(r)\psi(s) \\
&= \eta(r + K)\eta(s + K).
\end{aligned}$$

∎

Theorem 16.32. Second Isomorphism Theorem. Let I be a subring of a ring R and J an ideal of R. Then $I \cap J$ is an ideal of I and

$$I/I \cap J \cong (I + J)/J.$$

Theorem 16.33. Third Isomorphism Theorem. Let R be a ring and I and J be ideals of R where $J \subset I$. Then

$$R/I \cong \frac{R/J}{I/J}.$$

Theorem 16.34. Correspondence Theorem. Let I be an ideal of a ring R. Then $S \mapsto S/I$ is a one-to-one correspondence between the set of subrings S containing I and the set of subrings of R/I. Furthermore, the ideals of R containing I correspond to ideals of R/I.

16.4 Maximal and Prime Ideals

In this particular section we are especially interested in certain ideals of commutative rings. These ideals give us special types of factor rings. More specifically, we would like to characterize those ideals I of a commutative ring R such that R/I is an integral domain or a field.

A proper ideal M of a ring R is a *maximal ideal* of R if the ideal M is not a proper subset of any ideal of R except R itself. That is, M is a maximal ideal if for any ideal I properly containing M, $I = R$. The following theorem completely characterizes maximal ideals for commutative rings with identity in terms of their corresponding factor rings.

Theorem 16.35. Let R be a commutative ring with identity and M an ideal in R. Then M is a maximal ideal of R if and only if R/M is a field.

Proof. Let M be a maximal ideal in R. If R is a commutative ring, then R/M must also be a commutative ring. Clearly, $1 + M$ acts as an identity for R/M. We must also show that every nonzero element in R/M has an inverse. If $a + M$ is a nonzero element in R/M, then $a \notin M$. Define I to be the set $\{ra + m : r \in R \text{ and } m \in M\}$. We will show that I is an ideal in R. The set I is nonempty since $0a + 0 = 0$ is in I. If $r_1 a + m_1$ and $r_2 a + m_2$ are two elements in I, then

$$(r_1 a + m_1) - (r_2 a + m_2) = (r_1 - r_2)a + (m_1 - m_2)$$

is in I. Also, for any $r \in R$ it is true that $rI \subset I$; hence, I is closed under multiplication and satisfies the necessary conditions to be an ideal. Therefore, by Proposition 16.10 on page 244 and the definition of an ideal, I is an ideal properly containing M. Since M is a maximal ideal, $I = R$; consequently, by the definition of I there must be an m in M and an element b in R such that $1 = ab + m$. Therefore,

$$1 + M = ab + M = ba + M = (a + M)(b + M).$$

Conversely, suppose that M is an ideal and R/M is a field. Since R/M is a field, it must contain at least two elements: $0 + M = M$ and $1 + M$. Hence, M is a proper ideal of R. Let I be any ideal properly containing M. We need to show that $I = R$. Choose a in I but not in M. Since $a + M$ is a nonzero element in a field, there exists an element $b + M$ in R/M such that $(a + M)(b + M) = ab + M = 1 + M$. Consequently, there exists an element $m \in M$ such that $ab + m = 1$ and 1 is in I. Therefore, $r1 = r \in I$ for all $r \in R$. Consequently, $I = R$.   ∎

Example 16.36. Let $p\mathbb{Z}$ be an ideal in $\mathbb{Z}$, where p is prime. Then $p\mathbb{Z}$ is a maximal ideal since $\mathbb{Z}/p\mathbb{Z} \cong \mathbb{Z}_p$ is a field.   □

A proper ideal P in a commutative ring R is called a ***prime ideal*** if whenever $ab \in P$, then either $a \in P$ or $b \in P$.[5]

Example 16.37. It is easy to check that the set $P = \{0, 2, 4, 6, 8, 10\}$ is an ideal in $\mathbb{Z}_{12}$. This ideal is prime. In fact, it is a maximal ideal.   □

Proposition 16.38. Let R be a commutative ring with identity 1, where $1 \neq 0$. Then P is a prime ideal in R if and only if R/P is an integral domain.

Proof. First let us assume that P is an ideal in R and R/P is an integral domain. Suppose that $ab \in P$. If $a + P$ and $b + P$ are two elements of R/P such that $(a + P)(b + P) = 0 + P = P$, then either $a + P = P$ or $b + P = P$. This means that either a is in P or b is in P, which shows that P must be prime.

Conversely, suppose that P is prime and

$$(a + P)(b + P) = ab + P = 0 + P = P.$$

Then $ab \in P$. If $a \notin P$, then b must be in P by the definition of a prime ideal; hence, $b + P = 0 + P$ and R/P is an integral domain.   ■

Example 16.39. Every ideal in $\mathbb{Z}$ is of the form $n\mathbb{Z}$. The factor ring $\mathbb{Z}/n\mathbb{Z} \cong \mathbb{Z}_n$ is an integral domain only when n is prime. It is actually a field. Hence, the nonzero prime ideals in $\mathbb{Z}$ are the ideals $p\mathbb{Z}$, where p is prime. This example really justifies the use of the word "prime" in our definition of prime ideals.   □

Since every field is an integral domain, we have the following corollary.

Corollary 16.40. Every maximal ideal in a commutative ring with identity is also a prime ideal.

      ↫   Historical Note   ↬

Amalie Emmy Noether, one of the outstanding mathematicians of the twentieth century, was born in Erlangen, Germany in 1882. She was the daughter of Max Noether (1844–1921), a distinguished mathematician at the University of Erlangen. Together with Paul Gordon (1837–1912), Emmy Noether's father strongly influenced her early education. She entered the University of Erlangen at the age of 18. Although women had been admitted to universities in England, France, and Italy for decades, there

---

[5]It is possible to define prime ideals in a noncommutative ring. See [1] or [3].

was great resistance to their presence at universities in Germany. Noether was one of only two women among the university's 986 students. After completing her doctorate under Gordon in 1907, she continued to do research at Erlangen, occasionally lecturing when her father was ill.

Noether went to Göttingen to study in 1916. David Hilbert and Felix Klein tried unsuccessfully to secure her an appointment at Göttingen. Some of the faculty objected to women lecturers, saying, "What will our soldiers think when they return to the university and are expected to learn at the feet of a woman?" Hilbert, annoyed at the question, responded, "Meine Herren, I do not see that the sex of a candidate is an argument against her admission as a Privatdozent. After all, the Senate is not a bathhouse." At the end of World War I, attitudes changed and conditions greatly improved for women. After Noether passed her habilitation examination in 1919, she was given a title and was paid a small sum for her lectures.

In 1922, Noether became a Privatdozent at Göttingen. Over the next 11 years she used axiomatic methods to develop an abstract theory of rings and ideals. Though she was not good at lecturing, Noether was an inspiring teacher. One of her many students was B. L. van der Waerden, author of the first text treating abstract algebra from a modern point of view. Some of the other mathematicians Noether influenced or closely worked with were Alexandroff, Artin, Brauer, Courant, Hasse, Hopf, Pontryagin, von Neumann, and Weyl. One of the high points of her career was an invitation to address the International Congress of Mathematicians in Zurich in 1932. In spite of all the recognition she received from her colleagues, Noether's abilities were never recognized as they should have been during her lifetime. She was never promoted to full professor by the Prussian academic bureaucracy.

In 1933, Noether, who was Jewish, was banned from participation in all academic activities in Germany. She emigrated to the United States, took a position at Bryn Mawr College, and became a member of the Institute for Advanced Study at Princeton. Noether died suddenly on April 14, 1935. After her death she was eulogized by such notable scientists as Albert Einstein.

16.5 An Application to Software Design

The Chinese Remainder Theorem is a result from elementary number theory about the solution of systems of simultaneous congruences. The Chinese mathematician Sun-tsï wrote about the theorem in the first century A.D. This theorem has some interesting consequences in the design of software for parallel processors.

Lemma 16.41. Let m and n be positive integers such that $\gcd(m, n) = 1$. Then for $a, b \in \mathbb{Z}$ the system

$$x \equiv a \pmod{m}$$
$$x \equiv b \pmod{n}$$

has a solution. If x_1 and x_2 are two solutions of the system, then $x_1 \equiv x_2 \pmod{mn}$.

Proof. The equation $x \equiv a \pmod{m}$ has a solution since $a + km$ satisfies the equation for all $k \in \mathbb{Z}$. We must show that there exists an integer k_1 such that

$$a + k_1 m \equiv b \pmod{n}.$$

This is equivalent to showing that

$$k_1 m \equiv (b - a) \pmod{n}$$

has a solution for k_1. Since m and n are relatively prime, there exist integers s and t such that $ms + nt = 1$. Consequently,

$$(b - a)ms = (b - a) - (b - a)nt,$$

or

$$[(b - a)s]m \equiv (b - a) \pmod{n}.$$

Now let $k_1 = (b - a)s$.

To show that any two solutions are congruent modulo mn, let c_1 and c_2 be two solutions of the system. That is,

$$c_i \equiv a \pmod{m}$$
$$c_i \equiv b \pmod{n}$$

for $i = 1, 2$. Then

$$c_2 \equiv c_1 \pmod{m}$$
$$c_2 \equiv c_1 \pmod{n}.$$

Therefore, both m and n divide $c_1 - c_2$. Consequently, $c_2 \equiv c_1 \pmod{mn}$. ∎

Example 16.42. Let us solve the system

$$x \equiv 3 \pmod 4$$
$$x \equiv 4 \pmod 5.$$

Using the Euclidean algorithm, we can find integers s and t such that $4s + 5t = 1$. Two such integers are $s = 4$ and $t = -3$. Consequently,

$$x = a + k_1 m = 3 + 4k_1 = 3 + 4\left[(5-4)4\right] = 19.$$

$\square$

Theorem 16.43. Chinese Remainder Theorem. Let $n_1, n_2, \ldots, n_k$ be positive integers such that $\gcd(n_i, n_j) = 1$ for $i \neq j$. Then for any integers $a_1, \ldots, a_k$, the system

$$x \equiv a_1 \pmod{n_1}$$
$$x \equiv a_2 \pmod{n_2}$$
$$\vdots$$
$$x \equiv a_k \pmod{n_k}$$

has a solution. Furthermore, any two solutions of the system are congruent modulo $n_1 n_2 \cdots n_k$.

Proof. We will use mathematical induction on the number of equations in the system. If there are $k = 2$ equations, then the theorem is true by Lemma 16.41. Now suppose that the result is true for a system of k equations or less and that we wish to find a solution of

$$x \equiv a_1 \pmod{n_1}$$
$$x \equiv a_2 \pmod{n_2}$$
$$\vdots$$
$$x \equiv a_{k+1} \pmod{n_{k+1}}.$$

Considering the first k equations, there exists a solution that is unique modulo $n_1 \cdots n_k$, say a. Since $n_1 \cdots n_k$ and n_{k+1} are relatively prime, the system

$$x \equiv a \pmod{n_1 \cdots n_k}$$
$$x \equiv a_{k+1} \pmod{n_{k+1}}$$

has a solution that is unique modulo $n_1 \ldots n_{k+1}$ by the lemma. $\blacksquare$

Example 16.44. Let us solve the system

$$x \equiv 3 \pmod{4}$$
$$x \equiv 4 \pmod{5}$$
$$x \equiv 1 \pmod{9}$$
$$x \equiv 5 \pmod{7}.$$

From Example 16.42 on the previous page we know that 19 is a solution of the first two congruences and any other solution of the system is congruent to 19 (mod 20). Hence, we can reduce the system to a system of three congruences:

$$x \equiv 19 \pmod{20}$$
$$x \equiv 1 \pmod{9}$$
$$x \equiv 5 \pmod{7}.$$

Solving the next two equations, we can reduce the system to

$$x \equiv 19 \pmod{180}$$
$$x \equiv 5 \pmod{7}.$$

Solving this last system, we find that 19 is a solution for the system that is unique up to modulo 1260.   □

One interesting application of the Chinese Remainder Theorem in the design of computer software is that the theorem allows us to break up a calculation involving large integers into several less formidable calculations. A computer will handle integer calculations only up to a certain size due to the size of its processor chip, which is usually a 32 or 64-bit processor chip. For example, the largest integer available on a computer with a 64-bit processor chip is

$$2^{63} - 1 = 9{,}223{,}372{,}036{,}854{,}775{,}807.$$

Larger processors such as 128 or 256-bit have been proposed or are under development. There is even talk of a 512-bit processor chip. The largest integer that such a chip could store with be $2^{511} - 1$, which would be a 154 digit number. However, we would need to deal with much larger numbers to break sophisticated encryption schemes.

Special software is required for calculations involving larger integers which cannot be added directly by the machine. By using the Chinese Remainder Theorem we can break down large integer additions and multiplications into calculations that the computer can handle directly. This is especially useful on parallel processing computers which have the ability to run several programs concurrently.

Most computers have a single central processing unit (CPU) containing one processor chip and can only add two numbers at a time. To add a list of ten numbers, the CPU must do nine additions in sequence. However, a parallel processing computer has more than one CPU. A computer with 10 CPUs, for example, can perform 10 different additions at the same time. If we can take a large integer and break it down into parts, sending each part to a different CPU, then by performing several additions or multiplications simultaneously on those parts, we can work with an integer that the computer would not be able to handle as a whole.

Example 16.45. Suppose that we wish to multiply 2134 by 1531. We will use the integers 95, 97, 98, and 99 because they are relatively prime. We can break down each integer into four parts:

$$2134 \equiv 44 \pmod{95}$$
$$2134 \equiv 0 \pmod{97}$$
$$2134 \equiv 76 \pmod{98}$$
$$2134 \equiv 55 \pmod{99}$$

and

$$1531 \equiv 11 \pmod{95}$$
$$1531 \equiv 76 \pmod{97}$$
$$1531 \equiv 61 \pmod{98}$$
$$1531 \equiv 46 \pmod{99}.$$

Multiplying the corresponding equations, we obtain

$$2134 \cdot 1531 \equiv 44 \cdot 11 \equiv 9 \pmod{95}$$
$$2134 \cdot 1531 \equiv 0 \cdot 76 \equiv 0 \pmod{97}$$
$$2134 \cdot 1531 \equiv 76 \cdot 61 \equiv 30 \pmod{98}$$
$$2134 \cdot 1531 \equiv 55 \cdot 46 \equiv 55 \pmod{99}.$$

Each of these four computations can be sent to a different processor if our computer has several CPUs. By the above calculation, we know that $2134 \cdot 1531$ is a solution of the system

$$x \equiv 9 \pmod{95}$$
$$x \equiv 0 \pmod{97}$$
$$x \equiv 30 \pmod{98}$$
$$x \equiv 55 \pmod{99}.$$

The Chinese Remainder Theorem tells us that solutions are unique up to modulo $95 \cdot 97 \cdot 98 \cdot 99 = 89{,}403{,}930$. Solving this system of congruences for x tells us that $2134 \cdot 1531 = 3{,}267{,}154$.

The conversion of the computation into the four subcomputations will take some computing time. In addition, solving the system of congruences can also take considerable time. However, if we have many computations to be performed on a particular set of numbers, it makes sense to transform the problem as we have done above and to perform the necessary calculations simultaneously.   □

Sage.   Rings are at the heart of Sage's design, so you will find a wide range of possibilities for computing with rings and fields. Ideals, quotients, and homomorphisms are all available.

16.6 Reading Questions

1. What is the fundamental difference between groups and rings?

2. Give two characterizations of an integral domain.

3. Provide two examples of fields, one infinite, one finite.

4. Who was Emmy Noether?

5. Speculate on a computer program that might use the Chinese Remainder Theorem to speed up computations with large integers.

16.7 Exercises

1. Which of the following sets are rings with respect to the usual operations of addition and multiplication? If the set is a ring, is it also a field?

    (a) $7\mathbb{Z}$

    (b) $\mathbb{Z}_{18}$

    (c) $\mathbb{Q}(\sqrt{2}\,) = \{a + b\sqrt{2} : a, b \in \mathbb{Q}\}$

    (d) $\mathbb{Q}(\sqrt{2}, \sqrt{3}\,) = \{a + b\sqrt{2} + c\sqrt{3} + d\sqrt{6} : a, b, c, d \in \mathbb{Q}\}$

    (e) $\mathbb{Z}[\sqrt{3}\,] = \{a + b\sqrt{3} : a, b \in \mathbb{Z}\}$

(f) $R = \{a + b\sqrt[3]{3} : a, b \in \mathbb{Q}\}$

(g) $\mathbb{Z}[i] = \{a + bi : a, b \in \mathbb{Z} \text{ and } i^2 = -1\}$

(h) $\mathbb{Q}(\sqrt[3]{3}) = \{a + b\sqrt[3]{3} + c\sqrt[3]{9} : a, b, c \in \mathbb{Q}\}$

2. Let R be the ring of 2×2 matrices of the form

$$\begin{pmatrix} a & b \\ 0 & 0 \end{pmatrix},$$

where $a, b \in \mathbb{R}$. Show that although R is a ring that has no identity, we can find a subring S of R with an identity.

3. List or characterize all of the units in each of the following rings.

   (a) $\mathbb{Z}_{10}$

   (b) $\mathbb{Z}_{12}$

   (c) $\mathbb{Z}_7$

   (d) $\mathbb{M}_2(\mathbb{Z})$, the 2×2 matrices with entries in $\mathbb{Z}$

   (e) $\mathbb{M}_2(\mathbb{Z}_2)$, the 2×2 matrices with entries in $\mathbb{Z}_2$

4. Find all of the ideals in each of the following rings. Which of these ideals are maximal and which are prime?

   (a) $\mathbb{Z}_{18}$

   (b) $\mathbb{Z}_{25}$

   (c) $\mathbb{M}_2(\mathbb{R})$, the 2×2 matrices with entries in $\mathbb{R}$

   (d) $\mathbb{M}_2(\mathbb{Z})$, the 2×2 matrices with entries in $\mathbb{Z}$

   (e) $\mathbb{Q}$

5. For each of the following rings R with ideal I, give an addition table and a multiplication table for R/I.

   (a) $R = \mathbb{Z}$ and $I = 6\mathbb{Z}$

   (b) $R = \mathbb{Z}_{12}$ and $I = \{0, 3, 6, 9\}$

6. Find all homomorphisms $\phi : \mathbb{Z}/6\mathbb{Z} \to \mathbb{Z}/15\mathbb{Z}$.

7. Prove that $\mathbb{R}$ is not isomorphic to $\mathbb{C}$.

8. Prove or disprove: The ring $\mathbb{Q}(\sqrt{2}) = \{a + b\sqrt{2} : a, b \in \mathbb{Q}\}$ is isomorphic to the ring $\mathbb{Q}(\sqrt{3}) = \{a + b\sqrt{3} : a, b \in \mathbb{Q}\}$.

9. What is the characteristic of the field formed by the set of matrices

$$F = \left\{ \begin{pmatrix} 1 & 0 \\ 0 & 1 \end{pmatrix}, \begin{pmatrix} 1 & 1 \\ 1 & 0 \end{pmatrix}, \begin{pmatrix} 0 & 1 \\ 1 & 1 \end{pmatrix}, \begin{pmatrix} 0 & 0 \\ 0 & 0 \end{pmatrix} \right\}$$

with entries in $\mathbb{Z}_2$?

10. Define a map $\phi : \mathbb{C} \to \mathbb{M}_2(\mathbb{R})$ by

$$\phi(a + bi) = \begin{pmatrix} a & b \\ -b & a \end{pmatrix}.$$

Show that ϕ is an isomorphism of $\mathbb{C}$ with its image in $\mathbb{M}_2(\mathbb{R})$.

11. Prove that the Gaussian integers, $\mathbb{Z}[i]$, are an integral domain.

12. Prove that $\mathbb{Z}[\sqrt{3}\,i] = \{a + b\sqrt{3}\,i : a, b \in \mathbb{Z}\}$ is an integral domain.

13. Solve each of the following systems of congruences.

(a)

$$x \equiv 2 \pmod{5}$$
$$x \equiv 6 \pmod{11}$$

$$x \equiv 4 \pmod{7}$$
$$x \equiv 7 \pmod{9}$$
$$x \equiv 5 \pmod{11}$$

(b)

$$x \equiv 3 \pmod{7}$$
$$x \equiv 0 \pmod{8}$$
$$x \equiv 5 \pmod{15}$$

(d)

$$x \equiv 3 \pmod{5}$$
$$x \equiv 0 \pmod{8}$$
$$x \equiv 1 \pmod{11}$$
$$x \equiv 5 \pmod{13}$$

(c)

$$x \equiv 2 \pmod{4}$$

14. Use the method of parallel computation outlined in the text to calculate $2234 + 4121$ by dividing the calculation into four separate additions modulo 95, 97, 98, and 99.

15. Explain why the method of parallel computation outlined in the text fails for $2134 \cdot 1531$ if we attempt to break the calculation down into two smaller calculations modulo 98 and 99.

16. If R is a field, show that the only two ideals of R are $\{0\}$ and R itself.

17. Let a be any element in a ring R with identity. Show that $(-1)a = -a$.

18. Let $\phi : R \to S$ be a ring homomorphism. Prove each of the following statements.

    (a) If R is a commutative ring, then $\phi(R)$ is a commutative ring.

    (b) $\phi(0) = 0$.

    (c) Let 1_R and 1_S be the identities for R and S, respectively. If ϕ is onto, then $\phi(1_R) = 1_S$.

    (d) If R is a field and $\phi(R) \neq 0$, then $\phi(R)$ is a field.

19. Prove that the associative law for multiplication and the distributive laws hold in R/I.

20. Prove the Second Isomorphism Theorem for rings: Let I be a subring of a ring R and J an ideal in R. Then $I \cap J$ is an ideal in I and

$$I/I \cap J \cong I + J/J.$$

21. Prove the Third Isomorphism Theorem for rings: Let R be a ring and I and J be ideals of R, where $J \subset I$. Then

$$R/I \cong \frac{R/J}{I/J}.$$

22. Prove the Correspondence Theorem: Let I be an ideal of a ring R. Then $S \to S/I$ is a one-to-one correspondence between the set of subrings S containing I and the set of subrings of R/I. Furthermore, the ideals of R correspond to ideals of R/I.

23. Let R be a ring and S a subset of R. Show that S is a subring of R if and only if each of the following conditions is satisfied.

    (a) $S \neq \varnothing$.

    (b) $rs \in S$ for all $r, s \in S$.

    (c) $r - s \in S$ for all $r, s \in S$.

24. Let R be a ring with a collection of subrings $\{R_\alpha\}$. Prove that $\cap R_\alpha$ is a subring of R. Give an example to show that the union of two subrings is not necessarily a subring.

25. Let $\{I_\alpha\}_{\alpha \in A}$ be a collection of ideals in a ring R. Prove that $\cap_{\alpha \in A} I_\alpha$ is also an ideal in R. Give an example to show that if I_1 and I_2 are ideals in R, then $I_1 \cup I_2$ may not be an ideal.

26. Let R be an integral domain. Show that if the only ideals in R are $\{0\}$ and R itself, R must be a field.

27. Let R be a commutative ring. An element a in R is *nilpotent* if $a^n = 0$ for some positive integer n. Show that the set of all nilpotent elements forms an ideal in R.

28. A ring R is a **Boolean ring** if for every $a \in R$, $a^2 = a$. Show that every Boolean ring is a commutative ring.

29. Let R be a ring, where $a^3 = a$ for all $a \in R$. Prove that R must be a commutative ring.

30. Let R be a ring with identity 1_R and S a subring of R with identity 1_S. Prove or disprove that $1_R = 1_S$.

31. If we do not require the identity of a ring to be distinct from 0, we will not have a very interesting mathematical structure. Let R be a ring such that $1 = 0$. Prove that $R = \{0\}$.

32. Let R be a ring. Define the **center** of R to be

$$Z(R) = \{a \in R : ar = ra \text{ for all } r \in R\}.$$

Prove that $Z(R)$ is a commutative subring of R.

33. Let p be prime. Prove that

$$\mathbb{Z}_{(p)} = \{a/b : a, b \in \mathbb{Z} \text{ and } \gcd(b, p) = 1\}$$

is a ring. The ring $\mathbb{Z}_{(p)}$ is called the **ring of integers localized at** p.

34. Prove or disprove: Every finite integral domain is isomorphic to $\mathbb{Z}_p$.

35. Let R be a ring with identity.

   (a) Let u be a unit in R. Define a map $i_u : R \to R$ by $r \mapsto uru^{-1}$. Prove that i_u is an automorphism of R. Such an automorphism of R is called an inner automorphism of R. Denote the set of all inner automorphisms of R by $\mathrm{Inn}(R)$.

   (b) Denote the set of all automorphisms of R by $\mathrm{Aut}(R)$. Prove that $\mathrm{Inn}(R)$ is a normal subgroup of $\mathrm{Aut}(R)$.

   (c) Let $U(R)$ be the group of units in R. Prove that the map

$$\phi : U(R) \to \mathrm{Inn}(R)$$

   defined by $u \mapsto i_u$ is a homomorphism. Determine the kernel of ϕ.

   (d) Compute $\mathrm{Aut}(\mathbb{Z})$, $\mathrm{Inn}(\mathbb{Z})$, and $U(\mathbb{Z})$.

36. Let R and S be arbitrary rings. Show that their Cartesian product is a ring if we define addition and multiplication in $R \times S$ by

   (a) $(r, s) + (r', s') = (r + r', s + s')$

   (b) $(r, s)(r', s') = (rr', ss')$

37. An element x in a ring is called an *idempotent* if $x^2 = x$. Prove that the only idempotents in an integral domain are 0 and 1. Find a ring with a idempotent x not equal to 0 or 1.

38. Let $\gcd(a, n) = d$ and $\gcd(b, d) \neq 1$. Prove that $ax \equiv b \pmod{n}$ does not have a solution.

39. **The Chinese Remainder Theorem for Rings.** Let R be a ring and I and J be ideals in R such that $I + J = R$.

   (a) Show that for any r and s in R, the system of equations

$$x \equiv r \pmod{I}$$
$$x \equiv s \pmod{J}$$

   has a solution.

   (b) In addition, prove that any two solutions of the system are congruent modulo $I \cap J$.

   (c) Let I and J be ideals in a ring R such that $I + J = R$. Show that there exists a ring isomorphism

$$R/(I \cap J) \cong R/I \times R/J.$$

16.8 Programming Exercise

1. Write a computer program implementing fast addition and multiplication using the Chinese Remainder Theorem and the method outlined in the text.

16.9 References and Suggested Readings

[1] Anderson, F. W. and Fuller, K. R. *Rings and Categories of Modules.* 2nd ed. Springer, New York, 1992.

[2] Atiyah, M. F. and MacDonald, I. G. *Introduction to Commutative Algebra.* Westview Press, Boulder, CO, 1994.

[3] Herstein, I. N. *Noncommutative Rings.* Mathematical Association of America, Washington, DC, 1994.

[4] Kaplansky, I. *Commutative Rings.* Revised edition. University of Chicago Press, Chicago, 1974.

[5] Knuth, D. E. *The Art of Computer Programming: Semi-Numerical Algorithms,* vol. 2. 3rd ed. Addison-Wesley Professional, Boston, 1997.

[6] Lidl, R. and Pilz, G. *Applied Abstract Algebra*. 2nd ed. Springer, New York, 1998. A good source for applications.

[7] Mackiw, G. *Applications of Abstract Algebra*. Wiley, New York, 1985.

[8] McCoy, N. H. *Rings and Ideals*. Carus Monograph Series, No. 8. Mathematical Association of America, Washington, DC, 1968.

[9] McCoy, N. H. *The Theory of Rings*. Chelsea, New York, 1972.

[10] Zariski, O. and Samuel, P. *Commutative Algebra*, vols. I and II. Springer, New York, 1975, 1960.

Polynomials

$\mathcal{M}$ ost people are fairly familiar with polynomials by the time they begin to study abstract algebra. When we examine polynomial expressions such as

$$p(x) = x^3 - 3x + 2$$
$$q(x) = 3x^2 - 6x + 5,$$

we have a pretty good idea of what $p(x) + q(x)$ and $p(x)q(x)$ mean. We just add and multiply polynomials as functions; that is,

$$(p + q)(x) = p(x) + q(x)$$
$$= (x^3 - 3x + 2) + (3x^2 - 6x + 5)$$
$$= x^3 + 3x^2 - 9x + 7$$

and

$$(pq)(x) = p(x)q(x)$$
$$= (x^3 - 3x + 2)(3x^2 - 6x + 5)$$
$$= 3x^5 - 6x^4 - 4x^3 + 24x^2 - 27x + 10.$$

It is probably no surprise that polynomials form a ring. In this chapter we shall emphasize the algebraic structure of polynomials by studying polynomial rings. We can prove many results for polynomial rings that are similar to the theorems we proved for the integers. Analogs of prime numbers, the division algorithm, and the Euclidean algorithm exist for polynomials.

17.1 Polynomial Rings

Throughout this chapter we shall assume that R is a commutative ring with identity. Any expression of the form

$$f(x) = \sum_{i=0}^{n} a_i x^i = a_0 + a_1 x + a_2 x^2 + \cdots + a_n x^n,$$

where $a_i \in R$ and $a_n \neq 0$, is called a *polynomial over R* with **indeterminate** x. The elements $a_0, a_1, \ldots, a_n$ are called the *coefficients* of f. The coefficient a_n is called the *leading coefficient*. A polynomial is called *monic* if the leading coefficient is 1. If n is the largest nonnegative number for which $a_n \neq 0$, we say that the *degree* of f is n and write $\deg f(x) = n$. If no such n exists—that is, if $f = 0$ is the zero polynomial—then the degree of f is defined to be $-\infty$. We will denote the set of all polynomials with coefficients in a ring R by $R[x]$. Two polynomials are equal exactly when their corresponding coefficients are equal; that is, if we let

$$p(x) = a_0 + a_1 x + \cdots + a_n x^n$$
$$q(x) = b_0 + b_1 x + \cdots + b_m x^m,$$

then $p(x) = q(x)$ if and only if $a_i = b_i$ for all $i \geq 0$.

To show that the set of all polynomials forms a ring, we must first define addition and multiplication. We define the sum of two polynomials as follows. Let

$$p(x) = a_0 + a_1 x + \cdots + a_n x^n$$
$$q(x) = b_0 + b_1 x + \cdots + b_m x^m.$$

Then the sum of $p(x)$ and $q(x)$ is

$$p(x) + q(x) = c_0 + c_1 x + \cdots + c_k x^k,$$

where $c_i = a_i + b_i$ for each i. We define the product of $p(x)$ and $q(x)$ to be

$$p(x)q(x) = c_0 + c_1 x + \cdots + c_{m+n} x^{m+n},$$

where

$$c_i = \sum_{k=0}^{i} a_k b_{i-k} = a_0 b_i + a_1 b_{i-1} + \cdots + a_{i-1} b_1 + a_i b_0$$

for each i. Notice that in each case some of the coefficients may be zero.

Example 17.1. Suppose that

$$p(x) = 3 + 0x + 0x^2 + 2x^3 + 0x^4$$

and

$$q(x) = 2 + 0x - x^2 + 0x^3 + 4x^4$$

are polynomials in $\mathbb{Z}[x]$. If the coefficient of some term in a polynomial is zero, then we usually just omit that term. In this case we would write $p(x) = 3 + 2x^3$

and $q(x) = 2 - x^2 + 4x^4$. The sum of these two polynomials is

$$p(x) + q(x) = 5 - x^2 + 2x^3 + 4x^4.$$

The product,

$$p(x)q(x) = (3 + 2x^3)(2 - x^2 + 4x^4) = 6 - 3x^2 + 4x^3 + 12x^4 - 2x^5 + 8x^7,$$

can be calculated either by determining the c_is in the definition or by simply multiplying polynomials in the same way as we have always done.   □

Example 17.2. Let

$$p(x) = 3 + 3x^3 \qquad \text{and} \qquad q(x) = 4 + 4x^2 + 4x^4$$

be polynomials in $\mathbb{Z}_{12}[x]$. The sum of $p(x)$ and $q(x)$ is $7 + 4x^2 + 3x^3 + 4x^4$. The product of the two polynomials is the zero polynomial. This example tells us that we can not expect $R[x]$ to be an integral domain if R is not an integral domain. □

Theorem 17.3. Let R be a commutative ring with identity. Then $R[x]$ is a commutative ring with identity.

Proof. Our first task is to show that $R[x]$ is an abelian group under polynomial addition. The zero polynomial, $f(x) = 0$, is the additive identity. Given a polynomial $p(x) = \sum_{i=0}^{n} a_i x^i$, the inverse of $p(x)$ is easily verified to be $-p(x) = \sum_{i=0}^{n}(-a_i)x^i = -\sum_{i=0}^{n} a_i x^i$. Commutativity and associativity follow immediately from the definition of polynomial addition and from the fact that addition in R is both commutative and associative.

To show that polynomial multiplication is associative, let

$$p(x) = \sum_{i=0}^{m} a_i x^i,$$

$$q(x) = \sum_{i=0}^{n} b_i x^i,$$

$$r(x) = \sum_{i=0}^{p} c_i x^i.$$

Then

$$
\begin{aligned}
[p(x)q(x)]r(x) &= \left[\left(\sum_{i=0}^{m} a_i x^i\right)\left(\sum_{i=0}^{n} b_i x^i\right)\right]\left(\sum_{i=0}^{p} c_i x^i\right) \\
&= \left[\sum_{i=0}^{m+n}\left(\sum_{j=0}^{i} a_j b_{i-j}\right) x^i\right]\left(\sum_{i=0}^{p} c_i x^i\right) \\
&= \sum_{i=0}^{m+n+p}\left[\sum_{j=0}^{i}\left(\sum_{k=0}^{j} a_k b_{j-k}\right) c_{i-j}\right] x^i \\
&= \sum_{i=0}^{m+n+p}\left(\sum_{j+k+l=i} a_j b_k c_l\right) x^i \\
&= \sum_{i=0}^{m+n+p}\left[\sum_{j=0}^{i} a_j\left(\sum_{k=0}^{i-j} b_k c_{i-j-k}\right)\right] x^i \\
&= \left(\sum_{i=0}^{m} a_i x^i\right)\left[\sum_{i=0}^{n+p}\left(\sum_{j=0}^{i} b_j c_{i-j}\right) x^i\right] \\
&= \left(\sum_{i=0}^{m} a_i x^i\right)\left[\left(\sum_{i=0}^{n} b_i x^i\right)\left(\sum_{i=0}^{p} c_i x^i\right)\right] \\
&= p(x)[q(x)r(x)]
\end{aligned}
$$

The commutativity and distribution properties of polynomial multiplication are proved in a similar manner. We shall leave the proofs of these properties as an exercise.   ∎

Proposition 17.4. Let $p(x)$ and $q(x)$ be polynomials in $R[x]$, where R is an integral domain. Then $\deg p(x) + \deg q(x) = \deg(p(x)q(x))$. Furthermore, $R[x]$ is an integral domain.

Proof. Suppose that we have two nonzero polynomials

$$
p(x) = a_m x^m + \cdots + a_1 x + a_0
$$

and

$$
q(x) = b_n x^n + \cdots + b_1 x + b_0
$$

with $a_m \neq 0$ and $b_n \neq 0$. The degrees of $p(x)$ and $q(x)$ are m and n, respectively. The leading term of $p(x)q(x)$ is $a_m b_n x^{m+n}$, which cannot be zero since R is an integral domain; hence, the degree of $p(x)q(x)$ is $m+n$, and $p(x)q(x) \neq 0$. Since $p(x) \neq 0$ and $q(x) \neq 0$ imply that $p(x)q(x) \neq 0$, we know that $R[x]$ must also be an integral domain.   ∎

We also want to consider polynomials in two or more variables, such as $x^2 - 3xy + 2y^3$. Let R be a ring and suppose that we are given two indeterminates x and y. Certainly we can form the ring $(R[x])[y]$. It is straightforward but perhaps tedious to show that $(R[x])[y] \cong R([y])[x]$. We shall identify these two rings by this isomorphism and simply write $R[x, y]$. The ring $R[x, y]$ is called the *ring of polynomials in two indeterminates x and y with coefficients in R*. We can define the *ring of polynomials in n indeterminates with coefficients in R* similarly. We shall denote this ring by $R[x_1, x_2, \ldots, x_n]$.

Theorem 17.5. Let R be a commutative ring with identity and $\alpha \in R$. Then we have a ring homomorphism $\phi_\alpha : R[x] \to R$ defined by

$$\phi_\alpha(p(x)) = p(\alpha) = a_n \alpha^n + \cdots + a_1 \alpha + a_0,$$

where $p(x) = a_n x^n + \cdots + a_1 x + a_0$.

Proof. Let $p(x) = \sum_{i=0}^{n} a_i x^i$ and $q(x) = \sum_{i=0}^{m} b_i x^i$. It is easy to show that $\phi_\alpha(p(x) + q(x)) = \phi_\alpha(p(x)) + \phi_\alpha(q(x))$. To show that multiplication is preserved under the map ϕ_α, observe that

$$\phi_\alpha(p(x))\phi_\alpha(q(x)) = p(\alpha)q(\alpha)$$

$$= \left(\sum_{i=0}^{n} a_i \alpha^i \right) \left(\sum_{i=0}^{m} b_i \alpha^i \right)$$

$$= \sum_{i=0}^{m+n} \left(\sum_{k=0}^{i} a_k b_{i-k} \right) \alpha^i$$

$$= \phi_\alpha(p(x)q(x)).$$

∎

The map $\phi_\alpha : R[x] \to R$ is called the *evaluation homomorphism* at α.

17.2 The Division Algorithm

Recall that the division algorithm for integers (Theorem 2.9 on page 25) says that if a and b are integers with $b > 0$, then there exist unique integers q and r such that $a = bq + r$, where $0 \leq r < b$. The algorithm by which q and r are found is just long division. A similar theorem exists for polynomials. The division algorithm for polynomials has several important consequences. Since its proof is very similar to the corresponding proof for integers, it is worthwhile to review Theorem 2.9 on page 25 at this point.

Theorem 17.6. Division Algorithm. Let $f(x)$ and $g(x)$ be polynomials in $F[x]$, where F is a field and $g(x)$ is a nonzero polynomial. Then there exist unique polynomials $q(x), r(x) \in F[x]$ such that

$$f(x) = g(x)q(x) + r(x),$$

where either $\deg r(x) < \deg g(x)$ or $r(x)$ is the zero polynomial.

Proof. We will first consider the existence of $q(x)$ and $r(x)$. If $f(x)$ is the zero polynomial, then

$$0 = 0 \cdot g(x) + 0;$$

hence, both q and r must also be the zero polynomial. Now suppose that $f(x)$ is not the zero polynomial and that $\deg f(x) = n$ and $\deg g(x) = m$. If $m > n$, then we can let $q(x) = 0$ and $r(x) = f(x)$. Hence, we may assume that $m \le n$ and proceed by induction on n. If

$$f(x) = a_n x^n + a_{n-1} x^{n-1} + \cdots + a_1 x + a_0$$
$$g(x) = b_m x^m + b_{m-1} x^{m-1} + \cdots + b_1 x + b_0$$

the polynomial

$$f'(x) = f(x) - \frac{a_n}{b_m} x^{n-m} g(x)$$

has degree less than n or is the zero polynomial. By induction, there exist polynomials $q'(x)$ and $r(x)$ such that

$$f'(x) = q'(x)g(x) + r(x),$$

where $r(x) = 0$ or the degree of $r(x)$ is less than the degree of $g(x)$. Now let

$$q(x) = q'(x) + \frac{a_n}{b_m} x^{n-m}.$$

Then

$$f(x) = g(x)q(x) + r(x),$$

with $r(x)$ the zero polynomial or $\deg r(x) < \deg g(x)$.

To show that $q(x)$ and $r(x)$ are unique, suppose that there exist two other polynomials $q_1(x)$ and $r_1(x)$ such that $f(x) = g(x)q_1(x) + r_1(x)$ with $\deg r_1(x) < \deg g(x)$ or $r_1(x) = 0$, so that

$$f(x) = g(x)q(x) + r(x) = g(x)q_1(x) + r_1(x),$$

and

$$g(x)[q(x) - q_1(x)] = r_1(x) - r(x).$$

If $q(x) - q_1(x)$ is not the zero polynomial, then

$$\deg(g(x)[q(x) - q_1(x)]) = \deg(r_1(x) - r(x)) \geq \deg g(x).$$

However, the degrees of both $r(x)$ and $r_1(x)$ are strictly less than the degree of $g(x)$; therefore, $r(x) = r_1(x)$ and $q(x) = q_1(x)$. $\blacksquare$

Example 17.7. The division algorithm merely formalizes long division of polynomials, a task we have been familiar with since high school. For example, suppose that we divide $x^3 - x^2 + 2x - 3$ by $x - 2$.

$$
\begin{array}{r}
x^2 \quad + \quad x \quad + \quad 4 \\
x - 2 \enclose{longdiv}{x^3 \quad - \quad x^2 \quad + \quad 2x \quad - \quad 3} \\
\underline{x^3 \quad - \quad 2x^2} \\
x^2 \quad + \quad 2x \quad - \quad 3 \\
\underline{x^2 \quad - \quad 2x} \\
4x \quad - \quad 3 \\
\underline{4x \quad - \quad 8} \\
5
\end{array}
$$

Hence, $x^3 - x^2 + 2x - 3 = (x - 2)(x^2 + x + 4) + 5$. $\square$

Let $p(x)$ be a polynomial in $F[x]$ and $\alpha \in F$. We say that α is a *zero* or *root* of $p(x)$ if $p(x)$ is in the kernel of the evaluation homomorphism ϕ_α. All we are really saying here is that α is a zero of $p(x)$ if $p(\alpha) = 0$.

Corollary 17.8. Let F be a field. An element $\alpha \in F$ is a zero of $p(x) \in F[x]$ if and only if $x - \alpha$ is a factor of $p(x)$ in $F[x]$.

Proof. Suppose that $\alpha \in F$ and $p(\alpha) = 0$. By the division algorithm, there exist polynomials $q(x)$ and $r(x)$ such that

$$p(x) = (x - \alpha)q(x) + r(x)$$

and the degree of $r(x)$ must be less than the degree of $x - \alpha$. Since the degree of $r(x)$ is less than 1, $r(x) = a$ for $a \in F$; therefore,

$$p(x) = (x - \alpha)q(x) + a.$$

But

$$0 = p(\alpha) = 0 \cdot q(\alpha) + a = a;$$

consequently, $p(x) = (x - \alpha)q(x)$, and $x - \alpha$ is a factor of $p(x)$.

Conversely, suppose that $x - \alpha$ is a factor of $p(x)$; say $p(x) = (x - \alpha)q(x)$. Then $p(\alpha) = 0 \cdot q(\alpha) = 0$. $\blacksquare$

Corollary 17.9. Let F be a field. A nonzero polynomial $p(x)$ of degree n in $F[x]$ can have at most n distinct zeros in F.

Proof. We will use induction on the degree of $p(x)$. If $\deg p(x) = 0$, then $p(x)$ is a constant polynomial and has no zeros. Let $\deg p(x) = 1$. Then $p(x) = ax + b$ for some a and b in F. If α_1 and α_2 are zeros of $p(x)$, then $a\alpha_1 + b = a\alpha_2 + b$ or $\alpha_1 = \alpha_2$.

Now assume that $\deg p(x) > 1$. If $p(x)$ does not have a zero in F, then we are done. On the other hand, if α is a zero of $p(x)$, then $p(x) = (x - \alpha)q(x)$ for some $q(x) \in F[x]$ by Corollary 17.8 on the preceding page. The degree of $q(x)$ is $n - 1$ by Proposition 17.4 on page 268. Let β be some other zero of $p(x)$ that is distinct from α. Then $p(\beta) = (\beta - \alpha)q(\beta) = 0$. Since $\alpha \neq \beta$ and F is a field, $q(\beta) = 0$. By our induction hypothesis, $q(x)$ can have at most $n - 1$ zeros in F that are distinct from α. Therefore, $p(x)$ has at most n distinct zeros in F.   ∎

Let F be a field. A monic polynomial $d(x)$ is a ***greatest common divisor*** of polynomials $p(x), q(x) \in F[x]$ if $d(x)$ evenly divides both $p(x)$ and $q(x)$; and, if for any other polynomial $d'(x)$ dividing both $p(x)$ and $q(x)$, $d'(x) \mid d(x)$. We write $d(x) = \gcd(p(x), q(x))$. Two polynomials $p(x)$ and $q(x)$ are ***relatively prime*** if $\gcd(p(x), q(x)) = 1$.

Proposition 17.10. Let F be a field and suppose that $d(x)$ is a greatest common divisor of two polynomials $p(x)$ and $q(x)$ in $F[x]$. Then there exist polynomials $r(x)$ and $s(x)$ such that

$$d(x) = r(x)p(x) + s(x)q(x).$$

Furthermore, the greatest common divisor of two polynomials is unique.

Proof. Let $d(x)$ be the monic polynomial of smallest degree in the set

$$S = \{f(x)p(x) + g(x)q(x) : f(x), g(x) \in F[x]\}.$$

We can write $d(x) = r(x)p(x) + s(x)q(x)$ for two polynomials $r(x)$ and $s(x)$ in $F[x]$. We need to show that $d(x)$ divides both $p(x)$ and $q(x)$. We shall first show that $d(x)$ divides $p(x)$. By the division algorithm, there exist polynomials $a(x)$ and $b(x)$ such that $p(x) = a(x)d(x) + b(x)$, where $b(x)$ is either the zero polynomial or $\deg b(x) < \deg d(x)$. Therefore,

$$\begin{aligned}
b(x) &= p(x) - a(x)d(x) \\
&= p(x) - a(x)(r(x)p(x) + s(x)q(x)) \\
&= p(x) - a(x)r(x)p(x) - a(x)s(x)q(x) \\
&= p(x)(1 - a(x)r(x)) + q(x)(-a(x)s(x))
\end{aligned}$$

is a linear combination of $p(x)$ and $q(x)$ and therefore must be in S. However, $b(x)$ must be the zero polynomial since $d(x)$ was chosen to be of smallest degree; consequently, $d(x)$ divides $p(x)$. A symmetric argument shows that $d(x)$ must also divide $q(x)$; hence, $d(x)$ is a common divisor of $p(x)$ and $q(x)$.

To show that $d(x)$ is a greatest common divisor of $p(x)$ and $q(x)$, suppose that $d'(x)$ is another common divisor of $p(x)$ and $q(x)$. We will show that $d'(x) \mid d(x)$. Since $d'(x)$ is a common divisor of $p(x)$ and $q(x)$, there exist polynomials $u(x)$ and $v(x)$ such that $p(x) = u(x)d'(x)$ and $q(x) = v(x)d'(x)$. Therefore,

$$
\begin{aligned}
d(x) &= r(x)p(x) + s(x)q(x) \\
&= r(x)u(x)d'(x) + s(x)v(x)d'(x) \\
&= d'(x)[r(x)u(x) + s(x)v(x)].
\end{aligned}
$$

Since $d'(x) \mid d(x)$, $d(x)$ is a greatest common divisor of $p(x)$ and $q(x)$.

Finally, we must show that the greatest common divisor of $p(x)$ and $q(x)$ is unique. Suppose that $d'(x)$ is another greatest common divisor of $p(x)$ and $q(x)$. We have just shown that there exist polynomials $u(x)$ and $v(x)$ in $F[x]$ such that $d(x) = d'(x)[r(x)u(x) + s(x)v(x)]$. Since

$$
\deg d(x) = \deg d'(x) + \deg[r(x)u(x) + s(x)v(x)]
$$

and $d(x)$ and $d'(x)$ are both greatest common divisors, $\deg d(x) = \deg d'(x)$. Since $d(x)$ and $d'(x)$ are both monic polynomials of the same degree, it must be the case that $d(x) = d'(x)$.   ∎

Notice the similarity between the proof of Proposition 17.10 and the proof of Theorem 2.10 on page 26.

17.3 Irreducible Polynomials

A nonconstant polynomial $f(x) \in F[x]$ is ***irreducible*** over a field F if $f(x)$ cannot be expressed as a product of two polynomials $g(x)$ and $h(x)$ in $F[x]$, where the degrees of $g(x)$ and $h(x)$ are both smaller than the degree of $f(x)$. Irreducible polynomials function as the "prime numbers" of polynomial rings.

Example 17.11. The polynomial $x^2 - 2 \in \mathbb{Q}[x]$ is irreducible since it cannot be factored any further over the rational numbers. Similarly, $x^2 + 1$ is irreducible over the real numbers.   □

Example 17.12. The polynomial $p(x) = x^3 + x^2 + 2$ is irreducible over $\mathbb{Z}_3[x]$. Suppose that this polynomial was reducible over $\mathbb{Z}_3[x]$. By the division algorithm there would have to be a factor of the form $x - a$, where a is some element in

$\mathbb{Z}_3[x]$. Hence, it would have to be true that $p(a) = 0$. However,

$$p(0) = 2$$
$$p(1) = 1$$
$$p(2) = 2.$$

Therefore, $p(x)$ has no zeros in $\mathbb{Z}_3$ and must be irreducible.   □

Lemma 17.13. Let $p(x) \in \mathbb{Q}[x]$. Then

$$p(x) = \frac{r}{s}(a_0 + a_1 x + \cdots + a_n x^n),$$

where $r, s, a_0, \ldots, a_n$ are integers, the a_i's are relatively prime, and r and s are relatively prime.

Proof. Suppose that

$$p(x) = \frac{b_0}{c_0} + \frac{b_1}{c_1}x + \cdots + \frac{b_n}{c_n}x^n,$$

where the b_i's and the c_i's are integers. We can rewrite $p(x)$ as

$$p(x) = \frac{1}{c_0 \cdots c_n}(d_0 + d_1 x + \cdots + d_n x^n),$$

where $d_0, \ldots, d_n$ are integers. Let d be the greatest common divisor of $d_0, \ldots, d_n$. Then

$$p(x) = \frac{d}{c_0 \cdots c_n}(a_0 + a_1 x + \cdots + a_n x^n),$$

where $d_i = d a_i$ and the a_i's are relatively prime. Reducing $d/(c_0 \cdots c_n)$ to its lowest terms, we can write

$$p(x) = \frac{r}{s}(a_0 + a_1 x + \cdots + a_n x^n),$$

where $\gcd(r, s) = 1$.   ■

Theorem 17.14. Gauss's Lemma. Let $p(x) \in \mathbb{Z}[x]$ be a monic polynomial such that $p(x)$ factors into a product of two polynomials $\alpha(x)$ and $\beta(x)$ in $\mathbb{Q}[x]$, where the degrees of both $\alpha(x)$ and $\beta(x)$ are less than the degree of $p(x)$. Then $p(x) = a(x)b(x)$, where $a(x)$ and $b(x)$ are monic polynomials in $\mathbb{Z}[x]$ with $\deg \alpha(x) = \deg a(x)$ and $\deg \beta(x) = \deg b(x)$.

Proof. By Lemma 17.13, we can assume that

$$\alpha(x) = \frac{c_1}{d_1}(a_0 + a_1 x + \cdots + a_m x^m) = \frac{c_1}{d_1}\alpha_1(x)$$

$$\beta(x) = \frac{c_2}{d_2}(b_0 + b_1 x + \cdots + b_n x^n) = \frac{c_2}{d_2}\beta_1(x),$$

where the a_i's are relatively prime and the b_i's are relatively prime. Consequently,

$$p(x) = \alpha(x)\beta(x) = \frac{c_1 c_2}{d_1 d_2}\alpha_1(x)\beta_1(x) = \frac{c}{d}\alpha_1(x)\beta_1(x),$$

where c/d is the product of c_1/d_1 and c_2/d_2 expressed in lowest terms. Hence, $dp(x) = c\alpha_1(x)\beta_1(x)$.

If $d = 1$, then $c a_m b_n = 1$ since $p(x)$ is a monic polynomial. Hence, either $c = 1$ or $c = -1$. If $c = 1$, then either $a_m = b_n = 1$ or $a_m = b_n = -1$. In the first case $p(x) = \alpha_1(x)\beta_1(x)$, where $\alpha_1(x)$ and $\beta_1(x)$ are monic polynomials with $\deg \alpha(x) = \deg \alpha_1(x)$ and $\deg \beta(x) = \deg \beta_1(x)$. In the second case $a(x) = -\alpha_1(x)$ and $b(x) = -\beta_1(x)$ are the correct monic polynomials since $p(x) = (-\alpha_1(x))(-\beta_1(x)) = a(x)b(x)$. The case in which $c = -1$ can be handled similarly.

Now suppose that $d \neq 1$. Since $\gcd(c, d) = 1$, there exists a prime p such that $p \mid d$ and $p \nmid c$. Also, since the coefficients of $\alpha_1(x)$ are relatively prime, there exists a coefficient a_i such that $p \nmid a_i$. Similarly, there exists a coefficient b_j of $\beta_1(x)$ such that $p \nmid b_j$. Let $\alpha_1'(x)$ and $\beta_1'(x)$ be the polynomials in $\mathbb{Z}_p[x]$ obtained by reducing the coefficients of $\alpha_1(x)$ and $\beta_1(x)$ modulo p. Since $p \mid d$, $\alpha_1'(x)\beta_1'(x) = 0$ in $\mathbb{Z}_p[x]$. However, this is impossible since neither $\alpha_1'(x)$ nor $\beta_1'(x)$ is the zero polynomial and $\mathbb{Z}_p[x]$ is an integral domain. Therefore, $d = 1$ and the theorem is proven. ∎

Corollary 17.15. Let $p(x) = x^n + a_{n-1}x^{n-1} + \cdots + a_0$ be a polynomial with coefficients in $\mathbb{Z}$ and $a_0 \neq 0$. If $p(x)$ has a zero in $\mathbb{Q}$, then $p(x)$ also has a zero α in $\mathbb{Z}$. Furthermore, α divides a_0.

Proof. Let $p(x)$ have a zero $a \in \mathbb{Q}$. Then $p(x)$ must have a linear factor $x - a$. By Gauss's Lemma, $p(x)$ has a factorization with a linear factor in $\mathbb{Z}[x]$. Hence, for some $\alpha \in \mathbb{Z}$

$$p(x) = (x - \alpha)(x^{n-1} + \cdots - a_0/\alpha).$$

Thus $a_0/\alpha \in \mathbb{Z}$ and so $\alpha \mid a_0$. ∎

Example 17.16. Let $p(x) = x^4 - 2x^3 + x + 1$. We shall show that $p(x)$ is irreducible over $\mathbb{Q}[x]$. Assume that $p(x)$ is reducible. Then either $p(x)$ has a linear factor, say $p(x) = (x - \alpha)q(x)$, where $q(x)$ is a polynomial of degree three, or $p(x)$ has two quadratic factors.

If $p(x)$ has a linear factor in $\mathbb{Q}[x]$, then it has a zero in $\mathbb{Z}$. By Corollary 17.15 on the previous page, any zero must divide 1 and therefore must be ± 1; however, $p(1) = 1$ and $p(-1) = 3$. Consequently, we have eliminated the possibility that $p(x)$ has any linear factors.

Therefore, if $p(x)$ is reducible it must factor into two quadratic polynomials, say

$$p(x) = (x^2 + ax + b)(x^2 + cx + d)$$
$$= x^4 + (a + c)x^3 + (ac + b + d)x^2 + (ad + bc)x + bd,$$

where each factor is in $\mathbb{Z}[x]$ by Gauss's Lemma. Hence,

$$a + c = -2$$
$$ac + b + d = 0$$
$$ad + bc = 1$$
$$bd = 1.$$

Since $bd = 1$, either $b = d = 1$ or $b = d = -1$. In either case $b = d$ and so

$$ad + bc = b(a + c) = 1.$$

Since $a + c = -2$, we know that $-2b = 1$. This is impossible since b is an integer. Therefore, $p(x)$ must be irreducible over $\mathbb{Q}$.   □

Theorem 17.17. Eisenstein's Criterion. Let p be a prime and suppose that

$$f(x) = a_n x^n + \cdots + a_0 \in \mathbb{Z}[x].$$

If $p \mid a_i$ for $i = 0, 1, \ldots, n - 1$, but $p \nmid a_n$ and $p^2 \nmid a_0$, then $f(x)$ is irreducible over $\mathbb{Q}$.

Proof. By Gauss's Lemma, we need only show that $f(x)$ does not factor into polynomials of lower degree in $\mathbb{Z}[x]$. Let

$$f(x) = (b_r x^r + \cdots + b_0)(c_s x^s + \cdots + c_0)$$

be a factorization in $\mathbb{Z}[x]$, with b_r and c_s not equal to zero and $r, s < n$. Since p^2 does not divide $a_0 = b_0 c_0$, either b_0 or c_0 is not divisible by p. Suppose that $p \nmid b_0$ and $p \mid c_0$. Since $p \nmid a_n$ and $a_n = b_r c_s$, neither b_r nor c_s is divisible by p. Let m be the smallest value of k such that $p \nmid c_k$. Then

$$a_m = b_0 c_m + b_1 c_{m-1} + \cdots + b_m c_0$$

is not divisible by p, since each term on the right-hand side of the equation is divisible by p except for $b_0 c_m$. Therefore, $m = n$ since a_i is divisible by p for $m < n$. Hence, $f(x)$ cannot be factored into polynomials of lower degree and therefore must be irreducible. ∎

Example 17.18. The polynomial

$$f(x) = 16x^5 - 9x^4 + 3x^2 + 6x - 21$$

is easily seen to be irreducible over $\mathbb{Q}$ by Eisenstein's Criterion if we let $p = 3$. □

Eisenstein's Criterion is more useful in constructing irreducible polynomials of a certain degree over $\mathbb{Q}$ than in determining the irreducibility of an arbitrary polynomial in $\mathbb{Q}[x]$: given an arbitrary polynomial, it is not very likely that we can apply Eisenstein's Criterion. The real value of Theorem 17.17 is that we now have an easy method of generating irreducible polynomials of any degree.

Ideals in $F[x]$

Let F be a field. Recall that a principal ideal in $F[x]$ is an ideal $\langle p(x) \rangle$ generated by some polynomial $p(x)$; that is,

$$\langle p(x) \rangle = \{ p(x)q(x) : q(x) \in F[x] \}.$$

Example 17.19. The polynomial x^2 in $F[x]$ generates the ideal $\langle x^2 \rangle$ consisting of all polynomials with no constant term or term of degree 1. □

Theorem 17.20. *If F is a field, then every ideal in $F[x]$ is a principal ideal.*

Proof. Let I be an ideal of $F[x]$. If I is the zero ideal, the theorem is easily true. Suppose that I is a nontrivial ideal in $F[x]$, and let $p(x) \in I$ be a nonzero element of minimal degree. If $\deg p(x) = 0$, then $p(x)$ is a nonzero constant and 1 must be in I. Since 1 generates all of $F[x]$, $\langle 1 \rangle = I = F[x]$ and I is again a principal ideal.

Now assume that $\deg p(x) \geq 1$ and let $f(x)$ be any element in I. By the division algorithm there exist $q(x)$ and $r(x)$ in $F[x]$ such that $f(x) = p(x)q(x) + r(x)$ and $\deg r(x) < \deg p(x)$. Since $f(x), p(x) \in I$ and I is an ideal, $r(x) = f(x) - p(x)q(x)$ is also in I. However, since we chose $p(x)$ to be of minimal degree, $r(x)$ must be the zero polynomial. Since we can write any element $f(x)$ in I as $p(x)q(x)$ for some $q(x) \in F[x]$, it must be the case that $I = \langle p(x) \rangle$. ∎

Example 17.21. It is not the case that every ideal in the ring $F[x, y]$ is a principal ideal. Consider the ideal of $F[x, y]$ generated by the polynomials x and y. This is the ideal of $F[x, y]$ consisting of all polynomials with no constant term. Since both x and y are in the ideal, no single polynomial can generate the entire ideal. □

Theorem 17.22. Let F be a field and suppose that $p(x) \in F[x]$. Then the ideal generated by $p(x)$ is maximal if and only if $p(x)$ is irreducible.

Proof. Suppose that $p(x)$ generates a maximal ideal of $F[x]$. Then $\langle p(x) \rangle$ is also a prime ideal of $F[x]$. Since a maximal ideal must be properly contained inside $F[x]$, $p(x)$ cannot be a constant polynomial. Let us assume that $p(x)$ factors into two polynomials of lesser degree, say $p(x) = f(x)g(x)$. Since $\langle p(x) \rangle$ is a prime ideal one of these factors, say $f(x)$, is in $\langle p(x) \rangle$ and therefore be a multiple of $p(x)$. But this would imply that $\langle p(x) \rangle \subset \langle f(x) \rangle$, which is impossible since $\langle p(x) \rangle$ is maximal.

Conversely, suppose that $p(x)$ is irreducible over $F[x]$. Let I be an ideal in $F[x]$ containing $\langle p(x) \rangle$. By Theorem 17.20 on the previous page, I is a principal ideal; hence, $I = \langle f(x) \rangle$ for some $f(x) \in F[x]$. Since $p(x) \in I$, it must be the case that $p(x) = f(x)g(x)$ for some $g(x) \in F[x]$. However, $p(x)$ is irreducible; hence, either $f(x)$ or $g(x)$ is a constant polynomial. If $f(x)$ is constant, then $I = F[x]$ and we are done. If $g(x)$ is constant, then $f(x)$ is a constant multiple of I and $I = \langle p(x) \rangle$. Thus, there are no proper ideals of $F[x]$ that properly contain $\langle p(x) \rangle$.   ∎

Sage.   Polynomial rings are very important for computational approaches to algebra, and so Sage makes it very easy to compute with polynomials, over rings, or over fields. And it is trivial to check if a polynomial is irreducible.

⤚   Historical Note   ⤙

Throughout history, the solution of polynomial equations has been a challenging problem. The Babylonians knew how to solve the equation $ax^2 + bx + c = 0$. Omar Khayyam (1048–1131) devised methods of solving cubic equations through the use of geometric constructions and conic sections. The algebraic solution of the general cubic equation $ax^3 + bx^2 + cx + d = 0$ was not discovered until the sixteenth century. An Italian mathematician, Luca Pacioli (ca. 1445–1509), wrote in *Summa de Arithmetica* that the solution of the cubic was impossible. This was taken as a challenge by the rest of the mathematical community.

Scipione del Ferro (1465–1526), of the University of Bologna, solved the "depressed cubic,"

$$ax^3 + cx + d = 0.$$

He kept his solution an absolute secret. This may seem surprising today, when mathematicians are usually very eager to publish their results, but in the days of the Italian Renaissance secrecy was customary. Academic appointments were not easy to secure and depended on the ability to prevail in public contests. Such challenges could be issued at any time. Consequently, any major new discovery was a valuable weapon in such a contest. If an opponent presented a list of problems to be solved, del Ferro could in turn present a list of depressed cubics. He kept the secret of his discovery throughout his life, passing it on only on his deathbed to his student Antonio Fior (ca. 1506–?).

Although Fior was not the equal of his teacher, he immediately issued a challenge to Niccolo Fontana (1499–1557). Fontana was known as Tartaglia (the Stammerer). As a youth he had suffered a blow from the sword of a French soldier during an attack on his village. He survived the savage wound, but his speech was permanently impaired. Tartaglia sent Fior a list of 30 various mathematical problems; Fior countered by sending Tartaglia a list of 30 depressed cubics. Tartaglia would either solve all 30 of the problems or absolutely fail. After much effort Tartaglia finally succeeded in solving the depressed cubic and defeated Fior, who faded into obscurity.

At this point another mathematician, Gerolamo Cardano (1501–1576), entered the story. Cardano wrote to Tartaglia, begging him for the solution to the depressed cubic. Tartaglia refused several of his requests, then finally revealed the solution to Cardano after the latter swore an oath not to publish the secret or to pass it on to anyone else. Using the knowledge that he had obtained from Tartaglia, Cardano eventually solved the general cubic

$$ax^3 + bx^2 + cx + d = 0.$$

Cardano shared the secret with his student, Ludovico Ferrari (1522–1565), who solved the general quartic equation,

$$ax^4 + bx^3 + cx^2 + dx + e = 0.$$

In 1543, Cardano and Ferrari examined del Ferro's papers and discovered that he had also solved the depressed cubic. Cardano felt that this relieved him of his obligation to Tartaglia, so he proceeded to publish the solutions in *Ars Magna* (1545), in which he gave credit to del Ferro for solving the special case of the cubic. This resulted in a bitter dispute between Cardano and Tartaglia, who published the story of the oath a year later.

17.4 Reading Questions

1. Suppose $p(x)$ is a polynomial of degree n with coefficients from any field. How many roots can $p(x)$ have? How does this generalize your high school algebra experience?

2. What is the definition of an irreducible polynomial?

3. Find the remainder upon division of $8x^5 - 18x^4 + 20x^3 - 25x^2 + 20$ by $4x^2 - x - 2$.

4. A single theorem in this chapter connects many of the ideas of this chapter to many of the ideas of the previous chapter. State a paraphrased version of this theorem.

5. Early in this chapter, we say, "We can prove many results for polynomial rings that are similar to the theorems we proved for the integers." Write a short essay (or a very long paragraph) justifying this assertion.

17.5 Exercises

1. List all of the polynomials of degree 3 or less in $\mathbb{Z}_2[x]$.

2. Compute each of the following.

   (a) $(5x^2 + 3x - 4) + (4x^2 - x + 9)$ in $\mathbb{Z}_{12}[x]$

   (b) $(5x^2 + 3x - 4)(4x^2 - x + 9)$ in $\mathbb{Z}_{12}[x]$

   (c) $(7x^3 + 3x^2 - x) + (6x^2 - 8x + 4)$ in $\mathbb{Z}_9[x]$

   (d) $(3x^2 + 2x - 4) + (4x^2 + 2)$ in $\mathbb{Z}_5[x]$

   (e) $(3x^2 + 2x - 4)(4x^2 + 2)$ in $\mathbb{Z}_5[x]$

   (f) $(5x^2 + 3x - 2)^2$ in $\mathbb{Z}_{12}[x]$

3. Use the division algorithm to find $q(x)$ and $r(x)$ such that $a(x) = q(x)b(x) + r(x)$ with $\deg r(x) < \deg b(x)$ for each of the following pairs of polynomials.

   (a) $a(x) = 5x^3 + 6x^2 - 3x + 4$ and $b(x) = x - 2$ in $\mathbb{Z}_7[x]$

   (b) $a(x) = 6x^4 - 2x^3 + x^2 - 3x + 1$ and $b(x) = x^2 + x - 2$ in $\mathbb{Z}_7[x]$

   (c) $a(x) = 4x^5 - x^3 + x^2 + 4$ and $b(x) = x^3 - 2$ in $\mathbb{Z}_5[x]$

   (d) $a(x) = x^5 + x^3 - x^2 - x$ and $b(x) = x^3 + x$ in $\mathbb{Z}_2[x]$

4. Find the greatest common divisor of each of the following pairs $p(x)$ and $q(x)$ of polynomials. If $d(x) = \gcd(p(x), q(x))$, find two polynomials $a(x)$ and $b(x)$ such that $a(x)p(x) + b(x)q(x) = d(x)$.

(a) $p(x) = x^3 - 6x^2 + 14x - 15$ and $q(x) = x^3 - 8x^2 + 21x - 18$, where $p(x), q(x) \in \mathbb{Q}[x]$

(b) $p(x) = x^3 + x^2 - x + 1$ and $q(x) = x^3 + x - 1$, where $p(x), q(x) \in \mathbb{Z}_2[x]$

(c) $p(x) = x^3 + x^2 - 4x + 4$ and $q(x) = x^3 + 3x - 2$, where $p(x), q(x) \in \mathbb{Z}_5[x]$

(d) $p(x) = x^3 - 2x + 4$ and $q(x) = 4x^3 + x + 3$, where $p(x), q(x) \in \mathbb{Q}[x]$

5. Find all of the zeros for each of the following polynomials.

   (a) $5x^3 + 4x^2 - x + 9$ in $\mathbb{Z}_{12}[x]$ (c) $5x^4 + 2x^2 - 3$ in $\mathbb{Z}_7[x]$

   (b) $3x^3 - 4x^2 - x + 4$ in $\mathbb{Z}_5[x]$ (d) $x^3 + x + 1$ in $\mathbb{Z}_2[x]$

6. Find all of the units in $\mathbb{Z}[x]$.

7. Find a unit $p(x)$ in $\mathbb{Z}_4[x]$ such that $\deg p(x) > 1$.

8. Which of the following polynomials are irreducible over $\mathbb{Q}[x]$?

   (a) $x^4 - 2x^3 + 2x^2 + x + 4$ (c) $3x^5 - 4x^3 - 6x^2 + 6$

   (b) $x^4 - 5x^3 + 3x - 2$ (d) $5x^5 - 6x^4 - 3x^2 + 9x - 15$

9. Find all of the irreducible polynomials of degrees 2 and 3 in $\mathbb{Z}_2[x]$.

10. Give two different factorizations of $x^2 + x + 8$ in $\mathbb{Z}_{10}[x]$.

11. Prove or disprove: There exists a polynomial $p(x)$ in $\mathbb{Z}_6[x]$ of degree n with more than n distinct zeros.

12. If F is a field, show that $F[x_1, \ldots, x_n]$ is an integral domain.

13. Show that the division algorithm does not hold for $\mathbb{Z}[x]$. Why does it fail?

14. Prove or disprove: $x^p + a$ is irreducible for any $a \in \mathbb{Z}_p$, where p is prime.

15. Let $f(x)$ be irreducible in $F[x]$, where F is a field. If $f(x) \mid p(x)q(x)$, prove that either $f(x) \mid p(x)$ or $f(x) \mid q(x)$.

16. Suppose that R and S are isomorphic rings. Prove that $R[x] \cong S[x]$.

17. Let F be a field and $a \in F$. If $p(x) \in F[x]$, show that $p(a)$ is the remainder obtained when $p(x)$ is divided by $x - a$.

18. **The Rational Root Theorem.** Let

$$p(x) = a_n x^n + a_{n-1} x^{n-1} + \cdots + a_0 \in \mathbb{Z}[x],$$

where $a_n \neq 0$. Prove that if $p(r/s) = 0$, where $\gcd(r, s) = 1$, then $r \mid a_0$ and $s \mid a_n$.

19. Let $\mathbb{Q}^*$ be the multiplicative group of positive rational numbers. Prove that $\mathbb{Q}^*$ is isomorphic to $(\mathbb{Z}[x], +)$.

20. **Cyclotomic Polynomials.** The polynomial

$$\Phi_n(x) = \frac{x^n - 1}{x - 1} = x^{n-1} + x^{n-2} + \cdots + x + 1$$

is called the *cyclotomic polynomial*. Show that $\Phi_p(x)$ is irreducible over $\mathbb{Q}$ for any prime p.

21. If F is a field, show that there are infinitely many irreducible polynomials in $F[x]$.

22. Let R be a commutative ring with identity. Prove that multiplication is commutative in $R[x]$.

23. Let R be a commutative ring with identity. Prove that multiplication is distributive in $R[x]$.

24. Show that $x^p - x$ has p distinct zeros in $\mathbb{Z}_p$, for any prime p. Conclude that

$$x^p - x = x(x - 1)(x - 2)\cdots(x - (p - 1)).$$

25. Let F be a field and $f(x) = a_0 + a_1 x + \cdots + a_n x^n$ be in $F[x]$. Define $f'(x) = a_1 + 2a_2 x + \cdots + na_n x^{n-1}$ to be the *derivative* of $f(x)$.

    (a) Prove that

    $$(f + g)'(x) = f'(x) + g'(x).$$

    Conclude that we can define a homomorphism of abelian groups $D : F[x] \to F[x]$ by $D(f(x)) = f'(x)$.

    (b) Calculate the kernel of D if char $F = 0$.

    (c) Calculate the kernel of D if char $F = p$.

    (d) Prove that

    $$(fg)'(x) = f'(x)g(x) + f(x)g'(x).$$

    (e) Suppose that we can factor a polynomial $f(x) \in F[x]$ into linear factors, say

    $$f(x) = a(x - a_1)(x - a_2)\cdots(x - a_n).$$

    Prove that $f(x)$ has no repeated factors if and only if $f(x)$ and $f'(x)$ are relatively prime.

26. Let F be a field. Show that $F[x]$ is never a field.

27. Let R be an integral domain. Prove that $R[x_1, \ldots, x_n]$ is an integral domain.

28. Let R be a commutative ring with identity. Show that $R[x]$ has a subring R' isomorphic to R.

29. Let $p(x)$ and $q(x)$ be polynomials in $R[x]$, where R is a commutative ring with identity. Prove that $\deg(p(x) + q(x)) \leq \max(\deg p(x), \deg q(x))$.

17.6 Additional Exercises: Solving the Cubic and Quartic Equations

1. Solve the general quadratic equation

$$ax^2 + bx + c = 0$$

to obtain

$$x = \frac{-b \pm \sqrt{b^2 - 4ac}}{2a}.$$

The *discriminant* of the quadratic equation $\Delta = b^2 - 4ac$ determines the nature of the solutions of the equation. If $\Delta > 0$, the equation has two distinct real solutions. If $\Delta = 0$, the equation has a single repeated real root. If $\Delta < 0$, there are two distinct imaginary solutions.

2. Show that any cubic equation of the form

$$x^3 + bx^2 + cx + d = 0$$

can be reduced to the form $y^3 + py + q = 0$ by making the substitution $x = y - b/3$.

3. Prove that the cube roots of 1 are given by

$$\omega = \frac{-1 + i\sqrt{3}}{2}$$

$$\omega^2 = \frac{-1 - i\sqrt{3}}{2}$$

$$\omega^3 = 1.$$

4. Make the substitution

$$y = z - \frac{p}{3z}$$

for y in the equation $y^3 + py + q = 0$ and obtain two solutions A and B for z^3.

5. Show that the product of the solutions obtained in (4) is $-p^3/27$, deducing that $\sqrt[3]{AB} = -p/3$.

6. Prove that the possible solutions for z in (4) are given by

$$\sqrt[3]{A}, \quad \omega\sqrt[3]{A}, \quad \omega^2\sqrt[3]{A}, \quad \sqrt[3]{B}, \quad \omega\sqrt[3]{B}, \quad \omega^2\sqrt[3]{B}$$

and use this result to show that the three possible solutions for y are

$$\omega^i \sqrt[3]{-\frac{q}{2} + \sqrt{\frac{p^3}{27} + \frac{q^2}{4}}} + \omega^{2i} \sqrt[3]{-\frac{q}{2} - \sqrt{\frac{p^3}{27} + \frac{q^2}{4}}},$$

where $i = 0, 1, 2$.

7. The *discriminant* of the cubic equation is

$$\Delta = \frac{p^3}{27} + \frac{q^2}{4}.$$

Show that $y^3 + py + q = 0$

(a) has three real roots, at least two of which are equal, if $\Delta = 0$.

(b) has one real root and two conjugate imaginary roots if $\Delta > 0$.

(c) has three distinct real roots if $\Delta < 0$.

8. Solve the following cubic equations.

(a) $x^3 - 4x^2 + 11x + 30 = 0$

(b) $x^3 - 3x + 5 = 0$

(c) $x^3 - 3x + 2 = 0$

(d) $x^3 + x + 3 = 0$

9. Show that the general quartic equation

$$x^4 + ax^3 + bx^2 + cx + d = 0$$

can be reduced to

$$y^4 + py^2 + qy + r = 0$$

by using the substitution $x = y - a/4$.

10. Show that

$$\left(y^2 + \frac{1}{2}z\right)^2 = (z - p)y^2 - qy + \left(\frac{1}{4}z^2 - r\right).$$

11. Show that the right-hand side of Exercise 17.6.10 can be put in the form $(my + k)^2$ if and only if

$$q^2 - 4(z - p)\left(\frac{1}{4}z^2 - r\right) = 0.$$

12. From Exercise 17.6.11 obtain the *resolvent cubic equation*

$$z^3 - pz^2 - 4rz + (4pr - q^2) = 0.$$

Solving the resolvent cubic equation, put the equation found in Exercise 17.6.10 in the form

$$\left(y^2 + \frac{1}{2}z\right)^2 = (my + k)^2$$

to obtain the solution of the quartic equation.

13. Use this method to solve the following quartic equations.

(a) $x^4 - x^2 - 3x + 2 = 0$

(b) $x^4 + x^3 - 7x^2 - x + 6 = 0$

(c) $x^4 - 2x^2 + 4x - 3 = 0$

(d) $x^4 - 4x^3 + 3x^2 - 5x + 2 = 0$

18

Integral Domains

One of the most important we study is the ring of integers. It was our first example of an algebraic structure: the first polynomial ring that we examined was $\mathbb{Z}[x]$. We also know that the integers sit naturally inside the field of rational numbers, $\mathbb{Q}$. The ring of integers is the model for all integral domains. In this chapter we will examine integral domains in general, answering questions about the ideal structure of integral domains, polynomial rings over integral domains, and whether or not an integral domain can be embedded in a field.

18.1 Fields of Fractions

Every field is also an integral domain; however, there are many integral domains that are not fields. For example, the integers $\mathbb{Z}$ form an integral domain but not a field. A question that naturally arises is how we might associate an integral domain with a field. There is a natural way to construct the rationals $\mathbb{Q}$ from the integers: the rationals can be represented as formal quotients of two integers. The rational numbers are certainly a field. In fact, it can be shown that the rationals are the smallest field that contains the integers. Given an integral domain D, our question now becomes how to construct a smallest field F containing D. We will do this in the same way as we constructed the rationals from the integers.

An element $p/q \in \mathbb{Q}$ is the quotient of two integers p and q; however, different pairs of integers can represent the same rational number. For instance, $1/2 = 2/4 = 3/6$. We know that

$$\frac{a}{b} = \frac{c}{d}$$

if and only if $ad = bc$. A more formal way of considering this problem is to examine fractions in terms of equivalence relations. We can think of elements in $\mathbb{Q}$ as ordered pairs in $\mathbb{Z} \times \mathbb{Z}$. A quotient p/q can be written as (p, q). For instance, $(3, 7)$ would represent the fraction $3/7$. However, there are problems if we consider all possible pairs in $\mathbb{Z} \times \mathbb{Z}$. There is no fraction $5/0$ corresponding to the pair $(5, 0)$. Also, the pairs $(3, 6)$ and $(2, 4)$ both represent the fraction $1/2$. The first problem is easily solved if we require the second coordinate to be nonzero. The second problem is solved by considering two pairs (a, b) and (c, d) to be equivalent if $ad = bc$.

If we use the approach of ordered pairs instead of fractions, then we can study integral domains in general. Let D be any integral domain and let

$$S = \{(a, b) : a, b \in D \text{ and } b \neq 0\}.$$

Define a relation on S by $(a, b) \sim (c, d)$ if $ad = bc$.

Lemma 18.1. *The relation $\sim$ between elements of S is an equivalence relation.*

Proof. Since D is commutative, $ab = ba$; hence, $\sim$ is reflexive on D. Now suppose that $(a, b) \sim (c, d)$. Then $ad = bc$ or $cb = da$. Therefore, $(c, d) \sim (a, b)$ and the relation is symmetric. Finally, to show that the relation is transitive, let $(a, b) \sim (c, d)$ and $(c, d) \sim (e, f)$. In this case $ad = bc$ and $cf = de$. Multiplying both sides of $ad = bc$ by f yields

$$afd = adf = bcf = bde = bed.$$

Since D is an integral domain, we can deduce that $af = be$ or $(a, b) \sim (e, f)$. ∎

We will denote the set of equivalence classes on S by F_D. We now need to define the operations of addition and multiplication on F_D. Recall how fractions are added and multiplied in $\mathbb{Q}$:

$$\frac{a}{b} + \frac{c}{d} = \frac{ad + bc}{bd};$$
$$\frac{a}{b} \cdot \frac{c}{d} = \frac{ac}{bd}.$$

It seems reasonable to define the operations of addition and multiplication on F_D in a similar manner. If we denote the equivalence class of $(a, b) \in S$ by $[a, b]$, then we are led to define the operations of addition and multiplication on F_D by

$$[a, b] + [c, d] = [ad + bc, bd]$$

and

$$[a, b] \cdot [c, d] = [ac, bd],$$

respectively. The next lemma demonstrates that these operations are independent of the choice of representatives from each equivalence class.

Lemma 18.2. *The operations of addition and multiplication on F_D are well-defined.*

Proof. We will prove that the operation of addition is well-defined. The proof that multiplication is well-defined is left as an exercise. Let $[a_1, b_1] = [a_2, b_2]$ and $[c_1, d_1] = [c_2, d_2]$. We must show that

$$[a_1 d_1 + b_1 c_1, b_1 d_1] = [a_2 d_2 + b_2 c_2, b_2 d_2]$$

or, equivalently, that

$$(a_1 d_1 + b_1 c_1)(b_2 d_2) = (b_1 d_1)(a_2 d_2 + b_2 c_2).$$

Since $[a_1, b_1] = [a_2, b_2]$ and $[c_1, d_1] = [c_2, d_2]$, we know that $a_1 b_2 = b_1 a_2$ and $c_1 d_2 = d_1 c_2$. Therefore,

$$\begin{aligned}
(a_1 d_1 + b_1 c_1)(b_2 d_2) &= a_1 d_1 b_2 d_2 + b_1 c_1 b_2 d_2 \\
&= a_1 b_2 d_1 d_2 + b_1 b_2 c_1 d_2 \\
&= b_1 a_2 d_1 d_2 + b_1 b_2 d_1 c_2 \\
&= (b_1 d_1)(a_2 d_2 + b_2 c_2).
\end{aligned}$$

∎

Lemma 18.3. The set of equivalence classes of S, F_D, under the equivalence relation $\sim$, together with the operations of addition and multiplication defined by

$$[a, b] + [c, d] = [ad + bc, bd]$$
$$[a, b] \cdot [c, d] = [ac, bd],$$

is a field.

Proof. The additive and multiplicative identities are $[0, 1]$ and $[1, 1]$, respectively. To show that $[0, 1]$ is the additive identity, observe that

$$[a, b] + [0, 1] = [a1 + b0, b1] = [a, b].$$

It is easy to show that $[1, 1]$ is the multiplicative identity. Let $[a, b] \in F_D$ such that $a \neq 0$. Then $[b, a]$ is also in F_D and $[a, b] \cdot [b, a] = [1, 1]$; hence, $[b, a]$ is the multiplicative inverse for $[a, b]$. Similarly, $[-a, b]$ is the additive inverse of $[a, b]$. We leave as exercises the verification of the associative and commutative properties of multiplication in F_D. We also leave it to the reader to show that F_D is an abelian group under addition.

It remains to show that the distributive property holds in F_D; however,

$$\begin{aligned}
[a, b][e, f] + [c, d][e, f] &= [ae, bf] + [ce, df] \\
&= [aedf + bfce, bdf^2] \\
&= [aed + bce, bdf]
\end{aligned}$$

$$= [ade + bce, bdf]$$
$$= ([a, b] + [c, d])[e, f]$$

and the lemma is proved.   ∎

The field F_D in Lemma 18.3 is called the *field of fractions* or *field of quotients* of the integral domain D.

Theorem 18.4.   Let D be an integral domain. Then D can be embedded in a field of fractions F_D, where any element in F_D can be expressed as the quotient of two elements in D. Furthermore, the field of fractions F_D is unique in the sense that if E is any field containing D, then there exists a map $\psi : F_D \to E$ giving an isomorphism with a subfield of E such that $\psi(a) = a$ for all elements $a \in D$, where we identify a with its image in F_D.

Proof. We will first demonstrate that D can be embedded in the field F_D. Define a map $\phi : D \to F_D$ by $\phi(a) = [a, 1]$. Then for a and b in D,

$$\phi(a + b) = [a + b, 1] = [a, 1] + [b, 1] = \phi(a) + \phi(b)$$

and

$$\phi(ab) = [ab, 1] = [a, 1][b, 1] = \phi(a)\phi(b);$$

hence, ϕ is a homomorphism. To show that ϕ is one-to-one, suppose that $\phi(a) = \phi(b)$. Then $[a, 1] = [b, 1]$, or $a = a1 = 1b = b$. Finally, any element of F_D can be expressed as the quotient of two elements in D, since

$$\phi(a)[\phi(b)]^{-1} = [a, 1][b, 1]^{-1} = [a, 1] \cdot [1, b] = [a, b].$$

Now let E be a field containing D and define a map $\psi : F_D \to E$ by $\psi([a, b]) = ab^{-1}$. To show that ψ is well-defined, let $[a_1, b_1] = [a_2, b_2]$. Then $a_1 b_2 = b_1 a_2$. Therefore, $a_1 b_1^{-1} = a_2 b_2^{-1}$ and $\psi([a_1, b_1]) = \psi([a_2, b_2])$.

If $[a, b]$ and $[c, d]$ are in F_D, then

$$\psi([a, b] + [c, d]) = \psi([ad + bc, bd])$$
$$= (ad + bc)(bd)^{-1}$$
$$= ab^{-1} + cd^{-1}$$
$$= \psi([a, b]) + \psi([c, d])$$

and

$$\psi([a,b] \cdot [c,d]) = \psi([ac,bd])$$
$$= (ac)(bd)^{-1}$$
$$= ab^{-1}cd^{-1}$$
$$= \psi([a,b])\psi([c,d]).$$

Therefore, ψ is a homomorphism.

To complete the proof of the theorem, we need to show that ψ is one-to-one. Suppose that $\psi([a,b]) = ab^{-1} = 0$. Then $a = 0b = 0$ and $[a,b] = [0,b]$. Therefore, the kernel of ψ is the zero element $[0,b]$ in F_D, and ψ is injective.   ■

Example 18.5. Since $\mathbb{Q}$ is a field, $\mathbb{Q}[x]$ is an integral domain. The field of fractions of $\mathbb{Q}[x]$ is the set of all rational expressions $p(x)/q(x)$, where $p(x)$ and $q(x)$ are polynomials over the rationals and $q(x)$ is not the zero polynomial. We will denote this field by $\mathbb{Q}(x)$.   □

We will leave the proofs of the following corollaries of Theorem 18.4 on the preceding page as exercises.

Corollary 18.6. Let F be a field of characteristic zero. Then F contains a subfield isomorphic to $\mathbb{Q}$.

Corollary 18.7. Let F be a field of characteristic p. Then F contains a subfield isomorphic to $\mathbb{Z}_p$.

18.2 Factorization in Integral Domains

The building blocks of the integers are the prime numbers. If F is a field, then irreducible polynomials in $F[x]$ play a role that is very similar to that of the prime numbers in the ring of integers. Given an arbitrary integral domain, we are led to the following series of definitions.

Let R be a commutative ring with identity, and let a and b be elements in R. We say that a *divides* b, and write $a \mid b$, if there exists an element $c \in R$ such that $b = ac$. A *unit* in R is an element that has a multiplicative inverse. Two elements a and b in R are said to be *associates* if there exists a unit u in R such that $a = ub$.

Let D be an integral domain. A nonzero element $p \in D$ that is not a unit is said to be *irreducible* provided that whenever $p = ab$, either a or b is a unit. Furthermore, p is *prime* if whenever $p \mid ab$ either $p \mid a$ or $p \mid b$.

Example 18.8. It is important to notice that prime and irreducible elements do not always coincide. Let R be the subring (with identity) of $\mathbb{Q}[x,y]$ generated by x^2, y^2, and xy. Each of these elements is irreducible in R; however, xy is not prime, since xy divides $x^2 y^2$ but does not divide either x^2 or y^2. □

The Fundamental Theorem of Arithmetic states that every positive integer $n > 1$ can be factored into a product of prime numbers $p_1 \cdots p_k$, where the p_i's are not necessarily distinct. We also know that such factorizations are unique up to the order of the p_i's. We can easily extend this result to the integers. The question arises of whether or not such factorizations are possible in other rings. Generalizing this definition, we say an integral domain D is a *unique factorization domain*, or *UFD*, if D satisfies the following criteria.

1. Let $a \in D$ such that $a \neq 0$ and a is not a unit. Then a can be written as the product of irreducible elements in D.

2. Let $a = p_1 \cdots p_r = q_1 \cdots q_s$, where the p_i's and the q_i's are irreducible. Then $r = s$ and there is a $\pi \in S_r$ such that p_i and $q_{\pi(j)}$ are associates for $j = 1, \ldots, r$.

Example 18.9. The integers are a unique factorization domain by the Fundamental Theorem of Arithmetic. □

Example 18.10. Not every integral domain is a unique factorization domain. The subring $\mathbb{Z}[\sqrt{3}\,i] = \{a + b\sqrt{3}\,i\}$ of the complex numbers is an integral domain (Exercise 16.7.12 on page 260). Let $z = a + b\sqrt{3}\,i$ and define $v : \mathbb{Z}[\sqrt{3}\,i] \to \mathbb{N} \cup \{0\}$ by $v(z) = |z|^2 = a^2 + 3b^2$. It is clear that $v(z) \geq 0$ with equality when $z = 0$. Also, from our knowledge of complex numbers we know that $v(zw) = v(z)v(w)$. It is easy to show that if $v(z) = 1$, then z is a unit, and that the only units of $\mathbb{Z}[\sqrt{3}\,i]$ are 1 and -1.

We claim that 4 has two distinct factorizations into irreducible elements:

$$4 = 2 \cdot 2 = (1 - \sqrt{3}\,i)(1 + \sqrt{3}\,i).$$

We must show that each of these factors is an irreducible element in $\mathbb{Z}[\sqrt{3}\,i]$. If 2 is not irreducible, then $2 = zw$ for elements z, w in $\mathbb{Z}[\sqrt{3}\,i]$ where $v(z) = v(w) = 2$. However, there does not exist an element in z in $\mathbb{Z}[\sqrt{3}\,i]$ such that $v(z) = 2$ because the equation $a^2 + 3b^2 = 2$ has no integer solutions. Therefore, 2 must be irreducible. A similar argument shows that both $1 - \sqrt{3}\,i$ and $1 + \sqrt{3}\,i$ are irreducible. Since 2 is not a unit multiple of either $1 - \sqrt{3}\,i$ or $1 + \sqrt{3}\,i$, 4 has at least two distinct factorizations into irreducible elements. □

Principal Ideal Domains

Let R be a commutative ring with identity. Recall that a principal ideal generated by $a \in R$ is an ideal of the form $\langle a \rangle = \{ra : r \in R\}$. An integral domain in which every ideal is principal is called a *principal ideal domain*, or *PID*.

Lemma 18.11. Let D be an integral domain and let $a, b \in D$. Then

1. $a \mid b$ if and only if $\langle b \rangle \subset \langle a \rangle$.

2. a and b are associates if and only if $\langle b \rangle = \langle a \rangle$.

3. a is a unit in D if and only if $\langle a \rangle = D$.

Proof. (1) Suppose that $a \mid b$. Then $b = ax$ for some $x \in D$. Hence, for every r in D, $br = (ax)r = a(xr)$ and $\langle b \rangle \subset \langle a \rangle$. Conversely, suppose that $\langle b \rangle \subset \langle a \rangle$. Then $b \in \langle a \rangle$. Consequently, $b = ax$ for some $x \in D$. Thus, $a \mid b$.

(2) Since a and b are associates, there exists a unit u such that $a = ub$. Therefore, $b \mid a$ and $\langle a \rangle \subset \langle b \rangle$. Similarly, $\langle b \rangle \subset \langle a \rangle$. It follows that $\langle a \rangle = \langle b \rangle$. Conversely, suppose that $\langle a \rangle = \langle b \rangle$. By part (1), $a \mid b$ and $b \mid a$. Then $a = bx$ and $b = ay$ for some $x, y \in D$. Therefore, $a = bx = ayx$. Since D is an integral domain, $xy = 1$; that is, x and y are units and a and b are associates.

(3) An element $a \in D$ is a unit if and only if a is an associate of 1. However, a is an associate of 1 if and only if $\langle a \rangle = \langle 1 \rangle = D$. ∎

Theorem 18.12. Let D be a PID and $\langle p \rangle$ be a nonzero ideal in D. Then $\langle p \rangle$ is a maximal ideal if and only if p is irreducible.

Proof. Suppose that $\langle p \rangle$ is a maximal ideal. If some element a in D divides p, then $\langle p \rangle \subset \langle a \rangle$. Since $\langle p \rangle$ is maximal, either $D = \langle a \rangle$ or $\langle p \rangle = \langle a \rangle$. Consequently, either a and p are associates or a is a unit. Therefore, p is irreducible.

Conversely, let p be irreducible. If $\langle a \rangle$ is an ideal in D such that $\langle p \rangle \subset \langle a \rangle \subset D$, then $a \mid p$. Since p is irreducible, either a must be a unit or a and p are associates. Therefore, either $D = \langle a \rangle$ or $\langle p \rangle = \langle a \rangle$. Thus, $\langle p \rangle$ is a maximal ideal. ∎

Corollary 18.13. Let D be a PID. If p is irreducible, then p is prime.

Proof. Let p be irreducible and suppose that $p \mid ab$. Then $\langle ab \rangle \subset \langle p \rangle$. By Corollary 16.40 on page 252, since $\langle p \rangle$ is a maximal ideal, $\langle p \rangle$ must also be a prime ideal. Thus, either $a \in \langle p \rangle$ or $b \in \langle p \rangle$. Hence, either $p \mid a$ or $p \mid b$. ∎

Lemma 18.14. Let D be a PID. Let $I_1, I_2, \ldots$ be a set of ideals such that $I_1 \subset I_2 \subset \cdots$. Then there exists an integer N such that $I_n = I_N$ for all $n \geq N$.

Proof. We claim that $I = \bigcup_{i=1}^{\infty} I_i$ is an ideal of D. Certainly I is not empty, since $I_1 \subset I$ and $0 \in I$. If $a, b \in I$, then $a \in I_i$ and $b \in I_j$ for some i and j in $\mathbb{N}$. Without loss of generality we can assume that $i \leq j$. Hence, a and b are both in I_j and so $a - b$ is also in I_j. Now let $r \in D$ and $a \in I$. Again, we note that $a \in I_i$ for some positive integer i. Since I_i is an ideal, $ra \in I_i$ and hence must be in I. Therefore, we have shown that I is an ideal in D.

Since D is a principal ideal domain, there exists an element $\bar{a} \in D$ that generates I. Since $\bar{a}$ is in I_N for some $N \in \mathbb{N}$, we know that $I_N = I = \langle \bar{a} \rangle$. Consequently, $I_n = I_N$ for $n \geq N$. ∎

Any commutative ring satisfying the condition in Lemma 18.14 is said to satisfy the *ascending chain condition*, or *ACC*. Such rings are called *Noetherian rings*, after Emmy Noether.

Theorem 18.15. Every PID is a UFD.

Proof. Existence of a factorization. Let D be a PID and a be a nonzero element in D that is not a unit. If a is irreducible, then we are done. If not, then there exists a factorization $a = a_1 b_1$, where neither a_1 nor b_1 is a unit. Hence, $\langle a \rangle \subset \langle a_1 \rangle$. By Lemma 18.11, we know that $\langle a \rangle \neq \langle a_1 \rangle$; otherwise, a and a_1 would be associates and b_1 would be a unit, which would contradict our assumption. Now suppose that $a_1 = a_2 b_2$, where neither a_2 nor b_2 is a unit. By the same argument as before, $\langle a_1 \rangle \subset \langle a_2 \rangle$. We can continue with this construction to obtain an ascending chain of ideals

$$\langle a \rangle \subset \langle a_1 \rangle \subset \langle a_2 \rangle \subset \cdots.$$

By Lemma 18.14, there exists a positive integer N such that $\langle a_n \rangle = \langle a_N \rangle$ for all $n \geq N$. Consequently, a_N must be irreducible. We have now shown that a is the product of two elements, one of which must be irreducible.

Now suppose that $a = c_1 p_1$, where p_1 is irreducible. If c_1 is not a unit, we can repeat the preceding argument to conclude that $\langle a \rangle \subset \langle c_1 \rangle$. Either c_1 is irreducible or $c_1 = c_2 p_2$, where p_2 is irreducible and c_2 is not a unit. Continuing in this manner, we obtain another chain of ideals

$$\langle a \rangle \subset \langle c_1 \rangle \subset \langle c_2 \rangle \subset \cdots.$$

This chain must satisfy the ascending chain condition; therefore,

$$a = p_1 p_2 \cdots p_r$$

for irreducible elements $p_1, \ldots, p_r$.

Uniqueness of the factorization. To show uniqueness, let

$$a = p_1 p_2 \cdots p_r = q_1 q_2 \cdots q_s,$$

where each p_i and each q_i is irreducible. Without loss of generality, we can assume that $r < s$. Since p_1 divides $q_1 q_2 \cdots q_s$, by Corollary 18.13 it must divide some q_i. By rearranging the q_i's, we can assume that $p_1 \mid q_1$; hence, $q_1 = u_1 p_1$ for some unit u_1 in D. Therefore,

$$a = p_1 p_2 \cdots p_r = u_1 p_1 q_2 \cdots q_s$$

or

$$p_2 \cdots p_r = u_1 q_2 \cdots q_s.$$

Continuing in this manner, we can arrange the q_i's such that $p_2 = q_2, p_3 = q_3, \ldots, p_r = q_r$, to obtain

$$u_1 u_2 \cdots u_r q_{r+1} \cdots q_s = 1.$$

In this case $q_{r+1} \cdots q_s$ is a unit, which contradicts the fact that $q_{r+1}, \ldots, q_s$ are irreducibles. Therefore, $r = s$ and the factorization of a is unique. ∎

Corollary 18.16. Let F be a field. Then $F[x]$ is a UFD.

Example 18.17. Every PID is a UFD, but it is not the case that every UFD is a PID. In Corollary 18.31 on page 298, we will prove that $\mathbb{Z}[x]$ is a UFD. However, $\mathbb{Z}[x]$ is not a PID. Let $I = \{5f(x) + xg(x) : f(x), g(x) \in \mathbb{Z}[x]\}$. We can easily show that I is an ideal of $\mathbb{Z}[x]$. Suppose that $I = \langle p(x) \rangle$. Since $5 \in I, 5 = f(x)p(x)$. In this case $p(x) = p$ must be a constant. Since $x \in I, x = pg(x)$; consequently, $p = \pm 1$. However, it follows from this fact that $\langle p(x) \rangle = \mathbb{Z}[x]$. But this would mean that 3 is in I. Therefore, we can write $3 = 5f(x) + xg(x)$ for some $f(x)$ and $g(x)$ in $\mathbb{Z}[x]$. Examining the constant term of this polynomial, we see that $3 = 5f(x)$, which is impossible. □

Euclidean Domains

We have repeatedly used the division algorithm when proving results about either $\mathbb{Z}$ or $F[x]$, where F is a field. We should now ask when a division algorithm is available for an integral domain.

Let D be an integral domain such that there is a function $v : D \setminus \{0\} \to \mathbb{N}$ satisfying the following conditions.

1. If a and b are nonzero elements in D, then $v(a) \le v(ab)$.

2. Let $a, b \in D$ and suppose that $b \ne 0$. Then there exist elements $q, r \in D$ such that $a = bq + r$ and either $r = 0$ or $v(r) < v(b)$.

Then D is called a *Euclidean domain* and v is called a *Euclidean valuation*.

Example 18.18. Absolute value on $\mathbb{Z}$ is a Euclidean valuation. □

Example 18.19. Let F be a field. Then the degree of a polynomial in $F[x]$ is a Euclidean valuation. □

Example 18.20. Recall that the Gaussian integers in Example 16.12 on page 245 are defined by

$$\mathbb{Z}[i] = \{a + bi : a, b \in \mathbb{Z}\}.$$

We usually measure the size of a complex number $a + bi$ by its absolute value, $|a + bi| = \sqrt{a^2 + b^2}$; however, $\sqrt{a^2 + b^2}$ may not be an integer. For our valuation we will let $v(a + bi) = a^2 + b^2$ to ensure that we have an integer.

We claim that $v(a + bi) = a^2 + b^2$ is a Euclidean valuation on $\mathbb{Z}[i]$. Let $z, w \in \mathbb{Z}[i]$. Then $v(zw) = |zw|^2 = |z|^2|w|^2 = v(z)v(w)$. Since $v(z) \geq 1$ for every nonzero $z \in \mathbb{Z}[i]$, $v(z) \leq v(z)v(w)$.

Next, we must show that for any $z = a + bi$ and $w = c + di$ in $\mathbb{Z}[i]$ with $w \neq 0$, there exist elements q and r in $\mathbb{Z}[i]$ such that $z = qw + r$ with either $r = 0$ or $v(r) < v(w)$. We can view z and w as elements in $\mathbb{Q}(i) = \{p + qi : p, q \in \mathbb{Q}\}$, the field of fractions of $\mathbb{Z}[i]$. Observe that

$$
\begin{aligned}
zw^{-1} &= (a + bi)\frac{c - di}{c^2 + d^2} \\
&= \frac{ac + bd}{c^2 + d^2} + \frac{bc - ad}{c^2 + d^2}i \\
&= \left(m_1 + \frac{n_1}{c^2 + d^2}\right) + \left(m_2 + \frac{n_2}{c^2 + d^2}\right)i \\
&= (m_1 + m_2 i) + \left(\frac{n_1}{c^2 + d^2} + \frac{n_2}{c^2 + d^2}i\right) \\
&= (m_1 + m_2 i) + (s + ti)
\end{aligned}
$$

in $\mathbb{Q}(i)$. In the last steps we are writing the real and imaginary parts as an integer plus a proper fraction. That is, we take the closest integer m_i such that the fractional part satisfies $|n_i/(a^2 + b^2)| \leq 1/2$. For example, we write

$$
\begin{aligned}
\frac{9}{8} &= 1 + \frac{1}{8} \\
\frac{15}{8} &= 2 - \frac{1}{8}.
\end{aligned}
$$

Thus, s and t are the "fractional parts" of $zw^{-1} = (m_1 + m_2 i) + (s + ti)$. We also know that $s^2 + t^2 \leq 1/4 + 1/4 = 1/2$. Multiplying by w, we have

$$
z = zw^{-1}w = w(m_1 + m_2 i) + w(s + ti) = qw + r,
$$

where $q = m_1 + m_2 i$ and $r = w(s + ti)$. Since z and qw are in $\mathbb{Z}[i]$, r must be in $\mathbb{Z}[i]$. Finally, we need to show that either $r = 0$ or $v(r) < v(w)$. However,

$$
v(r) = v(w)v(s + ti) \leq \frac{1}{2}v(w) < v(w).
$$

$\square$

Theorem 18.21. Every Euclidean domain is a principal ideal domain.

Proof. Let D be a Euclidean domain and let v be a Euclidean valuation on D. Suppose I is a nontrivial ideal in D and choose a nonzero element $b \in I$ such that $v(b)$ is minimal for all $a \in I$. Since D is a Euclidean domain, there exist elements q and r in D such that $a = bq + r$ and either $r = 0$ or $v(r) < v(b)$. But $r = a - bq$ is in I since I is an ideal; therefore, $r = 0$ by the minimality of b. It follows that $a = bq$ and $I = \langle b \rangle$.   ∎

Corollary 18.22. Every Euclidean domain is a unique factorization domain.

Factorization in $D[x]$

One of the most important polynomial rings is $\mathbb{Z}[x]$. One of the first questions that come to mind about $\mathbb{Z}[x]$ is whether or not it is a UFD. We will prove a more general statement here. Our first task is to obtain a more general version of Gauss's Lemma (Theorem 17.14 on page 274).

Let D be a unique factorization domain and suppose that

$$p(x) = a_n x^n + \cdots + a_1 x + a_0$$

in $D[x]$. Then the *content* of $p(x)$ is the greatest common divisor of $a_0, \ldots, a_n$. We say that $p(x)$ is *primitive* if $\gcd(a_0, \ldots, a_n) = 1$.

Example 18.23. In $\mathbb{Z}[x]$ the polynomial $p(x) = 5x^4 - 3x^3 + x - 4$ is a primitive polynomial since the greatest common divisor of the coefficients is 1; however, the polynomial $q(x) = 4x^2 - 6x + 8$ is not primitive since the content of $q(x)$ is 2. □

Theorem 18.24. Gauss's Lemma. Let D be a UFD and let $f(x)$ and $g(x)$ be primitive polynomials in $D[x]$. Then $f(x)g(x)$ is primitive.

Proof. Let $f(x) = \sum_{i=0}^{m} a_i x^i$ and $g(x) = \sum_{i=0}^{n} b_i x^i$. Suppose that p is a prime dividing the coefficients of $f(x)g(x)$. Let r be the smallest integer such that $p \nmid a_r$ and s be the smallest integer such that $p \nmid b_s$. The coefficient of x^{r+s} in $f(x)g(x)$ is

$$c_{r+s} = a_0 b_{r+s} + a_1 b_{r+s-1} + \cdots + a_{r+s-1} b_1 + a_{r+s} b_0.$$

Since p divides $a_0, \ldots, a_{r-1}$ and $b_0, \ldots, b_{s-1}$, p divides every term of c_{r+s} except for the term $a_r b_s$. However, since $p \mid c_{r+s}$, either p divides a_r or p divides b_s. But this is impossible.   ∎

Lemma 18.25. Let D be a UFD, and let $p(x)$ and $q(x)$ be in $D[x]$. Then the content of $p(x)q(x)$ is equal to the product of the contents of $p(x)$ and $q(x)$.

Proof. Let $p(x) = cp_1(x)$ and $q(x) = dq_1(x)$, where c and d are the contents of $p(x)$ and $q(x)$, respectively. Then $p_1(x)$ and $q_1(x)$ are primitive. We can now write $p(x)q(x) = cdp_1(x)q_1(x)$. Since $p_1(x)q_1(x)$ is primitive, the content of $p(x)q(x)$ must be cd. ∎

Lemma 18.26. Let D be a UFD and F its field of fractions. Suppose that $p(x) \in D[x]$ and $p(x) = f(x)g(x)$, where $f(x)$ and $g(x)$ are in $F[x]$. Then $p(x) = f_1(x)g_1(x)$, where $f_1(x)$ and $g_1(x)$ are in $D[x]$. Further, $\deg f(x) = \deg f_1(x)$ and $\deg g(x) = \deg g_1(x)$.

Proof. Let a and b be nonzero elements of D such that $af(x), bg(x)$ are in $D[x]$. We can find $a_1, b_1 \in D$ such that $af(x) = a_1 f_1(x)$ and $bg(x) = b_1 g_1(x)$, where $f_1(x)$ and $g_1(x)$ are primitive polynomials in $D[x]$. Therefore, $abp(x) = (a_1 f_1(x))(b_1 g_1(x))$. Since $f_1(x)$ and $g_1(x)$ are primitive polynomials, it must be the case that $ab \mid a_1 b_1$ by Gauss's Lemma. Thus there exists a $c \in D$ such that $p(x) = cf_1(x)g_1(x)$. Clearly, $\deg f(x) = \deg f_1(x)$ and $\deg g(x) = \deg g_1(x)$. ∎

The following corollaries are direct consequences of Lemma 18.26.

Corollary 18.27. Let D be a UFD and F its field of fractions. A primitive polynomial $p(x)$ in $D[x]$ is irreducible in $F[x]$ if and only if it is irreducible in $D[x]$.

Corollary 18.28. Let D be a UFD and F its field of fractions. If $p(x)$ is a monic polynomial in $D[x]$ with $p(x) = f(x)g(x)$ in $F[x]$, then $p(x) = f_1(x)g_1(x)$, where $f_1(x)$ and $g_1(x)$ are in $D[x]$. Furthermore, $\deg f(x) = \deg f_1(x)$ and $\deg g(x) = \deg g_1(x)$.

Theorem 18.29. If D is a UFD, then $D[x]$ is a UFD.

Proof. Let $p(x)$ be a nonzero polynomial in $D[x]$. If $p(x)$ is a constant polynomial, then it must have a unique factorization since D is a UFD. Now suppose that $p(x)$ is a polynomial of positive degree in $D[x]$. Let F be the field of fractions of D, and let $p(x) = f_1(x)f_2(x)\cdots f_n(x)$ by a factorization of $p(x)$, where each $f_i(x)$ is irreducible. Choose $a_i \in D$ such that $a_i f_i(x)$ is in $D[x]$. There exist $b_1, \ldots, b_n \in D$ such that $a_i f_i(x) = b_i g_i(x)$, where $g_i(x)$ is a primitive polynomial in $D[x]$. By Corollary 18.27, each $g_i(x)$ is irreducible in $D[x]$. Consequently, we can write

$$a_1 \cdots a_n p(x) = b_1 \cdots b_n g_1(x) \cdots g_n(x).$$

Let $b = b_1 \cdots b_n$. Since $g_1(x) \cdots g_n(x)$ is primitive, $a_1 \cdots a_n$ divides b. Therefore, $p(x) = ag_1(x) \cdots g_n(x)$, where $a \in D$. Since D is a UFD, we can factor a as $uc_1 \cdots c_k$, where u is a unit and each of the c_i's is irreducible in D.

We will now show the uniqueness of this factorization. Let

$$p(x) = a_1 \cdots a_m f_1(x) \cdots f_n(x) = b_1 \cdots b_r g_1(x) \cdots g_s(x)$$

be two factorizations of $p(x)$, where all of the factors are irreducible in $D[x]$. By Corollary 18.27 on the preceding page, each of the f_i's and g_i's is irreducible in $F[x]$. The a_i's and the b_i's are units in F. Since $F[x]$ is a PID, it is a UFD; therefore, $n = s$. Now rearrange the $g_i(x)$'s so that $f_i(x)$ and $g_i(x)$ are associates for $i = 1, \ldots, n$. Then there exist $c_1, \ldots, c_n$ and $d_1, \ldots, d_n$ in D such that $(c_i/d_i)f_i(x) = g_i(x)$ or $c_i f_i(x) = d_i g_i(x)$. The polynomials $f_i(x)$ and $g_i(x)$ are primitive; hence, c_i and d_i are associates in D. Thus, $a_1 \cdots a_m = u b_1 \cdots b_r$ in D, where u is a unit in D. Since D is a unique factorization domain, $m = s$. Finally, we can reorder the b_i's so that a_i and b_i are associates for each i. This completes the uniqueness part of the proof.

∎

The theorem that we have just proven has several obvious but important corollaries.

Corollary 18.30. Let F be a field. Then $F[x]$ is a UFD.

Corollary 18.31. The ring of polynomials over the integers, $\mathbb{Z}[x]$, is a UFD.

Corollary 18.32. Let D be a UFD. Then $D[x_1, \ldots, x_n]$ is a UFD.

Remark 18.33. It is important to notice that every Euclidean domain is a PID and every PID is a UFD. However, as demonstrated by our examples, the converse of each of these statements fails. There are principal ideal domains that are not Euclidean domains, and there are unique factorization domains that are not principal ideal domains ($\mathbb{Z}[x]$).

Sage. Sage supports distinctions between "plain" rings, domains, principal ideal domains and fields. Support is often very good for constructions and computations with PID's, but sometimes problems get significantly harder (computationally) when a ring has less structure that that of a PID. So be aware when using Sage that some questions may go unanswered for rings with less structure.

⤴ Historical Note ⤶

Karl Friedrich Gauss, born in Brunswick, Germany on April 30, 1777, is considered to be one of the greatest mathematicians who ever lived. Gauss was truly a child prodigy. At the age of three he was able to detect errors in the books of his father's business. Gauss entered college at the age of 15. Before the age of 20, Gauss was able to construct a regular 17-sided polygon with a ruler and compass. This was the first new construction of a regular n-sided polygon since the time of the ancient Greeks. Gauss succeeded in showing that if $N = 2^{2^n} + 1$ was prime, then it was possible to construct a regular N-sided polygon.

Gauss obtained his Ph.D. in 1799 under the direction of Pfaff at the University of Helmstedt. In his dissertation he gave the first complete proof of the Fundamental Theorem of Algebra, which states that every polynomial with real coefficients can be factored into linear factors over the complex numbers. The acceptance of complex numbers was brought about by Gauss, who was the first person to use the notation of i for $\sqrt{-1}$.

Gauss then turned his attention toward number theory; in 1801, he published his famous book on number theory, *Disquisitiones Arithmeticae*. Throughout his life Gauss was intrigued with this branch of mathematics. He once wrote, "Mathematics is the queen of the sciences, and the theory of numbers is the queen of mathematics."

In 1807, Gauss was appointed director of the Observatory at the University of Göttingen, a position he held until his death. This position required him to study applications of mathematics to the sciences. He succeeded in making contributions to fields such as astronomy, mechanics, optics, geodesy, and magnetism. Along with Wilhelm Weber, he coinvented the first practical electric telegraph some years before a better version was invented by Samuel F. B. Morse.

Gauss was clearly the most prominent mathematician in the world in the early nineteenth century. His status naturally made his discoveries subject to intense scrutiny. Gauss's cold and distant personality many times led him to ignore the work of his contemporaries, making him many enemies. He did not enjoy teaching very much, and young mathematicians who sought him out for encouragement were often rebuffed. Nevertheless, he had many outstanding students, including Eisenstein, Riemann, Kummer, Dirichlet, and Dedekind. Gauss also offered a great deal of encouragement to Sophie Germain (1776–1831), who overcame the many obstacles facing women in her day to become a very prominent mathematician. Gauss died at the age of 78 in Göttingen on February 23, 1855.

18.3 Reading Questions

1. Integral domains are an abstraction of which two fundamental rings that we have already studied?

2. What are the various types of integral domains defined in this section?

3. The field of fractions of a ring abstracts what idea from basic mathematics?

4. In the previous chapter we had a theorem about irreducible polynomials generating maximal ideals. Which theorem in this chapter generalizes this previous result?

5. Describe an example which is a UFD, but not a PID.

18.4 Exercises

1. Let $z = a + b\sqrt{3}\,i$ be in $\mathbb{Z}[\sqrt{3}\,i]$. If $a^2 + 3b^2 = 1$, show that z must be a unit. Show that the only units of $\mathbb{Z}[\sqrt{3}\,i]$ are 1 and -1.

2. The Gaussian integers, $\mathbb{Z}[i]$, are a UFD. Factor each of the following elements in $\mathbb{Z}[i]$ into a product of irreducibles.

   (a) 5   (c) $6 + 8i$

   (b) $1 + 3i$ (d) 2

3. Let D be an integral domain.

   (a) Prove that F_D is an abelian group under the operation of addition.

   (b) Show that the operation of multiplication is well-defined in the field of fractions, F_D.

   (c) Verify the associative and commutative properties for multiplication in F_D.

4. Prove or disprove: Any subring of a field F containing 1 is an integral domain.

5. Prove or disprove: If D is an integral domain, then every prime element in D is also irreducible in D.

6. Let F be a field of characteristic zero. Prove that F contains a subfield isomorphic to $\mathbb{Q}$.

7. Let F be a field.

   (a) Prove that the field of fractions of $F[x]$, denoted by $F(x)$, is isomorphic to the set all rational expressions $p(x)/q(x)$, where $q(x)$ is not the zero polynomial.

   (b) Let $p(x_1, \ldots, x_n)$ and $q(x_1, \ldots, x_n)$ be polynomials in $F[x_1, \ldots, x_n]$. Show that the set of all rational expressions $p(x_1, \ldots, x_n)/q(x_1, \ldots, x_n)$ is isomorphic to the field of fractions of $F[x_1, \ldots, x_n]$. We denote the field of fractions of $F[x_1, \ldots, x_n]$ by $F(x_1, \ldots, x_n)$.

8. Let p be prime and denote the field of fractions of $\mathbb{Z}_p[x]$ by $\mathbb{Z}_p(x)$. Prove that $\mathbb{Z}_p(x)$ is an infinite field of characteristic p.

9. Prove that the field of fractions of the Gaussian integers, $\mathbb{Z}[i]$, is

$$\mathbb{Q}(i) = \{p + qi : p, q \in \mathbb{Q}\}.$$

10. A field F is called a *prime field* if it has no proper subfields. If E is a subfield of F and E is a prime field, then E is a *prime subfield* of F.

   (a) Prove that every field contains a unique prime subfield.

   (b) If F is a field of characteristic 0, prove that the prime subfield of F is isomorphic to the field of rational numbers, $\mathbb{Q}$.

   (c) If F is a field of characteristic p, prove that the prime subfield of F is isomorphic to $\mathbb{Z}_p$.

11. Let $\mathbb{Z}[\sqrt{2}] = \{a + b\sqrt{2} : a, b \in \mathbb{Z}\}$.

   (a) Prove that $\mathbb{Z}[\sqrt{2}]$ is an integral domain.

   (b) Find all of the units in $\mathbb{Z}[\sqrt{2}]$.

   (c) Determine the field of fractions of $\mathbb{Z}[\sqrt{2}]$.

   (d) Prove that $\mathbb{Z}[\sqrt{2}i]$ is a Euclidean domain under the Euclidean valuation $v(a + b\sqrt{2}i) = a^2 + 2b^2$.

12. Let D be a UFD. An element $d \in D$ is a *greatest common divisor of a and b in D* if $d \mid a$ and $d \mid b$ and d is divisible by any other element dividing both a and b.

   (a) If D is a PID and a and b are both nonzero elements of D, prove there exists a unique greatest common divisor of a and b up to associates. That is, if d and d' are both greatest common divisors of a and b, then d and d' are associates. We write $\gcd(a, b)$ for the greatest common divisor of a and b.

   (b) Let D be a PID and a and b be nonzero elements of D. Prove that there exist elements s and t in D such that $\gcd(a, b) = as + bt$.

13. Let D be an integral domain. Define a relation on D by $a \sim b$ if a and b are associates in D. Prove that $\sim$ is an equivalence relation on D.

14. Let D be a Euclidean domain with Euclidean valuation v. If u is a unit in D, show that $v(u) = v(1)$.

15. Let D be a Euclidean domain with Euclidean valuation v. If a and b are associates in D, prove that $v(a) = v(b)$.

16. Show that $\mathbb{Z}[\sqrt{5}i]$ is not a unique factorization domain.

17. Prove or disprove: Every subdomain of a UFD is also a UFD.

18. An ideal of a commutative ring R is said to be *finitely generated* if there exist elements $a_1, \ldots, a_n$ in R such that every element r in the ideal can be written as $a_1 r_1 + \cdots + a_n r_n$ for some $r_1, \ldots, r_n$ in R. Prove that R satisfies the ascending chain condition if and only if every ideal of R is finitely generated.

19. Let D be an integral domain with a descending chain of ideals $I_1 \supset I_2 \supset I_3 \supset \cdots$. Suppose that there exists an N such that $I_k = I_N$ for all $k \geq N$. A ring satisfying this condition is said to satisfy the *descending chain condition*, or **DCC**. Rings satisfying the DCC are called *Artinian rings*, after Emil Artin. Show that if D satisfies the descending chain condition, it must satisfy the ascending chain condition.

20. Let R be a commutative ring with identity. We define a *multiplicative subset* of R to be a subset S such that $1 \in S$ and $ab \in S$ if $a, b \in S$.

    (a) Define a relation $\sim$ on $R \times S$ by $(a, s) \sim (a', s')$ if there exists an $s^* \in S$ such that $s^*(s'a - sa') = 0$. Show that $\sim$ is an equivalence relation on $R \times S$.

    (b) Let a/s denote the equivalence class of $(a, s) \in R \times S$ and let $S^{-1}R$ be the set of all equivalence classes with respect to $\sim$. Define the operations of addition and multiplication on $S^{-1}R$ by

    $$\frac{a}{s} + \frac{b}{t} = \frac{at + bs}{st}$$
    $$\frac{a}{s}\frac{b}{t} = \frac{ab}{st},$$

    respectively. Prove that these operations are well-defined on $S^{-1}R$ and that $S^{-1}R$ is a ring with identity under these operations. The ring $S^{-1}R$ is called the *ring of quotients* of R with respect to S.

    (c) Show that the map $\psi : R \to S^{-1}R$ defined by $\psi(a) = a/1$ is a ring homomorphism.

    (d) If R has no zero divisors and $0 \notin S$, show that ψ is one-to-one.

    (e) Prove that P is a prime ideal of R if and only if $S = R \setminus P$ is a multiplicative subset of R.

    (f) If P is a prime ideal of R and $S = R \setminus P$, show that the ring of quotients $S^{-1}R$ has a unique maximal ideal. Any ring that has a unique maximal ideal is called a *local ring*.

18.5 References and Suggested Readings

[1] Atiyah, M. F. and MacDonald, I. G. *Introduction to Commutative Algebra.* Westview Press, Boulder, CO, 1994.

[2] Zariski, O. and Samuel, P. *Commutative Algebra*, vols. I and II. Springer, New York, 1975, 1960.

19

Lattices and Boolean Algebras

The axioms of a ring give structure to the operations of addition and multiplication on a set. However, we can construct algebraic structures, known
as lattices and Boolean algebras, that generalize other types of operations. For
example, the important operations on sets are inclusion, union, and intersection.
Lattices are generalizations of order relations on algebraic spaces, such as set inclu-
sion in set theory and inequality in the familiar number systems $\mathbb{N}$, $\mathbb{Z}$, $\mathbb{Q}$, and $\mathbb{R}$.
Boolean algebras generalize the operations of intersection and union. Lattices and
Boolean algebras have found applications in logic, circuit theory, and probability.

19.1 Lattices

Partially Ordered Sets

We begin the study of lattices and Boolean algebras by generalizing the idea of
inequality. Recall that a *relation* on a set X is a subset of $X \times X$. A relation P on
X is called a *partial order* of X if it satisfies the following axioms.

1. The relation is *reflexive*: $(a, a) \in P$ for all $a \in X$.

2. The relation is *antisymmetric*: if $(a, b) \in P$ and $(b, a) \in P$, then $a = b$.

3. The relation is *transitive*: if $(a, b) \in P$ and $(b, c) \in P$, then $(a, c) \in P$.

We will usually write $a \leq b$ to mean $(a, b) \in P$ unless some symbol is naturally
associated with a particular partial order, such as $a \leq b$ with integers a and b,
or $A \subset B$ with sets A and B. A set X together with a partial order $\leq$ is called a
partially ordered set, or *poset*.

Example 19.1. The set of integers (or rationals or reals) is a poset where $a \leq b$ has
the usual meaning for two integers a and b in $\mathbb{Z}$. □

Example 19.2. Let X be any set. We will define the *power set* of X to be the
set of all subsets of X. We denote the power set of X by $\mathcal{P}(X)$. For example, let
$X = \{a, b, c\}$. Then $\mathcal{P}(X)$ is the set of all subsets of the set $\{a, b, c\}$:

$$\emptyset \qquad \{a\} \qquad \{b\} \qquad \{c\}$$

$$\{a, b\} \qquad \{a, c\} \qquad \{b, c\} \qquad \{a, b, c\}.$$

On any power set of a set X, set inclusion, $\subset$, is a partial order. We can represent the order on $\{a, b, c\}$ schematically by a diagram such as the one in Figure 19.3.

□

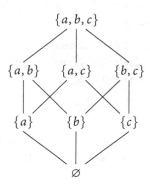

Figure 19.3. Partial order on $\mathcal{P}(\{a, b, c\})$

Example 19.4. Let G be a group. The set of subgroups of G is a poset, where the partial order is set inclusion. □

Example 19.5. There can be more than one partial order on a particular set. We can form a partial order on $\mathbb{N}$ by $a \preceq b$ if $a \mid b$. The relation is certainly reflexive since $a \mid a$ for all $a \in \mathbb{N}$. If $m \mid n$ and $n \mid m$, then $m = n$; hence, the relation is also antisymmetric. The relation is transitive, because if $m \mid n$ and $n \mid p$, then $m \mid p$. □

Example 19.6. Let $X = \{1, 2, 3, 4, 6, 8, 12, 24\}$ be the set of divisors of 24 with the partial order defined in Example 19.5. Figure 19.7 shows the partial order on X. □

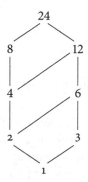

Figure 19.7. A partial order on the divisors of 24

Let Y be a subset of a poset X. An element u in X is an **upper bound** of Y if $a \leq u$ for every element $a \in Y$. If u is an upper bound of Y such that $u \leq v$ for every other upper bound v of Y, then u is called a **least upper bound** or **supremum** of Y. An element l in X is said to be a **lower bound** of Y if $l \leq a$ for all $a \in Y$. If l is a lower bound of Y such that $k \leq l$ for every other lower bound k of Y, then l is called a **greatest lower bound** or **infimum** of Y.

Example 19.8. Let $Y = \{2, 3, 4, 6\}$ be contained in the set X of Example 19.6 on the preceding page. Then Y has upper bounds 12 and 24, with 12 as a least upper bound. The only lower bound is 1; hence, it must be a greatest lower bound.   □

As it turns out, least upper bounds and greatest lower bounds are unique if they exist.

Theorem 19.9. Let Y be a nonempty subset of a poset X. If Y has a least upper bound, then Y has a unique least upper bound. If Y has a greatest lower bound, then Y has a unique greatest lower bound.

Proof. Let u_1 and u_2 be least upper bounds for Y. By the definition of the least upper bound, $u_1 \leq u$ for all upper bounds u of Y. In particular, $u_1 \leq u_2$. Similarly, $u_2 \leq u_1$. Therefore, $u_1 = u_2$ by antisymmetry. A similar argument show that the greatest lower bound is unique.   ∎

On many posets it is possible to define binary operations by using the greatest lower bound and the least upper bound of two elements. A **lattice** is a poset L such that every pair of elements in L has a least upper bound and a greatest lower bound. The least upper bound of $a, b \in L$ is called the **join** of a and b and is denoted by $a \vee b$. The greatest lower bound of $a, b \in L$ is called the **meet** of a and b and is denoted by $a \wedge b$.

Example 19.10. Let X be a set. Then the power set of X, $\mathcal{P}(X)$, is a lattice. For two sets A and B in $\mathcal{P}(X)$, the least upper bound of A and B is $A \cup B$. Certainly $A \cup B$ is an upper bound of A and B, since $A \subset A \cup B$ and $B \subset A \cup B$. If C is some other set containing both A and B, then C must contain $A \cup B$; hence, $A \cup B$ is the least upper bound of A and B. Similarly, the greatest lower bound of A and B is $A \cap B$.   □

Example 19.11. Let G be a group and suppose that X is the set of subgroups of G. Then X is a poset ordered by set-theoretic inclusion, $\subset$. The set of subgroups of G is also a lattice. If H and K are subgroups of G, the greatest lower bound of H and K is $H \cap K$. The set $H \cup K$ may not be a subgroup of G. We leave it as an exercise to show that the least upper bound of H and K is the subgroup generated by $H \cup K$.   □

In set theory we have certain duality conditions. For example, by De Morgan's laws, any statement about sets that is true about $(A \cup B)'$ must also be true about $A' \cap B'$. We also have a duality principle for lattices.

Axiom 19.12. Principle of Duality. Any statement that is true for all lattices remains true when $\leq$ is replaced by $\geq$ and $\vee$ and $\wedge$ are interchanged throughout the statement.

The following theorem tells us that a lattice is an algebraic structure with two binary operations that satisfy certain axioms.

Theorem 19.13. If L is a lattice, then the binary operations $\vee$ and $\wedge$ satisfy the following properties for $a, b, c \in L$.

1. Commutative laws: $a \vee b = b \vee a$ and $a \wedge b = b \wedge a$.

2. Associative laws: $a \vee (b \vee c) = (a \vee b) \vee c$ and $a \wedge (b \wedge c) = (a \wedge b) \wedge c$.

3. Idempotent laws: $a \vee a = a$ and $a \wedge a = a$.

4. Absorption laws: $a \vee (a \wedge b) = a$ and $a \wedge (a \vee b) = a$.

Proof. By the Principle of Duality, we need only prove the first statement in each part.

(1) By definition $a \vee b$ is the least upper bound of $\{a, b\}$, and $b \vee a$ is the least upper bound of $\{b, a\}$; however, $\{a, b\} = \{b, a\}$.

(2) We will show that $a \vee (b \vee c)$ and $(a \vee b) \vee c$ are both least upper bounds of $\{a, b, c\}$. Let $d = a \vee b$. Then $c \leq d \vee c = (a \vee b) \vee c$. We also know that

$$a \leq a \vee b = d \leq d \vee c = (a \vee b) \vee c.$$

A similar argument demonstrates that $b \leq (a \vee b) \vee c$. Therefore, $(a \vee b) \vee c$ is an upper bound of $\{a, b, c\}$. We now need to show that $(a \vee b) \vee c$ is the least upper bound of $\{a, b, c\}$. Let u be some other upper bound of $\{a, b, c\}$. Then $a \leq u$ and $b \leq u$; hence, $d = a \vee b \leq u$. Since $c \leq u$, it follows that $(a \vee b) \vee c = d \vee c \leq u$. Therefore, $(a \vee b) \vee c$ must be the least upper bound of $\{a, b, c\}$. The argument that shows $a \vee (b \vee c)$ is the least upper bound of $\{a, b, c\}$ is the same. Consequently, $a \vee (b \vee c) = (a \vee b) \vee c$.

(3) The join of a and a is the least upper bound of $\{a\}$; hence, $a \vee a = a$.

(4) Let $d = a \wedge b$. Then $a \leq a \vee d$. On the other hand, $d = a \wedge b \leq a$, and so $a \vee d \leq a$. Therefore, $a \vee (a \wedge b) = a$. ∎

Given any arbitrary set L with operations $\vee$ and $\wedge$, satisfying the conditions of the previous theorem, it is natural to ask whether or not this set comes from some lattice. The following theorem says that this is always the case.

Theorem 19.14. Let L be a nonempty set with two binary operations $\vee$ and $\wedge$ satisfying the commutative, associative, idempotent, and absorption laws. We can define a partial order on L by $a \leq b$ if $a \vee b = b$. Furthermore, L is a lattice with respect to $\leq$ if for all $a, b \in L$, we define the least upper bound and greatest lower bound of a and b by $a \vee b$ and $a \wedge b$, respectively.

Proof. We first show that L is a poset under $\leq$. Since $a \vee a = a$, $a \leq a$ and $\leq$ is reflexive. To show that $\leq$ is antisymmetric, let $a \leq b$ and $b \leq a$. Then $a \vee b = b$ and $b \vee a = a$. By the commutative law, $b = a \vee b = b \vee a = a$. Finally, we must show that $\leq$ is transitive. Let $a \leq b$ and $b \leq c$. Then $a \vee b = b$ and $b \vee c = c$. Thus,

$$a \vee c = a \vee (b \vee c) = (a \vee b) \vee c = b \vee c = c,$$

or $a \leq c$.

To show that L is a lattice, we must prove that $a \vee b$ and $a \wedge b$ are, respectively, the least upper and greatest lower bounds of a and b. Since $a = (a \vee b) \wedge a = a \wedge (a \vee b)$, it follows that $a \leq a \vee b$. Similarly, $b \leq a \vee b$. Therefore, $a \vee b$ is an upper bound for a and b. Let u be any other upper bound of both a and b. Then $a \leq u$ and $b \leq u$. But $a \vee b \leq u$ since

$$(a \vee b) \vee u = a \vee (b \vee u) = a \vee u = u.$$

The proof that $a \wedge b$ is the greatest lower bound of a and b is left as an exercise. ∎

19.2 Boolean Algebras

Let us investigate the example of the power set, $\mathcal{P}(X)$, of a set X more closely. The power set is a lattice that is ordered by inclusion. By the definition of the power set, the largest element in $\mathcal{P}(X)$ is X itself and the smallest element is $\varnothing$, the empty set. For any set A in $\mathcal{P}(X)$, we know that $A \cap X = A$ and $A \cup \varnothing = A$. This suggests the following definition for lattices. An element I in a poset X is a *largest element* if $a \leq I$ for all $a \in X$. An element O is a *smallest element* of X if $O \leq a$ for all $a \in X$.

Let A be in $\mathcal{P}(X)$. Recall that the complement of A is

$$A' = X \smallsetminus A = \{x : x \in X \text{ and } x \notin A\}.$$

We know that $A \cup A' = X$ and $A \cap A' = \varnothing$. We can generalize this example for lattices. A lattice L with a largest element I and a smallest element O is *complemented* if for each $a \in L$, there exists an a' such that $a \vee a' = I$ and $a \wedge a' = O$.

In a lattice L, the binary operations $\vee$ and $\wedge$ satisfy commutative and associative laws; however, they need not satisfy the distributive law

$$a \wedge (b \vee c) = (a \wedge b) \vee (a \wedge c);$$

however, in $\mathcal{P}(X)$ the distributive law is satisfied since

$$A \cap (B \cup C) = (A \cap B) \cup (A \cap C)$$

for $A, B, C \in \mathcal{P}(X)$. We will say that a lattice L is **distributive** if the following distributive law holds:

$$a \wedge (b \vee c) = (a \wedge b) \vee (a \wedge c)$$

for all $a, b, c \in L$.

Theorem 19.15. A lattice L is distributive if and only if

$$a \vee (b \wedge c) = (a \vee b) \wedge (a \vee c)$$

for all $a, b, c \in L$.

Proof. Let us assume that L is a distributive lattice.

$$
\begin{aligned}
a \vee (b \wedge c) &= [a \vee (a \wedge c)] \vee (b \wedge c) \\
&= a \vee [(a \wedge c) \vee (b \wedge c)] \\
&= a \vee [(c \wedge a) \vee (c \wedge b)] \\
&= a \vee [c \wedge (a \vee b)] \\
&= a \vee [(a \vee b) \wedge c] \\
&= [(a \vee b) \wedge a] \vee [(a \vee b) \wedge c] \\
&= (a \vee b) \wedge (a \vee c).
\end{aligned}
$$

The converse follows directly from the Duality Principle. ∎

A **Boolean algebra** is a lattice B with a greatest element I and a smallest element O such that B is both distributive and complemented. The power set of X, $\mathcal{P}(X)$, is our prototype for a Boolean algebra. As it turns out, it is also one of the most important Boolean algebras. The following theorem allows us to characterize Boolean algebras in terms of the binary relations $\vee$ and $\wedge$ without mention of the fact that a Boolean algebra is a poset.

Theorem 19.16. A set B is a Boolean algebra if and only if there exist binary operations $\vee$ and $\wedge$ on B satisfying the following axioms.

1. $a \vee b = b \vee a$ and $a \wedge b = b \wedge a$ for $a, b \in B$.

2. $a \vee (b \vee c) = (a \vee b) \vee c$ and $a \wedge (b \wedge c) = (a \wedge b) \wedge c$ for $a, b, c \in B$.

3. $a \wedge (b \vee c) = (a \wedge b) \vee (a \wedge c)$ and $a \vee (b \wedge c) = (a \vee b) \wedge (a \vee c)$ for $a, b, c \in B$.

4. There exist elements I and O such that $a \vee O = a$ and $a \wedge I = a$ for all $a \in B$.

5. For every $a \in B$ there exists an $a' \in B$ such that $a \vee a' = I$ and $a \wedge a' = O$.

Proof. Let B be a set satisfying (1)–(5) in the theorem. One of the idempotent laws is satisfied since

$$a = a \vee O$$
$$= a \vee (a \wedge a')$$
$$= (a \vee a) \wedge (a \vee a')$$
$$= (a \vee a) \wedge I$$
$$= a \vee a.$$

Observe that

$$I \vee b = (b \vee b') \vee b = (b' \vee b) \vee b = b' \vee (b \vee b) = b' \vee b = I.$$

Consequently, the first of the two absorption laws holds, since

$$a \vee (a \wedge b) = (a \wedge I) \vee (a \wedge b)$$
$$= a \wedge (I \vee b)$$
$$= a \wedge I$$
$$= a.$$

The other idempotent and absorption laws are proven similarly. Since B also satisfies (1)–(3), the conditions of Theorem 19.14 on page 308 are met; therefore, B must be a lattice. Condition (4) tells us that B is a distributive lattice.

For $a \in B$, $O \vee a = a$; hence, $O \leq a$ and O is the smallest element in B. To show that I is the largest element in B, we will first show that $a \vee b = b$ is equivalent to $a \wedge b = a$. Since $a \vee I = a$ for all $a \in B$, using the absorption laws we can determine that

$$a \vee I = (a \wedge I) \vee I = I \vee (I \wedge a) = I$$

or $a \leq I$ for all a in B. Finally, since we know that B is complemented by (5), B must be a Boolean algebra.

Conversely, suppose that B is a Boolean algebra. Let I and O be the greatest and least elements in B, respectively. If we define $a \vee b$ and $a \wedge b$ as least upper and greatest lower bounds of $\{a, b\}$, then B is a Boolean algebra by Theorem 19.14 on page 308, Theorem 19.15 on the previous page, and our hypothesis. ∎

Many other identities hold in Boolean algebras. Some of these identities are listed in the following theorem.

Theorem 19.17. Let B be a Boolean algebra. Then

1. $a \vee I = I$ and $a \wedge O = O$ for all $a \in B$.

2. If $a \vee b = a \vee c$ and $a \wedge b = a \wedge c$ for $a, b, c \in B$, then $b = c$.

3. If $a \vee b = I$ and $a \wedge b = O$, then $b = a'$.

4. $(a')' = a$ for all $a \in B$.

5. $I' = O$ and $O' = I$.

6. $(a \vee b)' = a' \wedge b'$ and $(a \wedge b)' = a' \vee b'$ (De Morgan's Laws).

Proof. We will prove only (2). The rest of the identities are left as exercises. For $a \vee b = a \vee c$ and $a \wedge b = a \wedge c$, we have

$$
\begin{aligned}
b &= b \vee (b \wedge a) \\
&= b \vee (a \wedge b) \\
&= b \vee (a \wedge c) \\
&= (b \vee a) \wedge (b \vee c) \\
&= (a \vee b) \wedge (b \vee c) \\
&= (a \vee c) \wedge (b \vee c) \\
&= (c \vee a) \wedge (c \vee b) \\
&= c \vee (a \wedge b) \\
&= c \vee (a \wedge c) \\
&= c \vee (c \wedge a) \\
&= c.
\end{aligned}
$$

■

Finite Boolean Algebras

A Boolean algebra is a *finite Boolean algebra* if it contains a finite number of elements as a set. Finite Boolean algebras are particularly nice since we can classify them up to isomorphism.

Let B and C be Boolean algebras. A bijective map $\phi : B \to C$ is an *isomorphism* of Boolean algebras if

$$
\begin{aligned}
\phi(a \vee b) &= \phi(a) \vee \phi(b) \\
\phi(a \wedge b) &= \phi(a) \wedge \phi(b)
\end{aligned}
$$

for all a and b in B.

We will show that any finite Boolean algebra is isomorphic to the Boolean algebra obtained by taking the power set of some finite set X. We will need a few lemmas and definitions before we prove this result. Let B be a finite Boolean algebra. An element $a \in B$ is an **atom** of B if $a \neq O$ and $a \wedge b = a$ for all $b \in B$ with $b \neq O$. Equivalently, a is an atom of B if there is no $b \in B$ with $b \neq O$ distinct from a such that $O \leq b \leq a$.

Lemma 19.18. Let B be a finite Boolean algebra. If b is a element of B with $b \neq O$, then there is an atom a in B such that $a \leq b$.

Proof. If b is an atom, let $a = b$. Otherwise, choose an element b_1, not equal to O or b, such that $b_1 \leq b$. We are guaranteed that this is possible since b is not an atom. If b_1 is an atom, then we are done. If not, choose b_2, not equal to O or b_1, such that $b_2 \leq b_1$. Again, if b_2 is an atom, let $a = b_2$. Continuing this process, we can obtain a chain

$$O \leq \cdots \leq b_3 \leq b_2 \leq b_1 \leq b.$$

Since B is a finite Boolean algebra, this chain must be finite. That is, for some k, b_k is an atom. Let $a = b_k$. ∎

Lemma 19.19. Let a and b be atoms in a finite Boolean algebra B such that $a \neq b$. Then $a \wedge b = O$.

Proof. Since $a \wedge b$ is the greatest lower bound of a and b, we know that $a \wedge b \leq a$. Hence, either $a \wedge b = a$ or $a \wedge b = O$. However, if $a \wedge b = a$, then either $a \leq b$ or $a = O$. In either case we have a contradiction because a and b are both atoms; therefore, $a \wedge b = O$. ∎

Lemma 19.20. Let B be a Boolean algebra and $a, b \in B$. The following statements are equivalent.

1. $a \leq b$.

2. $a \wedge b' = O$.

3. $a' \vee b = I$.

Proof. (1) $\Rightarrow$ (2). If $a \leq b$, then $a \vee b = b$. Therefore,

$$\begin{aligned}
a \wedge b' &= a \wedge (a \vee b)' \\
&= a \wedge (a' \wedge b') \\
&= (a \wedge a') \wedge b' \\
&= O \wedge b' \\
&= O.
\end{aligned}$$

(2) $\Rightarrow$ (3). If $a \wedge b' = O$, then $a' \vee b = (a \wedge b')' = O' = I$.

$(3) \Rightarrow (1)$. If $a' \vee b = I$, then

$$a = a \wedge (a' \vee b)$$
$$= (a \wedge a') \vee (a \wedge b)$$
$$= O \vee (a \wedge b)$$
$$= a \wedge b.$$

Thus, $a \leq b$.   ∎

Lemma 19.21. Let B be a Boolean algebra and b and c be elements in B such that $b \not\leq c$. Then there exists an atom $a \in B$ such that $a \leq b$ and $a \not\leq c$.

Proof. By Lemma 19.20, $b \wedge c' \neq O$. Hence, there exists an atom a such that $a \leq b \wedge c'$. Consequently, $a \leq b$ and $a \not\leq c$.   ∎

Lemma 19.22. Let $b \in B$ and $a_1, \ldots, a_n$ be the atoms of B such that $a_i \leq b$. Then $b = a_1 \vee \cdots \vee a_n$. Furthermore, if $a, a_1, \ldots, a_n$ are atoms of B such that $a \leq b$, $a_i \leq b$, and $b = a \vee a_1 \vee \cdots \vee a_n$, then $a = a_i$ for some $i = 1, \ldots, n$.

Proof. Let $b_1 = a_1 \vee \cdots \vee a_n$. Since $a_i \leq b$ for each i, we know that $b_1 \leq b$. If we can show that $b \leq b_1$, then the lemma is true by antisymmetry. Assume $b \not\leq b_1$. Then there exists an atom a such that $a \leq b$ and $a \not\leq b_1$. Since a is an atom and $a \leq b$, we can deduce that $a = a_i$ for some a_i. However, this is impossible since $a \leq b_1$. Therefore, $b \leq b_1$.

Now suppose that $b = a_1 \vee \cdots \vee a_n$. If a is an atom less than b,

$$a = a \wedge b = a \wedge (a_1 \vee \cdots \vee a_n) = (a \wedge a_1) \vee \cdots \vee (a \wedge a_n).$$

But each term is O or a with $a \wedge a_i$ occurring for only one a_i. Hence, by Lemma 19.19, $a = a_i$ for some i.   ∎

Theorem 19.23. Let B be a finite Boolean algebra. Then there exists a set X such that B is isomorphic to $\mathcal{P}(X)$.

Proof. We will show that B is isomorphic to $\mathcal{P}(X)$, where X is the set of atoms of B. Let $a \in B$. By Lemma 19.22, we can write a uniquely as $a = a_1 \vee \cdots \vee a_n$ for $a_1, \ldots, a_n \in X$. Consequently, we can define a map $\phi : B \to \mathcal{P}(X)$ by

$$\phi(a) = \phi(a_1 \vee \cdots \vee a_n) = \{a_1, \ldots, a_n\}.$$

Clearly, ϕ is onto.

Now let $a = a_1 \vee \cdots \vee a_n$ and $b = b_1 \vee \cdots \vee b_m$ be elements in B, where each a_i and each b_i is an atom. If $\phi(a) = \phi(b)$, then $\{a_1, \ldots, a_n\} = \{b_1, \ldots, b_m\}$ and $a = b$. Consequently, ϕ is injective.

The join of a and b is preserved by ϕ since

$$
\begin{aligned}
\phi(a \vee b) &= \phi(a_1 \vee \cdots \vee a_n \vee b_1 \vee \cdots \vee b_m) \\
&= \{a_1, \ldots, a_n, b_1, \ldots, b_m\} \\
&= \{a_1, \ldots, a_n\} \cup \{b_1, \ldots, b_m\} \\
&= \phi(a_1 \vee \cdots \vee a_n) \cup \phi(b_1 \wedge \cdots \vee b_m) \\
&= \phi(a) \cup \phi(b).
\end{aligned}
$$

Similarly, $\phi(a \wedge b) = \phi(a) \cap \phi(b)$. ∎

We leave the proof of the following corollary as an exercise.

Corollary 19.24. The order of any finite Boolean algebra must be 2^n for some positive integer n.

19.3 The Algebra of Electrical Circuits

The usefulness of Boolean algebras has become increasingly apparent over the past several decades with the development of the modern computer. The circuit design of computer chips can be expressed in terms of Boolean algebras. In this section we will develop the Boolean algebra of electrical circuits and switches; however, these results can easily be generalized to the design of integrated computer circuitry.

A *switch* is a device, located at some point in an electrical circuit, that controls the flow of current through the circuit. Each switch has two possible states: it can be *open*, and not allow the passage of current through the circuit, or a it can be *closed*, and allow the passage of current. These states are mutually exclusive. We require that every switch be in one state or the other—a switch cannot be open and closed at the same time. Also, if one switch is always in the same state as another, we will denote both by the same letter; that is, two switches that are both labeled with the same letter a will always be open at the same time and closed at the same time.

Given two switches, we can construct two fundamental types of circuits. Two switches a and b are in *series* if they make up a circuit of the type that is illustrated in Figure 19.25. Current can pass between the terminals A and B in a series circuit only if both of the switches a and b are closed. We will denote this combination of switches by $a \wedge b$. Two switches a and b are in *parallel* if they form a circuit of the type that appears in Figure 19.26. In the case of a parallel circuit, current can pass between A and B if either one of the switches is closed. We denote a parallel combination of circuits a and b by $a \vee b$.

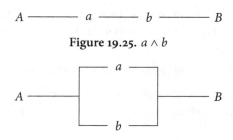

Figure 19.25. $a \wedge b$

Figure 19.26. $a \vee b$

We can build more complicated electrical circuits out of series and parallel circuits by replacing any switch in the circuit with one of these two fundamental types of circuits. Circuits constructed in this manner are called ***series-parallel circuits.***

We will consider two circuits equivalent if they act the same. That is, if we set the switches in equivalent circuits exactly the same we will obtain the same result. For example, in a series circuit $a \wedge b$ is exactly the same as $b \wedge a$. Notice that this is exactly the commutative law for Boolean algebras. In fact, the set of all series-parallel circuits forms a Boolean algebra under the operations of $\vee$ and $\wedge$. We can use diagrams to verify the different axioms of a Boolean algebra. The distributive law, $a \wedge (b \vee c) = (a \wedge b) \vee (a \wedge c)$, is illustrated in Figure 19.27. If a is a switch, then a' is the switch that is always open when a is closed and always closed when a is open. A circuit that is always closed is I in our algebra; a circuit that is always open is O. The laws for $a \wedge a' = O$ and $a \vee a' = I$ are shown in Figure 19.28.

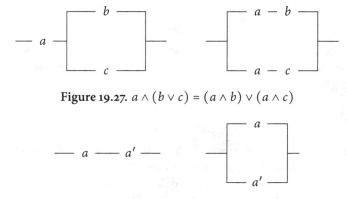

Figure 19.27. $a \wedge (b \vee c) = (a \wedge b) \vee (a \wedge c)$

Figure 19.28. $a \wedge a' = O$ and $a \vee a' = I$

Example 19.29. Every Boolean expression represents a switching circuit. For example, given the expression $(a \vee b) \wedge (a \vee b') \wedge (a \vee b)$, we can construct the circuit in Figure 19.32.   □

Theorem 19.30. The set of all circuits is a Boolean algebra.

We leave as an exercise the proof of this theorem for the Boolean algebra axioms not yet verified. We can now apply the techniques of Boolean algebras to switching theory.

Example 19.31. Given a complex circuit, we can now apply the techniques of Boolean algebra to reduce it to a simpler one. Consider the circuit in Figure 19.32. Since

$$(a \vee b) \wedge (a \vee b') \wedge (a \vee b) = (a \vee b) \wedge (a \vee b) \wedge (a \vee b')$$
$$= (a \vee b) \wedge (a \vee b')$$
$$= a \vee (b \wedge b')$$
$$= a \vee O$$
$$= a,$$

we can replace the more complicated circuit with a circuit containing the single switch a and achieve the same function.   □

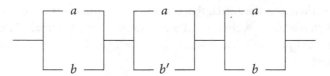

Figure 19.32. $(a \vee b) \wedge (a \vee b') \wedge (a \vee b)$

Sage.   Sage has a full suite of functionality for both posets and lattices, all as part of its excellent support for combinatorics. There is little in this chapter that cannot be investigated with Sage.

⤳ Historical Note ⤳

George Boole (1815–1864) was the first person to study lattices. In 1847, he published *The Investigation of the Laws of Thought*, a book in which he used lattices to formalize logic and the calculus of propositions. Boole believed that mathematics was the study of form rather than of content; that is, he was not so much concerned with what he was calculating as

with how he was calculating it. Boole's work was carried on by his friend
Augustus De Morgan (1806–1871). De Morgan observed that the principle
of duality often held in set theory, as is illustrated by De Morgan's laws for
set theory. He believed, as did Boole, that mathematics was the study of
symbols and abstract operations.

Set theory and logic were further advanced by such mathematicians
as Alfred North Whitehead (1861–1947), Bertrand Russell (1872–1970),
and David Hilbert (1862–1943). In *Principia Mathematica*, Whitehead
and Russell attempted to show the connection between mathematics and
logic by the deduction of the natural number system from the rules of
formal logic. If the natural numbers could be determined from logic itself,
then so could much of the rest of existing mathematics. Hilbert attempted
to build up mathematics by using symbolic logic in a way that would
prove the consistency of mathematics. His approach was dealt a mortal
blow by Kurt Gödel (1906–1978), who proved that there will always be
"undecidable" problems in any sufficiently rich axiomatic system; that is,
that in any mathematical system of any consequence, there will always be
statements that can never be proven either true or false.

As often occurs, this basic research in pure mathematics later became
indispensable in a wide variety of applications. Boolean algebras and
logic have become essential in the design of the large-scale integrated
circuitry found on today's computer chips. Sociologists have used lattices
and Boolean algebras to model social hierarchies; biologists have used
them to describe biosystems.

19.4 Reading Questions

1. Describe succinctly what a poset is. Do not just list the defining properties,
    but give a description that another student of algebra who has never seen a
    poset might understand. For example, part of your answer might include what
    type of common algebraic topics a poset generalizes, and your answer should
    be short on symbols.

2. How does a lattice differ from a poset? Answer this in the spirit of the previous
    question.

3. How does a Boolean algebra differ from a lattice? Again, answer this in the
    spirit of the previous two questions.

4. Give two (perhaps related) reasons why any discussion of finite Boolean alge-
    bras might center on the example of the power set of a finite set.

5. Describe a major innovation of the middle twentieth century made possible by Boolean algebra.

19.5 Exercises

1. Draw the lattice diagram for the power set of $X = \{a, b, c, d\}$ with the set inclusion relation, $\subseteq$.

2. Draw the diagram for the set of positive integers that are divisors of 30. Is this poset a Boolean algebra?

3. Draw a diagram of the lattice of subgroups of $\mathbb{Z}_{12}$.

4. Let B be the set of positive integers that are divisors of 210. Define an order on B by $a \leq b$ if $a \mid b$. Prove that B is a Boolean algebra. Find a set X such that B is isomorphic to $\mathcal{P}(X)$.

5. Prove or disprove: $\mathbb{Z}$ is a poset under the relation $a \leq b$ if $a \mid b$.

6. Draw the switching circuit for each of the following Boolean expressions.
    (a) $(a \vee b \vee a') \wedge a$ (c) $a \vee (a \wedge b)$
    (b) $(a \vee b)' \wedge (a \vee b)$ (d) $(c \vee a \vee b) \wedge c' \wedge (a \vee b)'$

7. Draw a circuit that will be closed exactly when only one of three switches a, b, and c are closed.

8. Prove or disprove that the two circuits shown are equivalent.

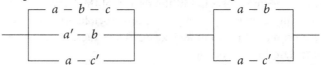

9. Let X be a finite set containing n elements. Prove that $|\mathcal{P}(X)| = 2^n$. Conclude that the order of any finite Boolean algebra must be 2^n for some $n \in \mathbb{N}$.

10. For each of the following circuits, write a Boolean expression. If the circuit can be replaced by one with fewer switches, give the Boolean expression and draw a diagram for the new circuit.

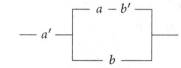

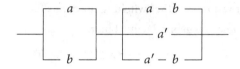

11. Prove or disprove: The set of all nonzero integers is a lattice, where $a \leq b$ is defined by $a \mid b$.

12. Let L be a nonempty set with two binary operations $\vee$ and $\wedge$ satisfying the commutative, associative, idempotent, and absorption laws. We can define a partial order on L, as in Theorem 19.14 on page 308, by $a \leq b$ if $a \vee b = b$. Prove that the greatest lower bound of a and b is $a \wedge b$.

13. Let G be a group and X be the set of subgroups of G ordered by set-theoretic inclusion. If H and K are subgroups of G, show that the least upper bound of H and K is the subgroup generated by $H \cup K$.

14. Let R be a ring and suppose that X is the set of ideals of R. Show that X is a poset ordered by set-theoretic inclusion, $\subset$. Define the meet of two ideals I and J in X by $I \cap J$ and the join of I and J by $I + J$. Prove that the set of ideals of R is a lattice under these operations.

15. Let B be a Boolean algebra. Prove each of the following identities.

    (a) $a \vee I = I$ and $a \wedge O = O$ for all $a \in B$.

    (b) If $a \vee b = I$ and $a \wedge b = O$, then $b = a'$.

    (c) $(a')' = a$ for all $a \in B$.

    (d) $I' = O$ and $O' = I$.

    (e) $(a \vee b)' = a' \wedge b'$ and $(a \wedge b)' = a' \vee b'$ (De Morgan's laws).

16. By drawing the appropriate diagrams, complete the proof of Theorem 19.30 on page 316 to show that the switching functions form a Boolean algebra.

17. Let B be a Boolean algebra. Define binary operations $+$ and $\cdot$ on B by

$$a + b = (a \wedge b') \vee (a' \wedge b)$$
$$a \cdot b = a \wedge b.$$

Prove that B is a commutative ring under these operations satisfying $a^2 = a$ for all $a \in B$.

18. Let X be a poset such that for every a and b in X, either $a \leq b$ or $b \leq a$. Then X is said to be a **totally ordered set**.

    (a) Is $a \mid b$ a total order on $\mathbb{N}$?

    (b) Prove that $\mathbb{N}$, $\mathbb{Z}$, $\mathbb{Q}$, and $\mathbb{R}$ are totally ordered sets under the usual ordering $\leq$.

19. Let X and Y be posets. A map $\phi : X \to Y$ is **order-preserving** if $a \leq b$ implies that $\phi(a) \leq \phi(b)$. Let L and M be lattices. A map $\psi : L \to M$ is a **lattice homomorphism** if $\psi(a \vee b) = \psi(a) \vee \psi(b)$ and $\psi(a \wedge b) = \psi(a) \wedge \psi(b)$. Show that every lattice homomorphism is order-preserving, but that it is not the case that every order-preserving homomorphism is a lattice homomorphism.

20. Let B be a Boolean algebra. Prove that $a = b$ if and only if $(a \wedge b') \vee (a' \wedge b) = O$ for $a, b \in B$.

21. Let B be a Boolean algebra. Prove that $a = O$ if and only if $(a \wedge b') \vee (a' \wedge b) = b$ for all $b \in B$.

22. Let L and M be lattices. Define an order relation on $L \times M$ by $(a, b) \leq (c, d)$ if $a \leq c$ and $b \leq d$. Show that $L \times M$ is a lattice under this partial order.

19.6 Programming Exercises

1. A *Boolean* or *switching function on n variables* is a map $f : \{O, I\}^n \to \{0, I\}$. A Boolean polynomial is a special type of Boolean function: it is any type of Boolean expression formed from a finite combination of variables $x_1, \ldots, x_n$ together with O and I, using the operations $\vee$, $\wedge$, and $'$. The values of the functions are defined in Table 19.33. Write a program to evaluate Boolean polynomials.

Table 19.33. Boolean polynomials

x	y	x'	$x \vee y$	$x \wedge y$
0	0	1	0	0
0	1	1	1	0
1	0	0	1	0
1	1	0	1	1

19.7 References and Suggested Readings

[1] Donnellan, T. *Lattice Theory* . Pergamon Press, Oxford, 1968.

[2] Halmos, P. R. "The Basic Concepts of Algebraic Logic," *American Mathematical Monthly* 53 (1956), 363–87.

[3] Hohn, F. "Some Mathematical Aspects of Switching," *American Mathematical Monthly* 62 (1955), 75–90.

[4] Hohn, F. *Applied Boolean Algebra*. 2nd ed. Macmillan, New York, 1966.

[5] Lidl, R. and Pilz, G. *Applied Abstract Algebra*. 2nd ed. Springer, New York, 1998.

[6] Whitesitt, J. *Boolean Algebra and Its Applications*. Dover, Mineola, NY, 2010.

20

Vector Spaces

$\mathcal{I}$n a physical system a quantity can often be described with a single number. For example, we need to know only a single number to describe temperature, mass, or volume. However, for some quantities, such as location, we need several numbers. To give the location of a point in space, we need x, y, and z coordinates. Temperature distribution over a solid object requires four numbers: three to identify each point within the object and a fourth to describe the temperature at that point. Often n-tuples of numbers, or vectors, also have certain algebraic properties, such as addition or scalar multiplication.

In this chapter we will examine mathematical structures called vector spaces. As with groups and rings, it is desirable to give a simple list of axioms that must be satisfied to make a set of vectors a structure worth studying.

20.1 Definitions and Examples

A *vector space* V over a field F is an abelian group with a *scalar product* $\alpha \cdot v$ or αv defined for all $\alpha \in F$ and all $v \in V$ satisfying the following axioms.

- $\alpha(\beta v) = (\alpha\beta)v$;
- $(\alpha + \beta)v = \alpha v + \beta v$;
- $\alpha(u + v) = \alpha u + \alpha v$;
- $1v = v$;

where $\alpha, \beta \in F$ and $u, v \in V$.

The elements of V are called *vectors*; the elements of F are called *scalars*. It is important to notice that in most cases two vectors cannot be multiplied. In general, it is only possible to multiply a vector with a scalar. To differentiate between the scalar zero and the vector zero, we will write them as o and **0**, respectively.

Let us examine several examples of vector spaces. Some of them will be quite familiar; others will seem less so.

Example 20.1. The n-tuples of real numbers, denoted by $\mathbb{R}^n$, form a vector space over $\mathbb{R}$. Given vectors $u = (u_1, \ldots, u_n)$ and $v = (v_1, \ldots, v_n)$ in $\mathbb{R}^n$ and α in $\mathbb{R}$, we can define vector addition by

$$u + v = (u_1, \ldots, u_n) + (v_1, \ldots, v_n) = (u_1 + v_1, \ldots, u_n + v_n)$$

and scalar multiplication by

$$\alpha u = \alpha(u_1, \ldots, u_n) = (\alpha u_1, \ldots, \alpha u_n).$$

□

Example 20.2. If F is a field, then $F[x]$ is a vector space over F. The vectors in $F[x]$ are simply polynomials, and vector addition is just polynomial addition. If $\alpha \in F$ and $p(x) \in F[x]$, then scalar multiplication is defined by $\alpha p(x)$.   □

Example 20.3. The set of all continuous real-valued functions on a closed interval $[a, b]$ is a vector space over $\mathbb{R}$. If $f(x)$ and $g(x)$ are continuous on $[a, b]$, then $(f + g)(x)$ is defined to be $f(x) + g(x)$. Scalar multiplication is defined by $(\alpha f)(x) = \alpha f(x)$ for $\alpha \in \mathbb{R}$. For example, if $f(x) = \sin x$ and $g(x) = x^2$, then $(2f + 5g)(x) = 2 \sin x + 5x^2$.   □

Example 20.4. Let $V = \mathbb{Q}(\sqrt{2}) = \{a + b\sqrt{2} : a, b \in \mathbb{Q}\}$. Then V is a vector space over $\mathbb{Q}$. If $u = a + b\sqrt{2}$ and $v = c + d\sqrt{2}$, then $u + v = (a + c) + (b + d)\sqrt{2}$ is again in V. Also, for $\alpha \in \mathbb{Q}$, αv is in V. We will leave it as an exercise to verify that all of the vector space axioms hold for V.   □

Proposition 20.5. Let V be a vector space over F. Then each of the following statements is true.

1. $0v = \mathbf{0}$ for all $v \in V$.

2. $\alpha \mathbf{0} = \mathbf{0}$ for all $\alpha \in F$.

3. If $\alpha v = \mathbf{0}$, then either $\alpha = 0$ or $v = \mathbf{0}$.

4. $(-1)v = -v$ for all $v \in V$.

5. $-(\alpha v) = (-\alpha)v = \alpha(-v)$ for all $\alpha \in F$ and all $v \in V$.

Proof. To prove (1), observe that

$$0v = (0 + 0)v = 0v + 0v;$$

consequently, $\mathbf{0} + 0v = 0v + 0v$. Since V is an abelian group, $\mathbf{0} = 0v$.

The proof of (2) is almost identical to the proof of (1). For (3), we are done if $\alpha = 0$. Suppose that $\alpha \neq 0$. Multiplying both sides of $\alpha v = \mathbf{0}$ by $1/\alpha$, we have $v = \mathbf{0}$.

To show (4), observe that

$$v + (-1)v = 1v + (-1)v = (1 - 1)v = 0v = \mathbf{0},$$

and so $-v = (-1)v$. We will leave the proof of (5) as an exercise.   ∎

20.2 Subspaces

Just as groups have subgroups and rings have subrings, vector spaces also have substructures. Let V be a vector space over a field F, and W a subset of V. Then W is a *subspace* of V if it is closed under vector addition and scalar multiplication; that is, if $u, v \in W$ and $\alpha \in F$, it will always be the case that $u + v$ and αv are also in W.

Example 20.6. Let W be the subspace of $\mathbb{R}^3$ defined by $W = \{(x_1, 2x_1 + x_2, x_1 - x_2) : x_1, x_2 \in \mathbb{R}\}$. We claim that W is a subspace of $\mathbb{R}^3$. Since

$$\alpha(x_1, 2x_1 + x_2, x_1 - x_2) = (\alpha x_1, \alpha(2x_1 + x_2), \alpha(x_1 - x_2))$$
$$= (\alpha x_1, 2(\alpha x_1) + \alpha x_2, \alpha x_1 - \alpha x_2),$$

W is closed under scalar multiplication. To show that W is closed under vector addition, let $u = (x_1, 2x_1 + x_2, x_1 - x_2)$ and $v = (y_1, 2y_1 + y_2, y_1 - y_2)$ be vectors in W. Then

$$u + v = (x_1 + y_1, 2(x_1 + y_1) + (x_2 + y_2), (x_1 + y_1) - (x_2 + y_2)).$$

□

Example 20.7. Let W be the subset of polynomials of $F[x]$ with no odd-power terms. If $p(x)$ and $q(x)$ have no odd-power terms, then neither will $p(x) + q(x)$. Also, $\alpha p(x) \in W$ for $\alpha \in F$ and $p(x) \in W$.   □

Let V be any vector space over a field F and suppose that $v_1, v_2, \ldots, v_n$ are vectors in V and $\alpha_1, \alpha_2, \ldots, \alpha_n$ are scalars in F. Any vector w in V of the form

$$w = \sum_{i=1}^{n} \alpha_i v_i = \alpha_1 v_1 + \alpha_2 v_2 + \cdots + \alpha_n v_n$$

is called a *linear combination* of the vectors $v_1, v_2, \ldots, v_n$. The *spanning set* of vectors $v_1, v_2, \ldots, v_n$ is the set of vectors obtained from all possible linear combinations of $v_1, v_2, \ldots, v_n$. If W is the spanning set of $v_1, v_2, \ldots, v_n$, then we say that W is *spanned* by $v_1, v_2, \ldots, v_n$.

Proposition 20.8. Let $S = \{v_1, v_2, \ldots, v_n\}$ be vectors in a vector space V. Then the span of S is a subspace of V.

Proof. Let u and v be in S. We can write both of these vectors as linear combinations of the v_i's:

$$u = \alpha_1 v_1 + \alpha_2 v_2 + \cdots + \alpha_n v_n$$

$$v = \beta_1 v_1 + \beta_2 v_2 + \cdots + \beta_n v_n.$$

Then

$$u + v = (\alpha_1 + \beta_1)v_1 + (\alpha_2 + \beta_2)v_2 + \cdots + (\alpha_n + \beta_n)v_n$$

is a linear combination of the v_i's. For $\alpha \in F$,

$$\alpha u = (\alpha \alpha_1)v_1 + (\alpha \alpha_2)v_2 + \cdots + (\alpha \alpha_n)v_n$$

is in the span of S. ∎

20.3 Linear Independence

Let $S = \{v_1, v_2, \ldots, v_n\}$ be a set of vectors in a vector space V. If there exist scalars $\alpha_1, \alpha_2 \ldots \alpha_n \in F$ such that not all of the α_i's are zero and

$$\alpha_1 v_1 + \alpha_2 v_2 + \cdots + \alpha_n v_n = 0,$$

then S is said to be *linearly dependent*. If the set S is not linearly dependent, then it is said to be *linearly independent*. More specifically, S is a linearly independent set if

$$\alpha_1 v_1 + \alpha_2 v_2 + \cdots + \alpha_n v_n = 0$$

implies that

$$\alpha_1 = \alpha_2 = \cdots = \alpha_n = 0$$

for any set of scalars $\{\alpha_1, \alpha_2 \ldots \alpha_n\}$.

Proposition 20.9. Let $\{v_1, v_2, \ldots, v_n\}$ be a set of linearly independent vectors in a vector space. Suppose that

$$v = \alpha_1 v_1 + \alpha_2 v_2 + \cdots + \alpha_n v_n = \beta_1 v_1 + \beta_2 v_2 + \cdots + \beta_n v_n.$$

Then $\alpha_1 = \beta_1, \alpha_2 = \beta_2, \ldots, \alpha_n = \beta_n$.

Proof. If

$$v = \alpha_1 v_1 + \alpha_2 v_2 + \cdots + \alpha_n v_n = \beta_1 v_1 + \beta_2 v_2 + \cdots + \beta_n v_n,$$

then

$$(\alpha_1 - \beta_1)v_1 + (\alpha_2 - \beta_2)v_2 + \cdots + (\alpha_n - \beta_n)v_n = 0.$$

Since $v_1, \ldots, v_n$ are linearly independent, $\alpha_i - \beta_i = 0$ for $i = 1, \ldots, n$. ∎

The definition of linear dependence makes more sense if we consider the following proposition.

Proposition 20.10. A set $\{v_1, v_2, \ldots, v_n\}$ of vectors in a vector space V is linearly dependent if and only if one of the v_i's is a linear combination of the rest.

Proof. Suppose that $\{v_1, v_2, \ldots, v_n\}$ is a set of linearly dependent vectors. Then there exist scalars $\alpha_1, \ldots, \alpha_n$ such that

$$\alpha_1 v_1 + \alpha_2 v_2 + \cdots + \alpha_n v_n = \mathbf{0},$$

with at least one of the α_i's not equal to zero. Suppose that $\alpha_k \neq 0$. Then

$$v_k = -\frac{\alpha_1}{\alpha_k} v_1 - \cdots - \frac{\alpha_{k-1}}{\alpha_k} v_{k-1} - \frac{\alpha_{k+1}}{\alpha_k} v_{k+1} - \cdots - \frac{\alpha_n}{\alpha_k} v_n.$$

Conversely, suppose that

$$v_k = \beta_1 v_1 + \cdots + \beta_{k-1} v_{k-1} + \beta_{k+1} v_{k+1} + \cdots + \beta_n v_n.$$

Then

$$\beta_1 v_1 + \cdots + \beta_{k-1} v_{k-1} - v_k + \beta_{k+1} v_{k+1} + \cdots + \beta_n v_n = \mathbf{0}.$$

■

The following proposition is a consequence of the fact that any system of homogeneous linear equations with more unknowns than equations will have a nontrivial solution. We leave the details of the proof for the end-of-chapter exercises.

Proposition 20.11. Suppose that a vector space V is spanned by n vectors. If $m > n$, then any set of m vectors in V must be linearly dependent.

A set $\{e_1, e_2, \ldots, e_n\}$ of vectors in a vector space V is called a ***basis*** for V if $\{e_1, e_2, \ldots, e_n\}$ is a linearly independent set that spans V.

Example 20.12. The vectors $e_1 = (1, 0, 0)$, $e_2 = (0, 1, 0)$, and $e_3 = (0, 0, 1)$ form a basis for $\mathbb{R}^3$. The set certainly spans $\mathbb{R}^3$, since any arbitrary vector (x_1, x_2, x_3) in $\mathbb{R}^3$ can be written as $x_1 e_1 + x_2 e_2 + x_3 e_3$. Also, none of the vectors e_1, e_2, e_3 can be written as a linear combination of the other two; hence, they are linearly independent. The vectors e_1, e_2, e_3 are not the only basis of $\mathbb{R}^3$: the set $\{(3, 2, 1), (3, 2, 0), (1, 1, 1)\}$ is also a basis for $\mathbb{R}^3$. □

Example 20.13. Let $\mathbb{Q}(\sqrt{2}) = \{a + b\sqrt{2} : a, b \in \mathbb{Q}\}$. The sets $\{1, \sqrt{2}\}$ and $\{1 + \sqrt{2}, 1 - \sqrt{2}\}$ are both bases of $\mathbb{Q}(\sqrt{2})$. □

From the last two examples it should be clear that a given vector space has several bases. In fact, there are an infinite number of bases for both of these examples. *In general, there is no unique basis for a vector space.* However, every basis of $\mathbb{R}^3$ consists of exactly three vectors, and every basis of $\mathbb{Q}(\sqrt{2})$ consists of exactly two vectors. This is a consequence of the next proposition.

Proposition 20.14. Let $\{e_1, e_2, \ldots, e_m\}$ and $\{f_1, f_2, \ldots, f_n\}$ be two bases for a vector space V. Then $m = n$.

Proof. Since $\{e_1, e_2, \ldots, e_m\}$ is a basis, it is a linearly independent set. By Proposition 20.11, $n \leq m$. Similarly, $\{f_1, f_2, \ldots, f_n\}$ is a linearly independent set, and the last proposition implies that $m \leq n$. Consequently, $m = n$. ∎

If $\{e_1, e_2, \ldots, e_n\}$ is a basis for a vector space V, then we say that the ***dimension*** of V is n and we write $\dim V = n$. We will leave the proof of the following theorem as an exercise.

Theorem 20.15. Let V be a vector space of dimension n.

1. If $S = \{v_1, \ldots, v_n\}$ is a set of linearly independent vectors for V, then S is a basis for V.

2. If $S = \{v_1, \ldots, v_n\}$ spans V, then S is a basis for V.

3. If $S = \{v_1, \ldots, v_k\}$ is a set of linearly independent vectors for V with $k < n$, then there exist vectors $v_{k+1}, \ldots, v_n$ such that

$$\{v_1, \ldots, v_k, v_{k+1}, \ldots, v_n\}$$

   is a basis for V.

Sage. Many of Sage's computations, in a wide variety of algebraic settings, come from solving problems in linear algebra. So you will find a wealth of linear algebra functionality. Further, you can use structures such as finite fields to find vector spaces in new settings.

20.4 Reading Questions

1. Why do the axioms of a vector space appear to only have four conditions, rather than the ten you may have seen the first time you saw an axiomatic definition?

2. The set $V = \mathbb{Q}(\sqrt{11}) = \{a + b\sqrt{11} \mid a, b \in \mathbb{Q}\}$ is a vector space. Carefully define the operations on this set that will make this possible. Describe the subspace spanned by $S = \{\mathbf{u}\}$, where $\mathbf{u} = 3 + \frac{2}{7}\sqrt{11} \in V$.

3. Write a long paragraph, or a short essay, on the importance of linear independence in linear algebra.

4. Write a long paragraph, or a short essay, on the importance of spanning sets in linear algebra.

5. "Linear algebra is all about linear combinations." Explain why you might say this.

20.5 Exercises

1. If F is a field, show that $F[x]$ is a vector space over F, where the vectors in $F[x]$ are polynomials. Vector addition is polynomial addition, and scalar multiplication is defined by $\alpha p(x)$ for $\alpha \in F$.

2. Prove that $\mathbb{Q}(\sqrt{2})$ is a vector space.

3. Let $\mathbb{Q}(\sqrt{2}, \sqrt{3})$ be the field generated by elements of the form $a + b\sqrt{2} + c\sqrt{3} + d\sqrt{6}$, where a, b, c, d are in $\mathbb{Q}$. Prove that $\mathbb{Q}(\sqrt{2}, \sqrt{3})$ is a vector space of dimension 4 over $\mathbb{Q}$. Find a basis for $\mathbb{Q}(\sqrt{2}, \sqrt{3})$.

4. Prove that the complex numbers are a vector space of dimension 2 over $\mathbb{R}$.

5. Prove that the set P_n of all polynomials of degree less than n form a subspace of the vector space $F[x]$. Find a basis for P_n and compute the dimension of P_n.

6. Let F be a field and denote the set of n-tuples of F by F^n. Given vectors $u = (u_1, \ldots, u_n)$ and $v = (v_1, \ldots, v_n)$ in F^n and α in F, define vector addition by

$$u + v = (u_1, \ldots, u_n) + (v_1, \ldots, v_n) = (u_1 + v_1, \ldots, u_n + v_n)$$

and scalar multiplication by

$$\alpha u = \alpha(u_1, \ldots, u_n) = (\alpha u_1, \ldots, \alpha u_n).$$

Prove that F^n is a vector space of dimension n under these operations.

7. Which of the following sets are subspaces of $\mathbb{R}^3$? If the set is indeed a subspace, find a basis for the subspace and compute its dimension.

   (a) $\{(x_1, x_2, x_3) : 3x_1 - 2x_2 + x_3 = 0\}$

   (b) $\{(x_1, x_2, x_3) : 3x_1 + 4x_3 = 0, 2x_1 - x_2 + x_3 = 0\}$

   (c) $\{(x_1, x_2, x_3) : x_1 - 2x_2 + 2x_3 = 2\}$

   (d) $\{(x_1, x_2, x_3) : 3x_1 - 2x_2^2 = 0\}$

8. Show that the set of all possible solutions $(x, y, z) \in \mathbb{R}^3$ of the equations

$$Ax + By + Cz = 0$$
$$Dx + Ey + Cz = 0$$

form a subspace of $\mathbb{R}^3$.

9. Let W be the subset of continuous functions on $[0,1]$ such that $f(0) = 0$. Prove that W is a subspace of $C[0,1]$.

10. Let V be a vector space over F. Prove that $-(\alpha v) = (-\alpha)v = \alpha(-v)$ for all $\alpha \in F$ and all $v \in V$.

11. Let V be a vector space of dimension n. Prove each of the following statements.

   (a) If $S = \{v_1, \ldots, v_n\}$ is a set of linearly independent vectors for V, then S is a basis for V.

   (b) If $S = \{v_1, \ldots, v_n\}$ spans V, then S is a basis for V.

   (c) If $S = \{v_1, \ldots, v_k\}$ is a set of linearly independent vectors for V with $k < n$, then there exist vectors $v_{k+1}, \ldots, v_n$ such that

   $$\{v_1, \ldots, v_k, v_{k+1}, \ldots, v_n\}$$

   is a basis for V.

12. Prove that any set of vectors containing $\mathbf{0}$ is linearly dependent.

13. Let V be a vector space. Show that $\{\mathbf{0}\}$ is a subspace of V of dimension zero.

14. If a vector space V is spanned by n vectors, show that any set of m vectors in V must be linearly dependent for $m > n$.

15. **Linear Transformations.** Let V and W be vector spaces over a field F, of dimensions m and n, respectively. If $T : V \to W$ is a map satisfying

   $$T(u + v) = T(u) + T(v)$$
   $$T(\alpha v) = \alpha T(v)$$

   for all $\alpha \in F$ and all $u, v \in V$, then T is called a *linear transformation* from V into W.

   (a) Prove that the *kernel* of T, $\ker(T) = \{v \in V : T(v) = \mathbf{0}\}$, is a subspace of V. The kernel of T is sometimes called the *null space* of T.

   (b) Prove that the *range* or *range space* of T, $R(V) = \{w \in W : T(v) = w \text{ for some } v \in V\}$, is a subspace of W.

   (c) Show that $T : V \to W$ is injective if and only if $\ker(T) = \{\mathbf{0}\}$.

   (d) Let $\{v_1, \ldots, v_k\}$ be a basis for the null space of T. We can extend this basis to be a basis $\{v_1, \ldots, v_k, v_{k+1}, \ldots, v_m\}$ of V. Why? Prove that $\{T(v_{k+1}), \ldots, T(v_m)\}$ is a basis for the range of T. Conclude that the range of T has dimension $m - k$.

   (e) Let $\dim V = \dim W$. Show that a linear transformation $T : V \to W$ is injective if and only if it is surjective.

16. Let V and W be finite dimensional vector spaces of dimension n over a field F. Suppose that $T : V \to W$ is a vector space isomorphism. If $\{v_1, \ldots, v_n\}$ is a basis of V, show that $\{T(v_1), \ldots, T(v_n)\}$ is a basis of W. Conclude that any vector space over a field F of dimension n is isomorphic to F^n.

17. **Direct Sums.** Let U and V be subspaces of a vector space W. The sum of U and V, denoted $U + V$, is defined to be the set of all vectors of the form $u + v$, where $u \in U$ and $v \in V$.

(a) Prove that $U + V$ and $U \cap V$ are subspaces of W.

(b) If $U + V = W$ and $U \cap V = \mathbf{0}$, then W is said to be the **direct sum.** In this case, we write $W = U \oplus V$. Show that every element $w \in W$ can be written uniquely as $w = u + v$, where $u \in U$ and $v \in V$.

(c) Let U be a subspace of dimension k of a vector space W of dimension n. Prove that there exists a subspace V of dimension $n - k$ such that $W = U \oplus V$. Is the subspace V unique?

(d) If U and V are arbitrary subspaces of a vector space W, show that

$$\dim(U + V) = \dim U + \dim V - \dim(U \cap V).$$

18. **Dual Spaces.** Let V and W be finite dimensional vector spaces over a field F.

(a) Show that the set of all linear transformations from V into W, denoted by $\mathrm{Hom}(V, W)$, is a vector space over F, where we define vector addition as follows:

$$(S + T)(v) = S(v) + T(v)$$
$$(\alpha S)(v) = \alpha S(v),$$

where $S, T \in \mathrm{Hom}(V, W)$, $\alpha \in F$, and $v \in V$.

(b) Let V be an F-vector space. Define the **dual space** of V to be $V^* = \mathrm{Hom}(V, F)$. Elements in the dual space of V are called **linear functionals.** Let $v_1, \ldots, v_n$ be an ordered basis for V. If $v = \alpha_1 v_1 + \cdots + \alpha_n v_n$ is any vector in V, define a linear functional $\phi_i : V \to F$ by $\phi_i(v) = \alpha_i$. Show that the ϕ_i's form a basis for V^*. This basis is called the **dual basis** of $v_1, \ldots, v_n$ (or simply the dual basis if the context makes the meaning clear).

(c) Consider the basis $\{(3,1), (2,-2)\}$ for $\mathbb{R}^2$. What is the dual basis for $(\mathbb{R}^2)^*$?

(d) Let V be a vector space of dimension n over a field F and let V^{**} be the dual space of V^*. Show that each element $v \in V$ gives rise to an element λ_v in V^{**} and that the map $v \mapsto \lambda_v$ is an isomorphism of V with V^{**}.

20.6 References and Suggested Readings

[1] Beezer, R. *A First Course in Linear Algebra* . Available online at http://linear.ups.edu/. 2004–2014.

[2] Bretscher, O. *Linear Algebra with Applications.* 4th ed. Pearson, Upper Saddle River, NJ, 2009.

[3] Curtis, C. W. *Linear Algebra: An Introductory Approach.* 4th ed. Springer, New York, 1984.

[4] Hoffman, K. and Kunze, R. *Linear Algebra.* 2nd ed. Prentice-Hall, Englewood Cliffs, NJ, 1971.

[5] Johnson, L. W., Riess, R. D., and Arnold, J. T. *Introduction to Linear Algebra.* 6th ed. Pearson, Upper Saddle River, NJ, 2011.

[6] Leon, S. J. *Linear Algebra with Applications.* 8th ed. Pearson, Upper Saddle River, NJ, 2010.

21

Fields

$\mathcal{I}$t is natural to ask whether or not some field F is contained in a larger field. We think of the rational numbers, which reside inside the real numbers, while in turn, the real numbers live inside the complex numbers. We can also study the fields between $\mathbb{Q}$ and $\mathbb{R}$ and inquire as to the nature of these fields.

More specifically if we are given a field F and a polynomial $p(x) \in F[x]$, we can ask whether or not we can find a field E containing F such that $p(x)$ factors into linear factors over $E[x]$. For example, if we consider the polynomial

$$p(x) = x^4 - 5x^2 + 6$$

in $\mathbb{Q}[x]$, then $p(x)$ factors as $(x^2 - 2)(x^2 - 3)$. However, both of these factors are irreducible in $\mathbb{Q}[x]$. If we wish to find a zero of $p(x)$, we must go to a larger field. Certainly the field of real numbers will work, since

$$p(x) = (x - \sqrt{2})(x + \sqrt{2})(x - \sqrt{3})(x + \sqrt{3}).$$

It is possible to find a smaller field in which $p(x)$ has a zero, namely

$$\mathbb{Q}(\sqrt{2}) = \{a + b\sqrt{2} : a, b \in \mathbb{Q}\}.$$

We wish to be able to compute and study such fields for arbitrary polynomials over a field F.

21.1 Extension Fields

A field E is an *extension field* of a field F if F is a subfield of E. The field F is called the *base field*. We write $F \subset E$.

Example 21.1. For example, let

$$F = \mathbb{Q}(\sqrt{2}) = \{a + b\sqrt{2} : a, b \in \mathbb{Q}\}$$

and let $E = \mathbb{Q}(\sqrt{2} + \sqrt{3})$ be the smallest field containing both $\mathbb{Q}$ and $\sqrt{2} + \sqrt{3}$. Both E and F are extension fields of the rational numbers. We claim that E is an extension field of F. To see this, we need only show that $\sqrt{2}$ is in E. Since $\sqrt{2} + \sqrt{3}$

is in E, $1/(\sqrt{2}+\sqrt{3}) = \sqrt{3}-\sqrt{2}$ must also be in E. Taking linear combinations of $\sqrt{2}+\sqrt{3}$ and $\sqrt{3}-\sqrt{2}$, we find that $\sqrt{2}$ and $\sqrt{3}$ must both be in E. $\square$

Example 21.2. Let $p(x) = x^2 + x + 1 \in \mathbb{Z}_2[x]$. Since neither 0 nor 1 is a root of this polynomial, we know that $p(x)$ is irreducible over $\mathbb{Z}_2$. We will construct a field extension of $\mathbb{Z}_2$ containing an element α such that $p(\alpha) = 0$. By Theorem 17.22 on page 278, the ideal $\langle p(x) \rangle$ generated by $p(x)$ is maximal; hence, $\mathbb{Z}_2[x]/\langle p(x) \rangle$ is a field. Let $f(x) + \langle p(x) \rangle$ be an arbitrary element of $\mathbb{Z}_2[x]/\langle p(x) \rangle$. By the division algorithm,

$$f(x) = (x^2 + x + 1)q(x) + r(x),$$

where the degree of $r(x)$ is less than the degree of $x^2 + x + 1$. Therefore,

$$f(x) + \langle x^2 + x + 1 \rangle = r(x) + \langle x^2 + x + 1 \rangle.$$

The only possibilities for $r(x)$ are then 0, 1, x, and $1 + x$. Consequently, $E = \mathbb{Z}_2[x]/\langle x^2 + x + 1 \rangle$ is a field with four elements and must be a field extension of $\mathbb{Z}_2$, containing a zero α of $p(x)$. The field $\mathbb{Z}_2(\alpha)$ consists of elements

$$0 + 0\alpha = 0$$
$$1 + 0\alpha = 1$$
$$0 + 1\alpha = \alpha$$
$$1 + 1\alpha = 1 + \alpha.$$

Notice that $\alpha^2 + \alpha + 1 = 0$; hence, if we compute $(1 + \alpha)^2$,

$$(1 + \alpha)(1 + \alpha) = 1 + \alpha + \alpha + (\alpha)^2 = \alpha.$$

Other calculations are accomplished in a similar manner. We summarize these computations in the following tables, which tell us how to add and multiply elements in E. $\square$

+	0	1	α	$1+\alpha$
0	0	1	α	$1+\alpha$
1	1	0	$1+\alpha$	α
α	α	$1+\alpha$	0	1
$1+\alpha$	$1+\alpha$	α	1	0

Figure 21.3. Addition Table for $\mathbb{Z}_2(\alpha)$

·	0	1	α	$1+\alpha$
0	0	0	0	0
1	0	1	α	$1+\alpha$
α	0	α	$1+\alpha$	1
$1+\alpha$	0	$1+\alpha$	1	α

Figure 21.4. Multiplication Table for $\mathbb{Z}_2(\alpha)$

The following theorem, due to Kronecker, is so important and so basic to our understanding of fields that it is often known as the Fundamental Theorem of Field Theory.

Theorem 21.5. Let F be a field and let $p(x)$ be a nonconstant polynomial in $F[x]$. Then there exists an extension field E of F and an element $\alpha \in E$ such that $p(\alpha) = 0$.

Proof. To prove this theorem, we will employ the method that we used to construct Example 21.2 on the preceding page. Clearly, we can assume that $p(x)$ is an irreducible polynomial. We wish to find an extension field E of F containing an element α such that $p(\alpha) = 0$. The ideal $\langle p(x) \rangle$ generated by $p(x)$ is a maximal ideal in $F[x]$ by Theorem 17.22 on page 278; hence, $F[x]/\langle p(x) \rangle$ is a field. We claim that $E = F[x]/\langle p(x) \rangle$ is the desired field.

We first show that E is a field extension of F. We can define a homomorphism of commutative rings by the map $\psi : F \to F[x]/\langle p(x) \rangle$, where $\psi(a) = a + \langle p(x) \rangle$ for $a \in F$. It is easy to check that ψ is indeed a ring homomorphism. Observe that

$$\psi(a) + \psi(b) = (a + \langle p(x) \rangle) + (b + \langle p(x) \rangle) = (a + b) + \langle p(x) \rangle = \psi(a + b)$$

and

$$\psi(a)\psi(b) = (a + \langle p(x) \rangle)(b + \langle p(x) \rangle) = ab + \langle p(x) \rangle = \psi(ab).$$

To prove that ψ is one-to-one, assume that

$$a + \langle p(x) \rangle = \psi(a) = \psi(b) = b + \langle p(x) \rangle.$$

Then $a - b$ is a multiple of $p(x)$, since it lives in the ideal $\langle p(x) \rangle$. Since $p(x)$ is a nonconstant polynomial, the only possibility is that $a - b = 0$. Consequently, $a = b$ and ψ is injective. Since ψ is one-to-one, we can identify F with the subfield $\{a + \langle p(x) \rangle : a \in F\}$ of E and view E as an extension field of F.

It remains for us to prove that $p(x)$ has a zero $\alpha \in E$. Set $\alpha = x + \langle p(x) \rangle$. Then α is in E. If $p(x) = a_0 + a_1 x + \cdots + a_n x^n$, then

$$
\begin{aligned}
p(\alpha) &= a_0 + a_1(x + \langle p(x) \rangle) + \cdots + a_n(x + \langle p(x) \rangle)^n \\
&= a_0 + (a_1 x + \langle p(x) \rangle) + \cdots + (a_n x^n + \langle p(x) \rangle) \\
&= a_0 + a_1 x + \cdots + a_n x^n + \langle p(x) \rangle \\
&= 0 + \langle p(x) \rangle.
\end{aligned}
$$

Therefore, we have found an element $\alpha \in E = F[x]/\langle p(x) \rangle$ such that α is a zero of $p(x)$.   ∎

Example 21.6. Let $p(x) = x^5 + x^4 + 1 \in \mathbb{Z}_2[x]$. Then $p(x)$ has irreducible factors $x^2 + x + 1$ and $x^3 + x + 1$. For a field extension E of $\mathbb{Z}_2$ such that $p(x)$ has a root in E, we can let E be either $\mathbb{Z}_2[x]/\langle x^2 + x + 1 \rangle$ or $\mathbb{Z}_2[x]/\langle x^3 + x + 1 \rangle$. We will leave it as an exercise to show that $\mathbb{Z}_2[x]/\langle x^3 + x + 1 \rangle$ is a field with $2^3 = 8$ elements. □

Algebraic Elements

An element α in an extension field E over F is **algebraic** over F if $f(\alpha) = 0$ for some nonzero polynomial $f(x) \in F[x]$. An element in E that is not algebraic over F is **transcendental** over F. An extension field E of a field F is an **algebraic extension** of F if every element in E is algebraic over F. If E is a field extension of F and $\alpha_1, \ldots, \alpha_n$ are contained in E, we denote the smallest field containing F and $\alpha_1, \ldots, \alpha_n$ by $F(\alpha_1, \ldots, \alpha_n)$. If $E = F(\alpha)$ for some $\alpha \in E$, then E is a **simple extension** of F.

Example 21.7. Both $\sqrt{2}$ and i are algebraic over $\mathbb{Q}$ since they are zeros of the polynomials $x^2 - 2$ and $x^2 + 1$, respectively. Clearly π and e are algebraic over the real numbers; however, it is a nontrivial fact that they are transcendental over $\mathbb{Q}$. Numbers in $\mathbb{R}$ that are algebraic over $\mathbb{Q}$ are in fact quite rare. Almost all real numbers are transcendental over $\mathbb{Q}$.[6] (In many cases we do not know whether or not a particular number is transcendental; for example, it is still not known whether $\pi + e$ is transcendental or algebraic.)   □

A complex number that is algebraic over $\mathbb{Q}$ is an **algebraic number**. A **transcendental number** is an element of $\mathbb{C}$ that is transcendental over $\mathbb{Q}$.

Example 21.8. We will show that $\sqrt{2 + \sqrt{3}}$ is algebraic over $\mathbb{Q}$. If $\alpha = \sqrt{2 + \sqrt{3}}$, then $\alpha^2 = 2 + \sqrt{3}$. Hence, $\alpha^2 - 2 = \sqrt{3}$ and $(\alpha^2 - 2)^2 = 3$. Since $\alpha^4 - 4\alpha^2 + 1 = 0$, it must be true that α is a zero of the polynomial $x^4 - 4x^2 + 1 \in \mathbb{Q}[x]$.   □

---

[6]The probability that a real number chosen at random from the interval $[0,1]$ will be transcendental over the rational numbers is one.

It is very easy to give an example of an extension field E over a field F, where E contains an element transcendental over F. The following theorem characterizes transcendental extensions.

Theorem 21.9. Let E be an extension field of F and $\alpha \in E$. Then α is transcendental over F if and only if $F(\alpha)$ is isomorphic to $F(x)$, the field of fractions of $F[x]$.

Proof. Let $\phi_\alpha : F[x] \to E$ be the evaluation homomorphism for α. Then α is transcendental over F if and only if $\phi_\alpha(p(x)) = p(\alpha) \neq 0$ for all nonconstant polynomials $p(x) \in F[x]$. This is true if and only if $\ker \phi_\alpha = \{0\}$; that is, it is true exactly when ϕ_α is one-to-one. Hence, E must contain a copy of $F[x]$. The smallest field containing $F[x]$ is the field of fractions $F(x)$. By Theorem 18.4 on page 289, E must contain a copy of this field. ∎

We have a more interesting situation in the case of algebraic extensions.

Theorem 21.10. Let E be an extension field of a field F and $\alpha \in E$ with α algebraic over F. Then there is a unique irreducible monic polynomial $p(x) \in F[x]$ of smallest degree such that $p(\alpha) = 0$. If $f(x)$ is another polynomial in $F[x]$ such that $f(\alpha) = 0$, then $p(x)$ divides $f(x)$.

Proof. Let $\phi_\alpha : F[x] \to E$ be the evaluation homomorphism. The kernel of ϕ_α is a principal ideal generated by some $p(x) \in F[x]$ with $\deg p(x) \geq 1$. We know that such a polynomial exists, since $F[x]$ is a principal ideal domain and α is algebraic. The ideal $\langle p(x) \rangle$ consists exactly of those elements of $F[x]$ having α as a zero. If $f(\alpha) = 0$ and $f(x)$ is not the zero polynomial, then $f(x) \in \langle p(x) \rangle$ and $p(x)$ divides $f(x)$. So $p(x)$ is a polynomial of minimal degree having α as a zero. Any other polynomial of the same degree having α as a zero must have the form $\beta p(x)$ for some $\beta \in F$.

Suppose now that $p(x) = r(x)s(x)$ is a factorization of $p(x)$ into polynomials of lower degree. Since $p(\alpha) = 0$, $r(\alpha)s(\alpha) = 0$; consequently, either $r(\alpha) = 0$ or $s(\alpha) = 0$, which contradicts the fact that p is of minimal degree. Therefore, $p(x)$ must be irreducible. ∎

Let E be an extension field of F and $\alpha \in E$ be algebraic over F. The unique monic polynomial $p(x)$ of the last theorem is called the ***minimal polynomial*** for α over F. The degree of $p(x)$ is the ***degree of α over*** F.

Example 21.11. Let $f(x) = x^2 - 2$ and $g(x) = x^4 - 4x^2 + 1$. These polynomials are the minimal polynomials of $\sqrt{2}$ and $\sqrt{2 + \sqrt{3}}$, respectively. □

Proposition 21.12. Let E be a field extension of F and $\alpha \in E$ be algebraic over F. Then $F(\alpha) \cong F[x]/\langle p(x) \rangle$, where $p(x)$ is the minimal polynomial of α over F.

Proof. Let $\phi_\alpha : F[x] \to E$ be the evaluation homomorphism. The kernel of this map is $\langle p(x) \rangle$, where $p(x)$ is the minimal polynomial of α. By the First Isomorphism Theorem for rings, the image of ϕ_α in E is isomorphic to $F(\alpha)$ since it contains both F and α. $\blacksquare$

Theorem 21.13. Let $E = F(\alpha)$ be a simple extension of F, where $\alpha \in E$ is algebraic over F. Suppose that the degree of α over F is n. Then every element $\beta \in E$ can be expressed uniquely in the form

$$\beta = b_0 + b_1 \alpha + \cdots + b_{n-1} \alpha^{n-1}$$

for $b_i \in F$.

Proof. Since $\phi_\alpha(F[x]) \cong F(\alpha)$, every element in $E = F(\alpha)$ must be of the form $\phi_\alpha(f(x)) = f(\alpha)$, where $f(\alpha)$ is a polynomial in α with coefficients in F. Let

$$p(x) = x^n + a_{n-1}x^{n-1} + \cdots + a_0$$

be the minimal polynomial of α. Then $p(\alpha) = 0$; hence,

$$\alpha^n = -a_{n-1}\alpha^{n-1} - \cdots - a_0.$$

Similarly,

$$\alpha^{n+1} = \alpha \alpha^n$$
$$= -a_{n-1}\alpha^n - a_{n-2}\alpha^{n-1} - \cdots - a_0 \alpha$$
$$= -a_{n-1}(-a_{n-1}\alpha^{n-1} - \cdots - a_0) - a_{n-2}\alpha^{n-1} - \cdots - a_0 \alpha.$$

Continuing in this manner, we can express every monomial α^m, $m \geq n$, as a linear combination of powers of α that are less than n. Hence, any $\beta \in F(\alpha)$ can be written as

$$\beta = b_0 + b_1 \alpha + \cdots + b_{n-1} \alpha^{n-1}.$$

To show uniqueness, suppose that

$$\beta = b_0 + b_1 \alpha + \cdots + b_{n-1} \alpha^{n-1} = c_0 + c_1 \alpha + \cdots + c_{n-1} \alpha^{n-1}$$

for b_i and c_i in F. Then

$$g(x) = (b_0 - c_0) + (b_1 - c_1)x + \cdots + (b_{n-1} - c_{n-1})x^{n-1}$$

is in $F[x]$ and $g(\alpha) = 0$. Since the degree of $g(x)$ is less than the degree of $p(x)$, the irreducible polynomial of α, $g(x)$ must be the zero polynomial. Consequently,

$$b_0 - c_0 = b_1 - c_1 = \cdots = b_{n-1} - c_{n-1} = 0,$$

or $b_i = c_i$ for $i = 0, 1, \ldots, n - 1$. Therefore, we have shown uniqueness. ∎

Example 21.14. Since $x^2 + 1$ is irreducible over $\mathbb{R}$, $\langle x^2 + 1 \rangle$ is a maximal ideal in $\mathbb{R}[x]$. So $E = \mathbb{R}[x]/\langle x^2 + 1 \rangle$ is a field extension of $\mathbb{R}$ that contains a root of $x^2 + 1$. Let $\alpha = x + \langle x^2 + 1 \rangle$. We can identify E with the complex numbers. By Proposition 21.12 on page 336, E is isomorphic to $\mathbb{R}(\alpha) = \{a + b\alpha : a, b \in \mathbb{R}\}$. We know that $\alpha^2 = -1$ in E, since

$$\begin{aligned}
\alpha^2 + 1 &= (x + \langle x^2 + 1 \rangle)^2 + (1 + \langle x^2 + 1 \rangle) \\
&= (x^2 + 1) + \langle x^2 + 1 \rangle \\
&= 0.
\end{aligned}$$

Hence, we have an isomorphism of $\mathbb{R}(\alpha)$ with $\mathbb{C}$ defined by the map that takes $a + b\alpha$ to $a + bi$. □

Let E be a field extension of a field F. If we regard E as a vector space over F, then we can bring the machinery of linear algebra to bear on the problems that we will encounter in our study of fields. The elements in the field E are vectors; the elements in the field F are scalars. We can think of addition in E as adding vectors. When we multiply an element in E by an element of F, we are multiplying a vector by a scalar. This view of field extensions is especially fruitful if a field extension E of F is a finite dimensional vector space over F, and Theorem 21.13 on the preceding page states that $E = F(\alpha)$ is finite dimensional vector space over F with basis $\{1, \alpha, \alpha^2, \ldots, \alpha^{n-1}\}$.

If an extension field E of a field F is a finite dimensional vector space over F of dimension n, then we say that E is a *finite extension of degree n over F*. We write

$$[E : F] = n.$$

to indicate the dimension of E over F.

Theorem 21.15. Every finite extension field E of a field F is an algebraic extension.

Proof. Let $\alpha \in E$. Since $[E : F] = n$, the elements

$$1, \alpha, \ldots, \alpha^n$$

cannot be linearly independent. Hence, there exist $a_i \in F$, not all zero, such that

$$a_n \alpha^n + a_{n-1} \alpha^{n-1} + \cdots + a_1 \alpha + a_0 = 0.$$

Therefore,

$$p(x) = a_n x^n + \cdots + a_0 \in F[x]$$

is a nonzero polynomial with $p(\alpha) = 0$. ∎

Remark 21.16. Theorem 21.15 says that every finite extension of a field F is an algebraic extension. The converse is false, however. We will leave it as an exercise to show that the set of all elements in $\mathbb{R}$ that are algebraic over $\mathbb{Q}$ forms an infinite field extension of $\mathbb{Q}$.

The next theorem is a counting theorem, similar to Lagrange's Theorem in group theory. Theorem 21.17 will prove to be an extremely useful tool in our investigation of finite field extensions.

Theorem 21.17. If E is a finite extension of F and K is a finite extension of E, then K is a finite extension of F and

$$[K : F] = [K : E][E : F].$$

Proof. Let $\{\alpha_1, \ldots, \alpha_n\}$ be a basis for E as a vector space over F and $\{\beta_1, \ldots, \beta_m\}$ be a basis for K as a vector space over E. We claim that $\{\alpha_i \beta_j\}$ is a basis for K over F. We will first show that these vectors span K. Let $u \in K$. Then $u = \sum_{j=1}^{m} b_j \beta_j$ and $b_j = \sum_{i=1}^{n} a_{ij} \alpha_i$, where $b_j \in E$ and $a_{ij} \in F$. Then

$$u = \sum_{j=1}^{m} \left(\sum_{i=1}^{n} a_{ij} \alpha_i \right) \beta_j = \sum_{i,j} a_{ij} (\alpha_i \beta_j).$$

So the mn vectors $\alpha_i \beta_j$ must span K over F.

We must show that $\{\alpha_i \beta_j\}$ are linearly independent. Recall that a set of vectors $v_1, v_2, \ldots, v_n$ in a vector space V are linearly independent if

$$c_1 v_1 + c_2 v_2 + \cdots + c_n v_n = 0$$

implies that

$$c_1 = c_2 = \cdots = c_n = 0.$$

Let

$$u = \sum_{i,j} c_{ij} (\alpha_i \beta_j) = 0$$

for $c_{ij} \in F$. We need to prove that all of the c_{ij}'s are zero. We can rewrite u as

$$\sum_{j=1}^{m} \left(\sum_{i=1}^{n} c_{ij} \alpha_i \right) \beta_j = 0,$$

where $\sum_i c_{ij} \alpha_i \in E$. Since the β_j's are linearly independent over E, it must be the case that

$$\sum_{i=1}^{n} c_{ij} \alpha_i = 0$$

for all j. However, the α_j are also linearly independent over F. Therefore, $c_{ij} = 0$ for all i and j, which completes the proof.   ∎

The following corollary is easily proved using mathematical induction.

Corollary 21.18. If F_i is a field for $i = 1, \ldots, k$ and F_{i+1} is a finite extension of F_i, then F_k is a finite extension of F_1 and

$$[F_k : F_1] = [F_k : F_{k-1}] \cdots [F_2 : F_1].$$

Corollary 21.19. Let E be an extension field of F. If $\alpha \in E$ is algebraic over F with minimal polynomial $p(x)$ and $\beta \in F(\alpha)$ with minimal polynomial $q(x)$, then $\deg q(x)$ divides $\deg p(x)$.

Proof. We know that $\deg p(x) = [F(\alpha) : F]$ and $\deg q(x) = [F(\beta) : F]$. Since $F \subset F(\beta) \subset F(\alpha)$,

$$[F(\alpha) : F] = [F(\alpha) : F(\beta)][F(\beta) : F].$$

∎

Example 21.20. Let us determine an extension field of $\mathbb{Q}$ containing $\sqrt{3} + \sqrt{5}$. It is easy to determine that the minimal polynomial of $\sqrt{3} + \sqrt{5}$ is $x^4 - 16x^2 + 4$. It follows that

$$[\mathbb{Q}(\sqrt{3} + \sqrt{5}) : \mathbb{Q}] = 4.$$

We know that $\{1, \sqrt{3}\}$ is a basis for $\mathbb{Q}(\sqrt{3})$ over $\mathbb{Q}$. Hence, $\sqrt{3} + \sqrt{5}$ cannot be in $\mathbb{Q}(\sqrt{3})$. It follows that $\sqrt{5}$ cannot be in $\mathbb{Q}(\sqrt{3})$ either. Therefore, $\{1, \sqrt{5}\}$ is a basis for $\mathbb{Q}(\sqrt{3}, \sqrt{5}) = (\mathbb{Q}(\sqrt{3}))(\sqrt{5})$ over $\mathbb{Q}(\sqrt{3})$ and $\{1, \sqrt{3}, \sqrt{5}, \sqrt{3}\sqrt{5} = \sqrt{15}\}$ is a basis for $\mathbb{Q}(\sqrt{3}, \sqrt{5}) = \mathbb{Q}(\sqrt{3} + \sqrt{5})$ over $\mathbb{Q}$. This example shows that it is possible that some extension $F(\alpha_1, \ldots, \alpha_n)$ is actually a simple extension of F even though $n > 1$.   □

Example 21.21. Let us compute a basis for $\mathbb{Q}(\sqrt[3]{5}, \sqrt{5}\, i)$, where $\sqrt{5}$ is the positive square root of 5 and $\sqrt[3]{5}$ is the real cube root of 5. We know that $\sqrt{5}\, i \notin \mathbb{Q}(\sqrt[3]{5})$, so

$$[\mathbb{Q}(\sqrt[3]{5}, \sqrt{5}\, i) : \mathbb{Q}(\sqrt[3]{5})] = 2.$$

It is easy to determine that $\{1, \sqrt{5}i\}$ is a basis for $\mathbb{Q}(\sqrt[3]{5}, \sqrt{5}\, i)$ over $\mathbb{Q}(\sqrt[3]{5})$. We also know that $\{1, \sqrt[3]{5}, (\sqrt[3]{5})^2\}$ is a basis for $\mathbb{Q}(\sqrt[3]{5})$ over $\mathbb{Q}$. Hence, a basis for $\mathbb{Q}(\sqrt[3]{5}, \sqrt{5}\, i)$ over $\mathbb{Q}$ is

$$\{1, \sqrt{5}\, i, \sqrt[3]{5}, (\sqrt[3]{5})^2, (\sqrt[6]{5})^5 i, (\sqrt[6]{5})^7 i = 5\sqrt[6]{5}\, i \text{ or } \sqrt[6]{5}\, i\}.$$

Notice that $\sqrt[6]{5}\,i$ is a zero of $x^6 + 5$. We can show that this polynomial is irreducible over $\mathbb{Q}$ using Eisenstein's Criterion, where we let $p = 5$. Consequently,

$$\mathbb{Q} \subset \mathbb{Q}(\sqrt[6]{5}\,i) \subset \mathbb{Q}(\sqrt[3]{5}, \sqrt{5}\,i).$$

But it must be the case that $\mathbb{Q}(\sqrt[6]{5}\,i) = \mathbb{Q}(\sqrt[3]{5}, \sqrt{5}\,i)$, since the degree of both of these extensions is 6.   □

Theorem 21.22. Let E be a field extension of F. Then the following statements are equivalent.

1. E is a finite extension of F.

2. There exists a finite number of algebraic elements $\alpha_1, \ldots, \alpha_n \in E$ such that $E = F(\alpha_1, \ldots, \alpha_n)$.

3. There exists a sequence of fields

$$E = F(\alpha_1, \ldots, \alpha_n) \supset F(\alpha_1, \ldots, \alpha_{n-1}) \supset \cdots \supset F(\alpha_1) \supset F,$$

   where each field $F(\alpha_1, \ldots, \alpha_i)$ is algebraic over $F(\alpha_1, \ldots, \alpha_{i-1})$.

Proof. $(1) \Rightarrow (2)$. Let E be a finite algebraic extension of F. Then E is a finite dimensional vector space over F and there exists a basis consisting of elements $\alpha_1, \ldots, \alpha_n$ in E such that $E = F(\alpha_1, \ldots, \alpha_n)$. Each α_i is algebraic over F by Theorem 21.15 on page 338.

$(2) \Rightarrow (3)$. Suppose that $E = F(\alpha_1, \ldots, \alpha_n)$, where every α_i is algebraic over F. Then

$$E = F(\alpha_1, \ldots, \alpha_n) \supset F(\alpha_1, \ldots, \alpha_{n-1}) \supset \cdots \supset F(\alpha_1) \supset F,$$

where each field $F(\alpha_1, \ldots, \alpha_i)$ is algebraic over $F(\alpha_1, \ldots, \alpha_{i-1})$.

$(3) \Rightarrow (1)$. Let

$$E = F(\alpha_1, \ldots, \alpha_n) \supset F(\alpha_1, \ldots, \alpha_{n-1}) \supset \cdots \supset F(\alpha_1) \supset F,$$

where each field $F(\alpha_1, \ldots, \alpha_i)$ is algebraic over $F(\alpha_1, \ldots, \alpha_{i-1})$. Since

$$F(\alpha_1, \ldots, \alpha_i) = F(\alpha_1, \ldots, \alpha_{i-1})(\alpha_i)$$

is simple extension and α_i is algebraic over $F(\alpha_1, \ldots, \alpha_{i-1})$, it follows that

$$[F(\alpha_1, \ldots, \alpha_i) : F(\alpha_1, \ldots, \alpha_{i-1})]$$

is finite for each i. Therefore, $[E : F]$ is finite.   ■

Algebraic Closure

Given a field F, the question arises as to whether or not we can find a field E such that every polynomial $p(x)$ has a root in E. This leads us to the following theorem.

Theorem 21.23. Let E be an extension field of F. The set of elements in E that are algebraic over F form a field.

Proof. Let $\alpha, \beta \in E$ be algebraic over F. Then $F(\alpha, \beta)$ is a finite extension of F. Since every element of $F(\alpha, \beta)$ is algebraic over F, $\alpha \pm \beta$, $\alpha\beta$, and α/β ($\beta \neq 0$) are all algebraic over F. Consequently, the set of elements in E that are algebraic over F form a field. ∎

Corollary 21.24. The set of all algebraic numbers forms a field; that is, the set of all complex numbers that are algebraic over $\mathbb{Q}$ makes up a field.

Let E be a field extension of a field F. We define the ***algebraic closure*** of a field F in E to be the field consisting of all elements in E that are algebraic over F. A field F is ***algebraically closed*** if every nonconstant polynomial in $F[x]$ has a root in F.

Theorem 21.25. A field F is algebraically closed if and only if every nonconstant polynomial in $F[x]$ factors into linear factors over $F[x]$.

Proof. Let F be an algebraically closed field. If $p(x) \in F[x]$ is a nonconstant polynomial, then $p(x)$ has a zero in F, say α. Therefore, $x - \alpha$ must be a factor of $p(x)$ and so $p(x) = (x - \alpha)q_1(x)$, where $\deg q_1(x) = \deg p(x) - 1$. Continue this process with $q_1(x)$ to find a factorization

$$p(x) = (x - \alpha)(x - \beta)q_2(x),$$

where $\deg q_2(x) = \deg p(x) - 2$. The process must eventually stop since the degree of $p(x)$ is finite.

Conversely, suppose that every nonconstant polynomial $p(x)$ in $F[x]$ factors into linear factors. Let $ax - b$ be such a factor. Then $p(b/a) = 0$. Consequently, F is algebraically closed. ∎

Corollary 21.26. An algebraically closed field F has no proper algebraic extension E.

Proof. Let E be an algebraic extension of F; then $F \subset E$. For $\alpha \in E$, the minimal polynomial of α is $x - \alpha$. Therefore, $\alpha \in F$ and $F = E$. ∎

Theorem 21.27. Every field F has a unique algebraic closure.

It is a nontrivial fact that every field has a unique algebraic closure. The proof is not extremely difficult, but requires some rather sophisticated set theory. We refer the reader to [3], [4], or [8] for a proof of this result.

We now state the Fundamental Theorem of Algebra, first proven by Gauss at the age of 22 in his doctoral thesis. This theorem states that every polynomial with coefficients in the complex numbers has a root in the complex numbers. The proof of this theorem will be given in Chapter 23.

Theorem 21.28. Fundamental Theorem of Algebra. The field of complex numbers is algebraically closed.

21.2 Splitting Fields

Let F be a field and $p(x)$ be a nonconstant polynomial in $F[x]$. We already know that we can find a field extension of F that contains a root of $p(x)$. However, we would like to know whether an extension E of F containing all of the roots of $p(x)$ exists. In other words, can we find a field extension of F such that $p(x)$ factors into a product of linear polynomials? What is the "smallest" extension containing all the roots of $p(x)$?

Let F be a field and $p(x) = a_0 + a_1 x + \cdots + a_n x^n$ be a nonconstant polynomial in $F[x]$. An extension field E of F is a *splitting field* of $p(x)$ if there exist elements $\alpha_1, \ldots, \alpha_n$ in E such that $E = F(\alpha_1, \ldots, \alpha_n)$ and

$$p(x) = (x - \alpha_1)(x - \alpha_2)\cdots(x - \alpha_n).$$

A polynomial $p(x) \in F[x]$ *splits* in E if it is the product of linear factors in $E[x]$.

Example 21.29. Let $p(x) = x^4 + 2x^2 - 8$ be in $\mathbb{Q}[x]$. Then $p(x)$ has irreducible factors $x^2 - 2$ and $x^2 + 4$. Therefore, the field $\mathbb{Q}(\sqrt{2}, i)$ is a splitting field for $p(x)$. $\square$

Example 21.30. Let $p(x) = x^3 - 3$ be in $\mathbb{Q}[x]$. Then $p(x)$ has a root in the field $\mathbb{Q}(\sqrt[3]{3})$. However, this field is not a splitting field for $p(x)$ since the complex cube roots of 3,

$$\frac{-\sqrt[3]{3} \pm (\sqrt[6]{3})^5 i}{2},$$

are not in $\mathbb{Q}(\sqrt[3]{3})$. $\square$

Theorem 21.31. Let $p(x) \in F[x]$ be a nonconstant polynomial. Then there exists a splitting field E for $p(x)$.

Proof. We will use mathematical induction on the degree of $p(x)$. If $\deg p(x) = 1$, then $p(x)$ is a linear polynomial and $E = F$. Assume that the theorem is true for all polynomials of degree k with $1 \leq k < n$ and let $\deg p(x) = n$. We can assume that $p(x)$ is irreducible; otherwise, by our induction hypothesis, we are done. By Theorem 21.5 on page 334, there exists a field K such that $p(x)$ has a zero α_1 in K. Hence, $p(x) = (x - \alpha_1)q(x)$, where $q(x) \in K[x]$. Since $\deg q(x) = n - 1$, there exists a splitting field $E \supset K$ of $q(x)$ that contains the zeros $\alpha_2, \ldots, \alpha_n$ of $p(x)$ by our induction hypothesis. Consequently,

$$E = K(\alpha_2, \ldots, \alpha_n) = F(\alpha_1, \ldots, \alpha_n)$$

is a splitting field of $p(x)$.   ∎

The question of uniqueness now arises for splitting fields. This question is answered in the affirmative. Given two splitting fields K and L of a polynomial $p(x) \in F[x]$, there exists a field isomorphism $\phi : K \to L$ that preserves F. In order to prove this result, we must first prove a lemma.

Lemma 21.32. Let $\phi : E \to F$ be an isomorphism of fields. Let K be an extension field of E and $\alpha \in K$ be algebraic over E with minimal polynomial $p(x)$. Suppose that L is an extension field of F such that β is root of the polynomial in $F[x]$ obtained from $p(x)$ under the image of ϕ. Then ϕ extends to a unique isomorphism $\overline{\phi} : E(\alpha) \to F(\beta)$ such that $\overline{\phi}(\alpha) = \beta$ and $\overline{\phi}$ agrees with ϕ on E.

Proof. If $p(x)$ has degree n, then by Theorem 21.13 on page 337 we can write any element in $E(\alpha)$ as a linear combination of $1, \alpha, \ldots, \alpha^{n-1}$. Therefore, the isomorphism that we are seeking must be

$$\overline{\phi}(a_0 + a_1\alpha + \cdots + a_{n-1}\alpha^{n-1}) = \phi(a_0) + \phi(a_1)\beta + \cdots + \phi(a_{n-1})\beta^{n-1},$$

where

$$a_0 + a_1\alpha + \cdots + a_{n-1}\alpha^{n-1}$$

is an element in $E(\alpha)$. The fact that $\overline{\phi}$ is an isomorphism could be checked by direct computation; however, it is easier to observe that $\overline{\phi}$ is a composition of maps that we already know to be isomorphisms.

We can extend ϕ to be an isomorphism from $E[x]$ to $F[x]$, which we will also denote by ϕ, by letting

$$\phi(a_0 + a_1x + \cdots + a_nx^n) = \phi(a_0) + \phi(a_1)x + \cdots + \phi(a_n)x^n.$$

This extension agrees with the original isomorphism $\phi : E \to F$, since constant polynomials get mapped to constant polynomials. By assumption, $\phi(p(x)) = q(x)$; hence, ϕ maps $\langle p(x) \rangle$ onto $\langle q(x) \rangle$. Consequently, we have an isomorphism $\psi : E[x]/\langle p(x) \rangle \to F[x]/\langle q(x) \rangle$. By Proposition 21.12 on page 336, we have isomorphisms $\sigma : E[x]/\langle p(x) \rangle \to E(\alpha)$ and $\tau : F[x]/\langle q(x) \rangle \to F(\beta)$, defined by evaluation at α and β, respectively. Therefore, $\bar{\phi} = \tau\psi\sigma^{-1}$ is the required isomorphism (see Figure 21.33).

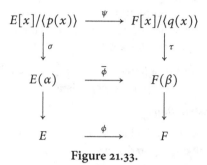

Figure 21.33.

We leave the proof of uniqueness as a exercise. ∎

Theorem 21.34. Let $\phi : E \to F$ be an isomorphism of fields and let $p(x)$ be a nonconstant polynomial in $E[x]$ and $q(x)$ the corresponding polynomial in $F[x]$ under the isomorphism. If K is a splitting field of $p(x)$ and L is a splitting field of $q(x)$, then ϕ extends to an isomorphism $\psi : K \to L$.

Proof. We will use mathematical induction on the degree of $p(x)$. We can assume that $p(x)$ is irreducible over E. Therefore, $q(x)$ is also irreducible over F. If $\deg p(x) = 1$, then by the definition of a splitting field, $K = E$ and $L = F$ and there is nothing to prove.

Assume that the theorem holds for all polynomials of degree less than n. Since K is a splitting field of $p(x)$, all of the roots of $p(x)$ are in K. Choose one of these roots, say α, such that $E \subset E(\alpha) \subset K$. Similarly, we can find a root β of $q(x)$ in L such that $F \subset F(\beta) \subset L$. By Lemma 21.32, there exists an isomorphism $\bar{\phi} : E(\alpha) \to F(\beta)$ such that $\bar{\phi}(\alpha) = \beta$ and $\bar{\phi}$ agrees with ϕ on E (see Figure 21.35).

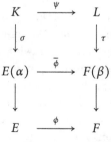

Figure 21.35.

Now write $p(x) = (x - \alpha)f(x)$ and $q(x) = (x - \beta)g(x)$, where the degrees of $f(x)$ and $g(x)$ are less than the degrees of $p(x)$ and $q(x)$, respectively. The field extension K is a splitting field for $f(x)$ over $E(\alpha)$, and L is a splitting field for $g(x)$ over $F(\beta)$. By our induction hypothesis there exists an isomorphism $\psi : K \to L$ such that ψ agrees with $\bar{\phi}$ on $E(\alpha)$. Hence, there exists an isomorphism $\psi : K \to L$ such that ψ agrees with ϕ on E.   ∎

Corollary 21.36. Let $p(x)$ be a polynomial in $F[x]$. Then there exists a splitting field K of $p(x)$ that is unique up to isomorphism.

21.3 Geometric Constructions

In ancient Greece, three classic problems were posed. These problems are geometric in nature and involve straightedge-and-compass constructions from what is now high school geometry; that is, we are allowed to use only a straightedge and compass to solve them. The problems can be stated as follows.

1. Given an arbitrary angle, can one trisect the angle into three equal subangles using only a straightedge and compass?

2. Given an arbitrary circle, can one construct a square with the same area using only a straightedge and compass?

3. Given a cube, can one construct the edge of another cube having twice the volume of the original? Again, we are only allowed to use a straightedge and compass to do the construction.

After puzzling mathematicians for over two thousand years, each of these constructions was finally shown to be impossible. We will use the theory of fields to provide a proof that the solutions do not exist. It is quite remarkable that the long-sought solution to each of these three geometric problems came from abstract algebra.

First, let us determine more specifically what we mean by a straightedge and compass, and also examine the nature of these problems in a bit more depth. To begin with, *a straightedge is not a ruler.* We cannot measure arbitrary lengths with a straightedge. It is merely a tool for drawing a line through two points. The statement that the trisection of an arbitrary angle is impossible means that there is at least one angle that is impossible to trisect with a straightedge-and-compass construction. Certainly it is possible to trisect an angle in special cases. We can construct a $30°$ angle; hence, it is possible to trisect a $90°$ angle. However, we will show that it is impossible to construct a $20°$ angle. Therefore, we cannot trisect a $60°$ angle.

Constructible Numbers

A real number α is *constructible* if we can construct a line segment of length $|\alpha|$ in a finite number of steps from a segment of unit length by using a straightedge and compass.

Theorem 21.37. The set of all constructible real numbers forms a subfield F of the field of real numbers.

Proof. Let α and β be constructible numbers. We must show that $\alpha + \beta$, $\alpha - \beta$, $\alpha\beta$, and α/β ($\beta \neq 0$) are also constructible numbers. We can assume that both α and β are positive with $\alpha > \beta$. It is quite obvious how to construct $\alpha + \beta$ and $\alpha - \beta$. To find a line segment with length $\alpha\beta$, we assume that $\beta > 1$ and construct the triangle in Figure 21.38 such that triangles $\triangle ABC$ and $\triangle ADE$ are similar. Since $\alpha/1 = x/\beta$, the line segment x has length $\alpha\beta$. A similar construction can be made if $\beta < 1$. We will leave it as an exercise to show that the same triangle can be used to construct α/β for $\beta \neq 0$.   ∎

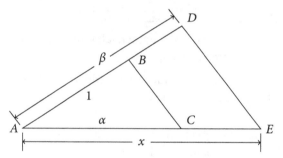

Figure 21.38. Construction of products

Lemma 21.39. If α is a constructible number, then $\sqrt{\alpha}$ is a constructible number.

Proof. In Figure 21.40 on the next page the triangles $\triangle ABD$, $\triangle BCD$, and $\triangle ABC$ are similar; hence, $1/x = x/\alpha$, or $x^2 = \alpha$.   ∎

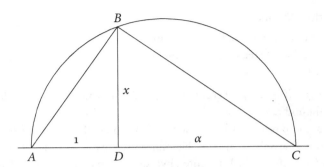

Figure 21.40. Construction of roots

By Theorem 21.37 on the preceding page, we can locate in the plane any point $P = (p, q)$ that has rational coordinates p and q. We need to know what other points can be constructed with a compass and straightedge from points with rational coordinates.

Lemma 21.41. Let F be a subfield of $\mathbb{R}$.

1. If a line contains two points in F, then it has the equation $ax + by + c = 0$, where a, b, and c are in F.

2. If a circle has a center at a point with coordinates in F and a radius that is also in F, then it has the equation $x^2 + y^2 + dx + ey + f = 0$, where d, e, and f are in F.

Proof. Let (x_1, y_1) and (x_2, y_2) be points on a line whose coordinates are in F. If $x_1 = x_2$, then the equation of the line through the two points is $x - x_1 = 0$, which has the form $ax + by + c = 0$. If $x_1 \neq x_2$, then the equation of the line through the two points is given by

$$y - y_1 = \left(\frac{y_2 - y_1}{x_2 - x_1} \right) (x - x_1),$$

which can also be put into the proper form.

To prove the second part of the lemma, suppose that (x_1, y_1) is the center of a circle of radius r. Then the circle has the equation

$$(x - x_1)^2 + (y - y_1)^2 - r^2 = 0.$$

This equation can easily be put into the appropriate form. ∎

Starting with a field of constructible numbers F, we have three possible ways of constructing additional points in $\mathbb{R}$ with a compass and straightedge.

1. To find possible new points in $\mathbb{R}$, we can take the intersection of two lines, each of which passes through two known points with coordinates in F.

2. The intersection of a line that passes through two points that have coordinates in F and a circle whose center has coordinates in F with radius of a length in F will give new points in $\mathbb{R}$.

3. We can obtain new points in $\mathbb{R}$ by intersecting two circles whose centers have coordinates in F and whose radii are of lengths in F.

The first case gives no new points in $\mathbb{R}$, since the solution of two equations of the form $ax + by + c = 0$ having coefficients in F will always be in F. The third case can be reduced to the second case. Let

$$x^2 + y^2 + d_1 x + e_1 y + f_1 = 0$$
$$x^2 + y^2 + d_2 x + e_2 y + f_2 = 0$$

be the equations of two circles, where d_i, e_i, and f_i are in F for $i = 1, 2$. These circles have the same intersection as the circle

$$x^2 + y^2 + d_1 x + e_1 x + f_1 = 0$$

and the line

$$(d_1 - d_2)x + b(e_2 - e_1)y + (f_2 - f_1) = 0.$$

The last equation is that of the chord passing through the intersection points of the two circles. Hence, the intersection of two circles can be reduced to the case of an intersection of a line with a circle.

Considering the case of the intersection of a line and a circle, we must determine the nature of the solutions of the equations

$$ax + by + c = 0$$
$$x^2 + y^2 + dx + ey + f = 0.$$

If we eliminate y from these equations, we obtain an equation of the form $Ax^2 + Bx + C = 0$, where A, B, and C are in F. The x coordinate of the intersection points is given by

$$x = \frac{-B \pm \sqrt{B^2 - 4AC}}{2A}$$

and is in $F(\sqrt{\alpha})$, where $\alpha = B^2 - 4AC > 0$. We have proven the following lemma.

Lemma 21.42. Let F be a field of constructible numbers. Then the points determined by the intersections of lines and circles in F lie in the field $F(\sqrt{\alpha})$ for some α in F.

Theorem 21.43. A real number α is a constructible number if and only if there exists a sequence of fields

$$\mathbb{Q} = F_0 \subset F_1 \subset \cdots \subset F_k$$

such that $F_i = F_{i-1}(\sqrt{\alpha_i})$ with $\alpha_i \in F_i$ and $\alpha \in F_k$. In particular, there exists an integer $k > 0$ such that $[\mathbb{Q}(\alpha) : \mathbb{Q}] = 2^k$.

Proof. The existence of the F_i's and the α_i's is a direct consequence of Lemma 21.42 and of the fact that

$$[F_k : \mathbb{Q}] = [F_k : F_{k-1}][F_{k-1} : F_{k-2}]\cdots[F_1 : \mathbb{Q}] = 2^k.$$

∎

Corollary 21.44. The field of all constructible numbers is an algebraic extension of $\mathbb{Q}$.

As we can see by the field of constructible numbers, not every algebraic extension of a field is a finite extension.

Doubling the Cube and Squaring the Circle

We are now ready to investigate the classical problems of doubling the cube and squaring the circle. We can use the field of constructible numbers to show exactly when a particular geometric construction can be accomplished.

Doubling the cube is impossible. Given the edge of the cube, it is impossible to construct with a straightedge and compass the edge of the cube that has twice the volume of the original cube. Let the original cube have an edge of length 1 and, therefore, a volume of 1. If we could construct a cube having a volume of 2, then this new cube would have an edge of length $\sqrt[3]{2}$. However, $\sqrt[3]{2}$ is a zero of the irreducible polynomial $x^3 - 2$ over $\mathbb{Q}$; hence,

$$[\mathbb{Q}(\sqrt[3]{2}) : \mathbb{Q}] = 3$$

This is impossible, since 3 is not a power of 2.

Squaring the circle. Suppose that we have a circle of radius 1. The area of the circle is π; therefore, we must be able to construct a square with side $\sqrt{\pi}$. This is impossible since π and consequently $\sqrt{\pi}$ are both transcendental. Therefore, using a straightedge and compass, it is not possible to construct a square with the same area as the circle.

Trisecting an Angle

Trisecting an arbitrary angle is impossible. We will show that it is impossible to construct a 20° angle. Consequently, a 60° angle cannot be trisected. We first need to calculate the triple-angle formula for the cosine:

$$
\begin{aligned}
\cos 3\theta &= \cos(2\theta + \theta) \\
&= \cos 2\theta \cos \theta - \sin 2\theta \sin \theta \\
&= (2\cos^2 \theta - 1)\cos \theta - 2\sin^2 \theta \cos \theta \\
&= (2\cos^2 \theta - 1)\cos \theta - 2(1 - \cos^2 \theta)\cos \theta \\
&= 4\cos^3 \theta - 3\cos \theta.
\end{aligned}
$$

The angle θ can be constructed if and only if $\alpha = \cos \theta$ is constructible. Let $\theta = 20°$. Then $\cos 3\theta = \cos 60° = 1/2$. By the triple-angle formula for the cosine,

$$
4\alpha^3 - 3\alpha = \frac{1}{2}.
$$

Therefore, α is a zero of $8x^3 - 6x - 1$. This polynomial has no factors in $\mathbb{Z}[x]$, and hence is irreducible over $\mathbb{Q}[x]$. Thus, $[\mathbb{Q}(\alpha) : \mathbb{Q}] = 3$. Consequently, α cannot be a constructible number.

Sage.   Extensions of the field of rational numbers are a central object of study in number theory, so with Sage's roots in this discipline, it is no surprise that there is extensive support for fields and for extensions of the rationals. Sage also contains an implementation of the entire field of algebraic numbers, with exact representations.

↪   Historical Note   ↩

Algebraic number theory uses the tools of algebra to solve problems in number theory. Modern algebraic number theory began with Pierre de Fermat (1601–1665). Certainly we can find many positive integers that satisfy the equation $x^2 + y^2 = z^2$; Fermat conjectured that the equation $x^n + y^n = z^n$ has no positive integer solutions for $n \geq 3$. He stated in the margin of his copy of the Latin translation of Diophantus' *Arithmetica* that he had found a marvelous proof of this theorem, but that the margin of the book was too narrow to contain it. Building on work of other mathematicians, it was Andrew Wiles who finally succeeded in proving Fermat's Last Theorem in the 1990s. Wiles's achievement was reported on the front page of the *New York Times*.

Attempts to prove Fermat's Last Theorem have led to important contributions to algebraic number theory by such notable mathematicians as Leonhard Euler (1707–1783). Significant advances in the understanding of Fermat's Last Theorem were made by Ernst Kummer (1810–1893). Kummer's student, Leopold Kronecker (1823–1891), became one of the leading algebraists of the nineteenth century. Kronecker's theory of ideals and his study of algebraic number theory added much to the understanding of fields.

David Hilbert (1862–1943) and Hermann Minkowski (1864–1909) were among the mathematicians who led the way in this subject at the beginning of the twentieth century. Hilbert and Minkowski were both mathematicians at Göttingen University in Germany. Göttingen was truly one the most important centers of mathematical research during the last two centuries. The large number of exceptional mathematicians who studied there included Gauss, Dirichlet, Riemann, Dedekind, Noether, and Weyl.

André Weil answered questions in number theory using algebraic geometry, a field of mathematics that studies geometry by studying commutative rings. From about 1955 to 1970, Alexander Grothendieck dominated the field of algebraic geometry. Pierre Deligne, a student of Grothendieck, solved several of Weil's number-theoretic conjectures. One of the most recent contributions to algebra and number theory is Gerd Falting's proof of the Mordell-Weil conjecture. This conjecture of Mordell and Weil essentially says that certain polynomials $p(x, y)$ in $\mathbb{Z}[x, y]$ have only a finite number of integral solutions.

21.4 Reading Questions

1. What does it mean for an extension field E of a field F to be a simple extension of F?

2. What is the definition of a minimal polynomial of an element $\alpha \in E$, where E is an extension of F, and α is algebraic over F?

3. Describe how linear algebra enters into this chapter. What critical result relies on a proof that is almost entirely linear algebra?

4. What is the definition of an algebraically closed field?

5. What is a splitting field of a polynomial $p(x) \in F[x]$?

21.5 Exercises

1. Show that each of the following numbers is algebraic over $\mathbb{Q}$ by finding the minimal polynomial of the number over $\mathbb{Q}$.

    (a) $\sqrt{1/3 + \sqrt{7}}$

    (b) $\sqrt{3} + \sqrt[3]{5}$

    (c) $\sqrt{3} + \sqrt{2}\,i$

    (d) $\cos\theta + i\sin\theta$ for $\theta = 2\pi/n$ with $n \in \mathbb{N}$

    (e) $\sqrt{\sqrt[3]{2} - i}$

2. Find a basis for each of the following field extensions. What is the degree of each extension?

    (a) $\mathbb{Q}(\sqrt{3}, \sqrt{6})$ over $\mathbb{Q}$

    (b) $\mathbb{Q}(\sqrt[3]{2}, \sqrt[3]{3})$ over $\mathbb{Q}$

    (c) $\mathbb{Q}(\sqrt{2}, i)$ over $\mathbb{Q}$

    (d) $\mathbb{Q}(\sqrt{3}, \sqrt{5}, \sqrt{7})$ over $\mathbb{Q}$

    (e) $\mathbb{Q}(\sqrt{2}, \sqrt[3]{2})$ over $\mathbb{Q}$

    (f) $\mathbb{Q}(\sqrt{8})$ over $\mathbb{Q}(\sqrt{2})$

    (g) $\mathbb{Q}(i, \sqrt{2} + i, \sqrt{3} + i)$ over $\mathbb{Q}$

    (h) $\mathbb{Q}(\sqrt{2} + \sqrt{5})$ over $\mathbb{Q}(\sqrt{5})$

    (i) $\mathbb{Q}(\sqrt{2}, \sqrt{6} + \sqrt{10})$ over $\mathbb{Q}(\sqrt{3} + \sqrt{5})$

3. Find the splitting field for each of the following polynomials.

    (a) $x^4 - 10x^2 + 21$ over $\mathbb{Q}$ (c) $x^3 + 2x + 2$ over $\mathbb{Z}_3$

    (b) $x^4 + 1$ over $\mathbb{Q}$ (d) $x^3 - 3$ over $\mathbb{Q}$

4. Consider the field extension $\mathbb{Q}(\sqrt[4]{3}, i)$ over $\mathbb{Q}$.

    (a) Find a basis for the field extension $\mathbb{Q}(\sqrt[4]{3}, i)$ over $\mathbb{Q}$. Conclude that $[\mathbb{Q}(\sqrt[4]{3}, i) : \mathbb{Q}] = 8$.

    (b) Find all subfields F of $\mathbb{Q}(\sqrt[4]{3}, i)$ such that $[F : \mathbb{Q}] = 2$.

    (c) Find all subfields F of $\mathbb{Q}(\sqrt[4]{3}, i)$ such that $[F : \mathbb{Q}] = 4$.

5. Show that $\mathbb{Z}_2[x]/\langle x^3 + x + 1 \rangle$ is a field with eight elements. Construct a multiplication table for the multiplicative group of the field.

6. Show that the regular 9-gon is not constructible with a straightedge and compass, but that the regular 20-gon is constructible.

7. Prove that the cosine of one degree ($\cos 1°$) is algebraic over $\mathbb{Q}$ but not constructible.

8. Can a cube be constructed with three times the volume of a given cube?

9. Prove that $\mathbb{Q}(\sqrt{3}, \sqrt[4]{3}, \sqrt[8]{3}, \ldots)$ is an algebraic extension of $\mathbb{Q}$ but not a finite extension.

10. Prove or disprove: π is algebraic over $\mathbb{Q}(\pi^3)$.

11. Let $p(x)$ be a nonconstant polynomial of degree n in $F[x]$. Prove that there exists a splitting field E for $p(x)$ such that $[E : F] \leq n!$.

12. Prove or disprove: $\mathbb{Q}(\sqrt{2}) \cong \mathbb{Q}(\sqrt{3})$.

13. Prove that the fields $\mathbb{Q}(\sqrt[4]{3})$ and $\mathbb{Q}(\sqrt[4]{3}\,i)$ are isomorphic but not equal.

14. Let K be an algebraic extension of E, and E an algebraic extension of F. Prove that K is algebraic over F. [*Caution*: Do not assume that the extensions are finite.]

15. Prove or disprove: $\mathbb{Z}[x]/\langle x^3 - 2 \rangle$ is a field.

16. Let F be a field of characteristic p. Prove that $p(x) = x^p - a$ either is irreducible over F or splits in F.

17. Let E be the algebraic closure of a field F. Prove that every polynomial $p(x)$ in $F[x]$ splits in E.

18. If every irreducible polynomial $p(x)$ in $F[x]$ is linear, show that F is an algebraically closed field.

19. Prove that if α and β are constructible numbers such that $\beta \neq 0$, then so is α/β.

20. Show that the set of all elements in $\mathbb{R}$ that are algebraic over $\mathbb{Q}$ form a field extension of $\mathbb{Q}$ that is not finite.

21. Let E be an algebraic extension of a field F, and let σ be an automorphism of E leaving F fixed. Let $\alpha \in E$. Show that σ induces a permutation of the set of all zeros of the minimal polynomial of α that are in E.

22. Show that $\mathbb{Q}(\sqrt{3}, \sqrt{7}) = \mathbb{Q}(\sqrt{3} + \sqrt{7})$. Extend your proof to show that $\mathbb{Q}(\sqrt{a}, \sqrt{b}) = \mathbb{Q}(\sqrt{a} + \sqrt{b})$, where $a \neq b$ and neither a nor b is a perfect square.

23. Let E be a finite extension of a field F. If $[E : F] = 2$, show that E is a splitting field of F for some polynomial $f(x) \in F[x]$.

24. Prove or disprove: Given a polynomial $p(x)$ in $\mathbb{Z}_6[x]$, it is possible to construct a ring R such that $p(x)$ has a root in R.

25. Let E be a field extension of F and $\alpha \in E$. Determine $[F(\alpha) : F(\alpha^3)]$.

26. Let α, β be transcendental over $\mathbb{Q}$. Prove that either $\alpha\beta$ or $\alpha + \beta$ is also transcendental.

27. Let E be an extension field of F and $\alpha \in E$ be transcendental over F. Prove that every element in $F(\alpha)$ that is not in F is also transcendental over F.

28. Let α be a root of an irreducible monic polynomial $p(x) \in F[x]$, with $\deg p = n$. Prove that $[F(\alpha) : F] = n$.

21.6 References and Suggested Readings

[1]   Dean, R. A. *Elements of Abstract Algebra* . Wiley, New York, 1966.

[2]   Dudley, U. *A Budget of Trisections*. Springer-Verlag, New York, 1987. An interesting and entertaining account of how not to trisect an angle.

[3]   Fraleigh, J. B. *A First Course in Abstract Algebra*. 7th ed. Pearson, Upper Saddle River, NJ, 2003.

[4]   Kaplansky, I. *Fields and Rings*, 2nd ed. University of Chicago Press, Chicago, 1972.

[5]   Klein, F. *Famous Problems of Elementary Geometry*. Chelsea, New York, 1955.

[6]   Martin, G. *Geometric Constructions*. Springer, New York, 1998.

[7]   H. Pollard and H. G. Diamond. *Theory of Algebraic Numbers*, Dover, Mineola, NY, 2010.

[8]   Walker, E. A. *Introduction to Abstract Algebra*. Random House, New York, 1987. This work contains a proof showing that every field has an algebraic closure.

22

Finite Fields

---

*F*inite fields appear in many applications of algebra, including coding theory and cryptography. We already know one finite field, $\mathbb{Z}_p$, where p is prime. In this chapter we will show that a unique finite field of order p^n exists for every prime p, where n is a positive integer. Finite fields are also called Galois fields in honor of Évariste Galois, who was one of the first mathematicians to investigate them.

22.1 Structure of a Finite Field

Recall that a field F has *characteristic* p if p is the smallest positive integer such that for every nonzero element α in F, we have $p\alpha = 0$. If no such integer exists, then F has characteristic 0. From Theorem 16.19 on page 246 we know that p must be prime. Suppose that F is a finite field with n elements. Then $n\alpha = 0$ for all α in F. Consequently, the characteristic of F must be p, where p is a prime dividing n. This discussion is summarized in the following proposition.

Proposition 22.1. If F is a finite field, then the characteristic of F is p, where p is prime.

Throughout this chapter we will assume that p is a prime number unless otherwise stated.

Proposition 22.2. If F is a finite field of characteristic p, then the order of F is p^n for some $n \in \mathbb{N}$.

Proof. Let $\phi : \mathbb{Z} \to F$ be the ring homomorphism defined by $\phi(n) = n \cdot 1$. Since the characteristic of F is p, the kernel of ϕ must be $p\mathbb{Z}$ and the image of ϕ must be a subfield of F isomorphic to $\mathbb{Z}_p$. We will denote this subfield by K. Since F is a finite field, it must be a finite extension of K and, therefore, an algebraic extension of K. Suppose that $[F : K] = n$ is the dimension of F, where F is a K vector space. There must exist elements $\alpha_1, \ldots, \alpha_n \in F$ such that any element α in F can be written uniquely in the form

$$\alpha = a_1\alpha_1 + \cdots + a_n\alpha_n,$$

where the a_i's are in K. Since there are p elements in K, there are p^n possible linear combinations of the α_i's. Therefore, the order of F must be p^n. ∎

Lemma 22.3. Freshman's Dream. Let p be prime and D be an integral domain of characteristic p. Then

$$a^{p^n} + b^{p^n} = (a + b)^{p^n}$$

for all positive integers n.

Proof. We will prove this lemma using mathematical induction on n. We can use the binomial formula (see Example 2.4 on page 23) to verify the case for $n = 1$; that is,

$$(a + b)^p = \sum_{k=0}^{p} \binom{p}{k} a^k b^{p-k}.$$

If $0 < k < p$, then

$$\binom{p}{k} = \frac{p!}{k!(p - k)!}$$

must be divisible by p, since p cannot divide $k!(p - k)!$. Note that D is an integral domain of characteristic p, so all but the first and last terms in the sum must be zero. Therefore, $(a + b)^p = a^p + b^p$.

Now suppose that the result holds for all k, where $1 \le k \le n$. By the induction hypothesis,

$$(a + b)^{p^{n+1}} = ((a + b)^p)^{p^n} = (a^p + b^p)^{p^n} = (a^p)^{p^n} + (b^p)^{p^n} = a^{p^{n+1}} + b^{p^{n+1}}.$$

Therefore, the lemma is true for $n + 1$ and the proof is complete. ∎

Let F be a field. A polynomial $f(x) \in F[x]$ of degree n is *separable* if it has n distinct roots in the splitting field of $f(x)$; that is, $f(x)$ is separable when it factors into distinct linear factors over the splitting field of f. An extension E of F is a *separable extension* of F if every element in E is the root of a separable polynomial in $F[x]$.

Example 22.4. The polynomial $x^2 - 2$ is separable over $\mathbb{Q}$ since it factors as $(x - \sqrt{2})(x + \sqrt{2})$. In fact, $\mathbb{Q}(\sqrt{2})$ is a separable extension of $\mathbb{Q}$. Let $\alpha = a + b\sqrt{2}$ be any element in $\mathbb{Q}(\sqrt{2})$. If $b = 0$, then α is a root of $x - a$. If $b \ne 0$, then α is the root of the separable polynomial

$$x^2 - 2ax + a^2 - 2b^2 = (x - (a + b\sqrt{2}))(x - (a - b\sqrt{2})).$$

□

Fortunately, we have an easy test to determine the separability of any polynomial. Let

$$f(x) = a_0 + a_1 x + \cdots + a_n x^n$$

be any polynomial in $F[x]$. Define the *derivative* of $f(x)$ to be

$$f'(x) = a_1 + 2a_2 x + \cdots + na_n x^{n-1}.$$

Lemma 22.5. Let F be a field and $f(x) \in F[x]$. Then $f(x)$ is separable if and only if $f(x)$ and $f'(x)$ are relatively prime.

Proof. Let $f(x)$ be separable. Then $f(x)$ factors over some extension field of F as $f(x) = (x - \alpha_1)(x - \alpha_2)\cdots(x - \alpha_n)$, where $\alpha_i \neq \alpha_j$ for $i \neq j$. Taking the derivative of $f(x)$, we see that

$$\begin{aligned}
f'(x) = {} & (x - \alpha_2)\cdots(x - \alpha_n) \\
& + (x - \alpha_1)(x - \alpha_3)\cdots(x - \alpha_n) \\
& + \cdots + (x - \alpha_1)\cdots(x - \alpha_{n-1}).
\end{aligned}$$

Hence, $f(x)$ and $f'(x)$ can have no common factors.

To prove the converse, we will show that the contrapositive of the statement is true. Suppose that $f(x) = (x - \alpha)^k g(x)$, where $k > 1$. Differentiating, we have

$$f'(x) = k(x - \alpha)^{k-1} g(x) + (x - \alpha)^k g'(x).$$

Therefore, $f(x)$ and $f'(x)$ have a common factor. ∎

Theorem 22.6. For every prime p and every positive integer n, there exists a finite field F with p^n elements. Furthermore, any field of order p^n is isomorphic to the splitting field of $x^{p^n} - x$ over $\mathbb{Z}_p$.

Proof. Let $f(x) = x^{p^n} - x$ and let F be the splitting field of $f(x)$. Then by Lemma 22.5, $f(x)$ has p^n distinct zeros in F, since $f'(x) = p^n x^{p^n - 1} - 1 = -1$ is relatively prime to $f(x)$. We claim that the roots of $f(x)$ form a subfield of F. Certainly 0 and 1 are zeros of $f(x)$. If α and β are zeros of $f(x)$, then $\alpha + \beta$ and $\alpha\beta$ are also zeros of $f(x)$, since $\alpha^{p^n} + \beta^{p^n} = (\alpha + \beta)^{p^n}$ and $\alpha^{p^n} \beta^{p^n} = (\alpha\beta)^{p^n}$. We also need to show that the additive inverse and the multiplicative inverse of each root of $f(x)$ are roots of $f(x)$. For any zero α of $f(x)$, we know that $-\alpha$ is also a zero of $f(x)$, since

$$f(-\alpha) = (-\alpha)^{p^n} - (-\alpha) = -\alpha^{p^n} + \alpha = -(\alpha^{p^n} - \alpha) = 0,$$

provided p is odd. If $p = 2$, then

$$f(-\alpha) = (-\alpha)^{2^n} - (-\alpha) = \alpha + \alpha = 0.$$

If $\alpha \neq 0$, then $(\alpha^{-1})^{p^n} = (\alpha^{p^n})^{-1} = \alpha^{-1}$. Since the zeros of $f(x)$ form a subfield of F and $f(x)$ splits in this subfield, the subfield must be all of F.

Let E be any other field of order p^n. To show that E is isomorphic to F, we must show that every element in E is a root of $f(x)$. Certainly 0 is a root of $f(x)$. Let α be a nonzero element of E. The order of the multiplicative group of nonzero elements of E is $p^n - 1$; hence, $\alpha^{p^n-1} = 1$ or $\alpha^{p^n} - \alpha = 0$. Since E contains p^n elements, E must be a splitting field of $f(x)$; however, by Corollary 21.36 on page 346, the splitting field of any polynomial is unique up to isomorphism. ∎

The unique finite field with p^n elements is called the **Galois field** of order p^n. We will denote this field by $\mathrm{GF}(p^n)$.

Theorem 22.7. Every subfield of the Galois field $\mathrm{GF}(p^n)$ has p^m elements, where m divides n. Conversely, if $m \mid n$ for $m > 0$, then there exists a unique subfield of $\mathrm{GF}(p^n)$ isomorphic to $\mathrm{GF}(p^m)$.

Proof. Let F be a subfield of $E = \mathrm{GF}(p^n)$. Then F must be a field extension of K that contains p^m elements, where K is isomorphic to $\mathbb{Z}_p$. Then $m \mid n$, since $[E : K] = [E : F][F : K]$.

To prove the converse, suppose that $m \mid n$ for some $m > 0$. Then $p^m - 1$ divides $p^n - 1$. Consequently, $x^{p^m-1} - 1$ divides $x^{p^n-1} - 1$. Therefore, $x^{p^m} - x$ must divide $x^{p^n} - x$, and every zero of $x^{p^m} - x$ is also a zero of $x^{p^n} - x$. Thus, $\mathrm{GF}(p^n)$ contains, as a subfield, a splitting field of $x^{p^m} - x$, which must be isomorphic to $\mathrm{GF}(p^m)$. ∎

Example 22.8. The lattice of subfields of $\mathrm{GF}(p^{24})$ is given in Figure 22.9. □

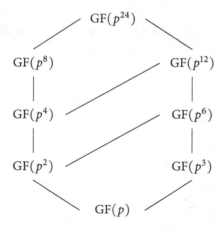

Figure 22.9. Subfields of $\mathrm{GF}(p^{24})$

With each field F we have a multiplicative group of nonzero elements of F which we will denote by F^*. The multiplicative group of any finite field is cyclic. This result follows from the more general result that we will prove in the next theorem.

Theorem 22.10. If G is a finite subgroup of F^*, the multiplicative group of nonzero elements of a field F, then G is cyclic.

Proof. Let G be a finite subgroup of F^* of order n. By the Fundamental Theorem of Finite Abelian Groups (Theorem 13.4 on page 200),

$$G \cong \mathbb{Z}_{p_1^{e_1}} \times \cdots \times \mathbb{Z}_{p_k^{e_k}},$$

where $n = p_1^{e_1} \cdots p_k^{e_k}$ and the $p_1, \ldots, p_k$ are (not necessarily distinct) primes. Let m be the least common multiple of $p_1^{e_1}, \ldots, p_k^{e_k}$. Then G contains an element of order m. Since every α in G satisfies $x^r - 1$ for some r dividing m, α must also be a root of $x^m - 1$. Since $x^m - 1$ has at most m roots in F, $n \leq m$. On the other hand, we know that $m \leq |G|$; therefore, $m = n$. Thus, G contains an element of order n and must be cyclic. ∎

Corollary 22.11. The multiplicative group of all nonzero elements of a finite field is cyclic.

Corollary 22.12. Every finite extension E of a finite field F is a simple extension of F.

Proof. Let α be a generator for the cyclic group E^* of nonzero elements of E. Then $E = F(\alpha)$. ∎

Example 22.13. The finite field $\mathrm{GF}(2^4)$ is isomorphic to the field $\mathbb{Z}_2/\langle 1 + x + x^4 \rangle$. Therefore, the elements of $\mathrm{GF}(2^4)$ can be taken to be

$$\{a_0 + a_1\alpha + a_2\alpha^2 + a_3\alpha^3 : a_i \in \mathbb{Z}_2 \text{ and } 1 + \alpha + \alpha^4 = 0\}.$$

Remembering that $1 + \alpha + \alpha^4 = 0$, we add and multiply elements of $\mathrm{GF}(2^4)$ exactly as we add and multiply polynomials. The multiplicative group of $\mathrm{GF}(2^4)$ is isomorphic to $\mathbb{Z}_{15}$ with generator α:

$\alpha^1 = \alpha$	$\alpha^6 = \alpha^2 + \alpha^3$	$\alpha^{11} = \alpha + \alpha^2 + \alpha^3$
$\alpha^2 = \alpha^2$	$\alpha^7 = 1 + \alpha + \alpha^3$	$\alpha^{12} = 1 + \alpha + \alpha^2 + \alpha^3$
$\alpha^3 = \alpha^3$	$\alpha^8 = 1 + \alpha^2$	$\alpha^{13} = 1 + \alpha^2 + \alpha^3$
$\alpha^4 = 1 + \alpha$	$\alpha^9 = \alpha + \alpha^3$	$\alpha^{14} = 1 + \alpha^3$
$\alpha^5 = \alpha + \alpha^2$	$\alpha^{10} = 1 + \alpha + \alpha^2$	$\alpha^{15} = 1.$

□

22.2 Polynomial Codes

With knowledge of polynomial rings and finite fields, it is now possible to derive more sophisticated codes than those of Chapter 8. First let us recall that an (n, k)-block code consists of a one-to-one encoding function $E : \mathbb{Z}_2^k \to \mathbb{Z}_2^n$ and a decoding function $D : \mathbb{Z}_2^n \to \mathbb{Z}_2^k$. The code is error-correcting if D is onto. A code is a linear code if it is the null space of a matrix $H \in \mathbb{M}_{k \times n}(\mathbb{Z}_2)$.

 We are interested in a class of codes known as cyclic codes. Let $\phi : \mathbb{Z}_2^k \to \mathbb{Z}_2^n$ be a binary (n, k)-block code. Then ϕ is a *cyclic code* if for every codeword $(a_1, a_2, \ldots, a_n)$, the cyclically shifted n-tuple $(a_n, a_1, a_2, \ldots, a_{n-1})$ is also a codeword. Cyclic codes are particularly easy to implement on a computer using shift registers [2, 3].

Example 22.14. Consider the $(6, 3)$-linear codes generated by the two matrices

$$
G_1 = \begin{pmatrix} 1 & 0 & 0 \\ 0 & 1 & 0 \\ 0 & 0 & 1 \\ 1 & 0 & 0 \\ 0 & 1 & 0 \\ 0 & 0 & 1 \end{pmatrix} \quad \text{and} \quad G_2 = \begin{pmatrix} 1 & 0 & 0 \\ 1 & 1 & 0 \\ 1 & 1 & 1 \\ 1 & 1 & 1 \\ 0 & 1 & 1 \\ 0 & 0 & 1 \end{pmatrix}.
$$

Messages in the first code are encoded as follows:

$$
\begin{array}{llll}
(000) & \mapsto & (000000) & \qquad (100) \mapsto (100100) \\
(001) & \mapsto & (001001) & \qquad (101) \mapsto (101101) \\
(010) & \mapsto & (010010) & \qquad (110) \mapsto (110110) \\
(011) & \mapsto & (011011) & \qquad (111) \mapsto (111111).
\end{array}
$$

It is easy to see that the codewords form a cyclic code. In the second code, 3-tuples are encoded in the following manner:

$$
\begin{array}{llll}
(000) & \mapsto & (000000) & \qquad (100) \mapsto (111100) \\
(001) & \mapsto & (001111) & \qquad (101) \mapsto (110011) \\
(010) & \mapsto & (011110) & \qquad (110) \mapsto (100010) \\
(011) & \mapsto & (010001) & \qquad (111) \mapsto (101101).
\end{array}
$$

This code cannot be cyclic, since (101101) is a codeword but (011011) is not a codeword. □

Polynomial Codes

We would like to find an easy method of obtaining cyclic linear codes. To accomplish this, we can use our knowledge of finite fields and polynomial rings over $\mathbb{Z}_2$. Any binary n-tuple can be interpreted as a polynomial in $\mathbb{Z}_2[x]$. Stated another way, the n-tuple $(a_0, a_1, \ldots, a_{n-1})$ corresponds to the polynomial

$$f(x) = a_0 + a_1 x + \cdots + a_{n-1} x^{n-1},$$

where the degree of $f(x)$ is at most $n - 1$. For example, the polynomial corresponding to the 5-tuple (10011) is

$$1 + 0x + 0x^2 + 1x^3 + 1x^4 = 1 + x^3 + x^4.$$

Conversely, with any polynomial $f(x) \in \mathbb{Z}_2[x]$ with $\deg f(x) < n$ we can associate a binary n-tuple. The polynomial $x + x^2 + x^4$ corresponds to the 5-tuple (01101).

Let us fix a nonconstant polynomial $g(x)$ in $\mathbb{Z}_2[x]$ of degree $n - k$. We can define an (n, k)-code C in the following manner. If $(a_0, \ldots, a_{k-1})$ is a k-tuple to be encoded, then $f(x) = a_0 + a_1 x + \cdots + a_{k-1} x^{k-1}$ is the corresponding polynomial in $\mathbb{Z}_2[x]$. To encode $f(x)$, we multiply by $g(x)$. The codewords in C are all those polynomials in $\mathbb{Z}_2[x]$ of degree less than n that are divisible by $g(x)$. Codes obtained in this manner are called *polynomial codes.*

Example 22.15. If we let $g(x) = 1 + x^3$, we can define a $(6, 3)$-code C as follows. To encode a 3-tuple (a_0, a_1, a_2), we multiply the corresponding polynomial $f(x) = a_0 + a_1 x + a_2 x^2$ by $1 + x^3$. We are defining a map $\phi : \mathbb{Z}_2^3 \to \mathbb{Z}_2^6$ by $\phi : f(x) \mapsto g(x)f(x)$. It is easy to check that this map is a group homomorphism. In fact, if we regard $\mathbb{Z}_2^n$ as a vector space over $\mathbb{Z}_2$, ϕ is a linear transformation of vector spaces (see Exercise 20.5.15 on page 329). Let us compute the kernel of ϕ. Observe that $\phi(a_0, a_1, a_2) = (000000)$ exactly when

$$0 + 0x + 0x^2 + 0x^3 + 0x^4 + 0x^5 = (1 + x^3)(a_0 + a_1 x + a_2 x^2)$$
$$= a_0 + a_1 x + a_2 x^2 + a_0 x^3 + a_1 x^4 + a_2 x^5.$$

Since the polynomials over a field form an integral domain, $a_0 + a_1 x + a_2 x^2$ must be the zero polynomial. Therefore, $\ker \phi = \{(000)\}$ and ϕ is one-to-one.

To calculate a generator matrix for C, we merely need to examine the way the polynomials 1, x, and x^2 are encoded:

$$(1 + x^3) \cdot 1 = 1 + x^3$$
$$(1 + x^3)x = x + x^4$$
$$(1 + x^3)x^2 = x^2 + x^5.$$

We obtain the code corresponding to the generator matrix G_1 in Example 22.14 on page 361. The parity-check matrix for this code is

$$H = \begin{pmatrix} 1 & 0 & 0 & 1 & 0 & 0 \\ 0 & 1 & 0 & 0 & 1 & 0 \\ 0 & 0 & 1 & 0 & 0 & 1 \end{pmatrix}.$$

Since the smallest weight of any nonzero codeword is 2, this code has the ability to detect all single errors. □

Rings of polynomials have a great deal of structure; therefore, our immediate goal is to establish a link between polynomial codes and ring theory. Recall that $x^n - 1 = (x-1)(x^{n-1} + \cdots + x + 1)$. The factor ring

$$R_n = \mathbb{Z}_2[x]/\langle x^n - 1 \rangle$$

can be considered to be the ring of polynomials of the form

$$f(t) = a_0 + a_1 t + \cdots + a_{n-1} t^{n-1}$$

that satisfy the condition $t^n = 1$. It is an easy exercise to show that $\mathbb{Z}_2^n$ and R_n are isomorphic as vector spaces. We will often identify elements in $\mathbb{Z}_2^n$ with elements in $\mathbb{Z}[x]/\langle x^n - 1 \rangle$. In this manner we can interpret a linear code as a subset of $\mathbb{Z}[x]/\langle x^n - 1 \rangle$.

The additional ring structure on polynomial codes is very powerful in describing cyclic codes. A cyclic shift of an n-tuple can be described by polynomial multiplication. If $f(t) = a_0 + a_1 t + \cdots + a_{n-1} t^{n-1}$ is a code polynomial in R_n, then

$$tf(t) = a_{n-1} + a_0 t + \cdots + a_{n-2} t^{n-1}$$

is the cyclically shifted word obtained from multiplying $f(t)$ by t. The following theorem gives a beautiful classification of cyclic codes in terms of the ideals of R_n.

Theorem 22.16. A linear code C in $\mathbb{Z}_2^n$ is cyclic if and only if it is an ideal in $R_n = \mathbb{Z}[x]/\langle x^n - 1 \rangle$.

Proof. Let C be a linear cyclic code and suppose that $f(t)$ is in C. Then $tf(t)$ must also be in C. Consequently, $t^k f(t)$ is in C for all $k \in \mathbb{N}$. Since C is a linear code, any linear combination of the codewords $f(t), tf(t), t^2 f(t), \ldots, t^{n-1} f(t)$ is also a codeword; therefore, for every polynomial $p(t)$, $p(t)f(t)$ is in C. Hence, C is an ideal.

Conversely, let C be an ideal in $\mathbb{Z}_2[x]/\langle x^n + 1 \rangle$. Suppose that $f(t) = a_0 + a_1 t + \cdots + a_{n-1} t^{n-1}$ is a codeword in C. Then $tf(t)$ is a codeword in C; that is, $(a_1, \ldots, a_{n-1}, a_0)$ is in C. ∎

Theorem 22.16 on the preceding page tells us that knowing the ideals of R_n is equivalent to knowing the linear cyclic codes in $\mathbb{Z}_2^n$. Fortunately, the ideals in R_n are easy to describe. The natural ring homomorphism $\phi : \mathbb{Z}_2[x] \to R_n$ defined by $\phi[f(x)] = f(t)$ is a surjective homomorphism. The kernel of ϕ is the ideal generated by $x^n - 1$. By Theorem 16.34 on page 250, every ideal C in R_n is of the form $\phi(I)$, where I is an ideal in $\mathbb{Z}_2[x]$ that contains $\langle x^n - 1\rangle$. By Theorem 17.20 on page 277, we know that every ideal I in $\mathbb{Z}_2[x]$ is a principal ideal, since $\mathbb{Z}_2$ is a field. Therefore, $I = \langle g(x)\rangle$ for some unique monic polynomial in $\mathbb{Z}_2[x]$. Since $\langle x^n - 1\rangle$ is contained in I, it must be the case that $g(x)$ divides $x^n - 1$. Consequently, every ideal C in R_n is of the form

$$C = \langle g(t)\rangle = \{f(t)g(t) : f(t) \in R_n \text{ and } g(x) \mid (x^n - 1) \text{ in } \mathbb{Z}_2[x]\}.$$

The unique monic polynomial of the smallest degree that generates C is called the *minimal generator polynomial* of C.

Example 22.17. If we factor $x^7 - 1$ into irreducible components, we have

$$x^7 - 1 = (1 + x)(1 + x + x^3)(1 + x^2 + x^3).$$

We see that $g(t) = (1 + t + t^3)$ generates an ideal C in R_7. This code is a $(7, 4)$-block code. As in Example 22.15 on page 362, it is easy to calculate a generator matrix by examining what $g(t)$ does to the polynomials 1, t, t^2, and t^3. A generator matrix for C is

$$G = \begin{pmatrix} 1 & 0 & 0 & 0 \\ 1 & 1 & 0 & 0 \\ 0 & 1 & 1 & 0 \\ 1 & 0 & 1 & 1 \\ 0 & 1 & 0 & 1 \\ 0 & 0 & 1 & 0 \\ 0 & 0 & 0 & 1 \end{pmatrix}.$$

□

In general, we can determine a generator matrix for an (n, k)-code C by the manner in which the elements t^k are encoded. Let $x^n - 1 = g(x)h(x)$ in $\mathbb{Z}_2[x]$. If $g(x) = g_0 + g_1 x + \cdots + g_{n-k}x^{n-k}$ and $h(x) = h_0 + h_1 x + \cdots + h_k x^k$, then the

$n \times k$ matrix

$$G = \begin{pmatrix} g_0 & 0 & \cdots & 0 \\ g_1 & g_0 & \cdots & 0 \\ \vdots & \vdots & \ddots & \vdots \\ g_{n-k} & g_{n-k-1} & \cdots & g_0 \\ 0 & g_{n-k} & \cdots & g_1 \\ \vdots & \vdots & \ddots & \vdots \\ 0 & 0 & \cdots & g_{n-k} \end{pmatrix}$$

is a generator matrix for the code C with generator polynomial $g(t)$. The parity-check matrix for C is the $(n-k) \times n$ matrix

$$H = \begin{pmatrix} 0 & \cdots & 0 & 0 & h_k & \cdots & h_0 \\ 0 & \cdots & 0 & h_k & \cdots & h_0 & 0 \\ \cdots & \cdots & \cdots & \cdots & \cdots & \cdots & \cdots \\ h_k & \cdots & h_0 & 0 & 0 & \cdots & 0 \end{pmatrix}.$$

We will leave the details of the proof of the following proposition as an exercise.

Proposition 22.18. Let $C = \langle g(t) \rangle$ be a cyclic code in R_n and suppose that $x^n - 1 = g(x)h(x)$. Then G and H are generator and parity-check matrices for C, respectively. Furthermore, $HG = 0$.

Example 22.19. In Example 22.17,

$$x^7 - 1 = g(x)h(x) = (1 + x + x^3)(1 + x + x^2 + x^4).$$

Therefore, a parity-check matrix for this code is

$$H = \begin{pmatrix} 0 & 0 & 1 & 0 & 1 & 1 & 1 \\ 0 & 1 & 0 & 1 & 1 & 1 & 0 \\ 1 & 0 & 1 & 1 & 1 & 0 & 0 \end{pmatrix}.$$

$\square$

To determine the error-detecting and error-correcting capabilities of a cyclic code, we need to know something about determinants. If $\alpha_1, \ldots, \alpha_n$ are elements in a field F, then the $n \times n$ matrix

$$\begin{pmatrix} 1 & 1 & \cdots & 1 \\ \alpha_1 & \alpha_2 & \cdots & \alpha_n \\ \alpha_1^2 & \alpha_2^2 & \cdots & \alpha_n^2 \\ \vdots & \vdots & \ddots & \vdots \\ \alpha_1^{n-1} & \alpha_2^{n-1} & \cdots & \alpha_n^{n-1} \end{pmatrix}$$

is called the **Vandermonde matrix.** The determinant of this matrix is called the **Vandermonde determinant.** We will need the following lemma in our investigation of cyclic codes.

Lemma 22.20. Let $\alpha_1, \ldots, \alpha_n$ be elements in a field F with $n \geq 2$. Then

$$\det \begin{pmatrix} 1 & 1 & \cdots & 1 \\ \alpha_1 & \alpha_2 & \cdots & \alpha_n \\ \alpha_1^2 & \alpha_2^2 & \cdots & \alpha_n^2 \\ \vdots & \vdots & \ddots & \vdots \\ \alpha_1^{n-1} & \alpha_2^{n-1} & \cdots & \alpha_n^{n-1} \end{pmatrix} = \prod_{1 \leq j < i \leq n} (\alpha_i - \alpha_j).$$

In particular, if the α_i's are distinct, then the determinant is nonzero.

Proof. We will induct on n. If $n = 2$, then the determinant is $\alpha_2 - \alpha_1$. Let us assume the result for $n - 1$ and consider the polynomial $p(x)$ defined by

$$p(x) = \det \begin{pmatrix} 1 & 1 & \cdots & 1 & 1 \\ \alpha_1 & \alpha_2 & \cdots & \alpha_{n-1} & x \\ \alpha_1^2 & \alpha_2^2 & \cdots & \alpha_{n-1}^2 & x^2 \\ \vdots & \vdots & \ddots & \vdots & \vdots \\ \alpha_1^{n-1} & \alpha_2^{n-1} & \cdots & \alpha_{n-1}^{n-1} & x^{n-1} \end{pmatrix}.$$

Expanding this determinant by cofactors on the last column, we see that $p(x)$ is a polynomial of at most degree $n - 1$. Moreover, the roots of $p(x)$ are $\alpha_1, \ldots, \alpha_{n-1}$, since the substitution of any one of these elements in the last column will produce a column identical to the last column in the matrix. Remember that the determinant of a matrix is zero if it has two identical columns. Therefore,

$$p(x) = (x - \alpha_1)(x - \alpha_2)\cdots(x - \alpha_{n-1})\beta,$$

where

$$\beta = (-1)^{n+n} \det \begin{pmatrix} 1 & 1 & \cdots & 1 \\ \alpha_1 & \alpha_2 & \cdots & \alpha_{n-1} \\ \alpha_1^2 & \alpha_2^2 & \cdots & \alpha_{n-1}^2 \\ \vdots & \vdots & \ddots & \vdots \\ \alpha_1^{n-2} & \alpha_2^{n-2} & \cdots & \alpha_{n-1}^{n-2} \end{pmatrix}.$$

By our induction hypothesis,

$$\beta = (-1)^{n+n} \prod_{1 \leq j < i \leq n-1} (\alpha_i - \alpha_j).$$

If we let $x = \alpha_n$, the result now follows immediately. ∎

The following theorem gives us an estimate on the error detection and correction capabilities for a particular generator polynomial.

Theorem 22.21. Let $C = \langle g(t) \rangle$ be a cyclic code in R_n and suppose that ω is a primitive nth root of unity over $\mathbb{Z}_2$. If s consecutive powers of ω are roots of $g(x)$, then the minimum distance of C is at least $s + 1$.

Proof. Suppose that

$$g(\omega^r) = g(\omega^{r+1}) = \cdots = g(\omega^{r+s-1}) = 0.$$

Let $f(x)$ be some polynomial in C with s or fewer nonzero coefficients. We can assume that

$$f(x) = a_{i_0} x^{i_0} + a_{i_1} x^{i_1} + \cdots + a_{i_{s-1}} x^{i_{s-1}}$$

be some polynomial in C. It will suffice to show that all of the a_i's must be 0. Since

$$g(\omega^r) = g(\omega^{r+1}) = \cdots = g(\omega^{r+s-1}) = 0$$

and $g(x)$ divides $f(x)$,

$$f(\omega^r) = f(\omega^{r+1}) = \cdots = f(\omega^{r+s-1}) = 0.$$

Equivalently, we have the following system of equations:

$$a_{i_0}(\omega^r)^{i_0} + a_{i_1}(\omega^r)^{i_1} + \cdots + a_{i_{s-1}}(\omega^r)^{i_{s-1}} = 0$$
$$a_{i_0}(\omega^{r+1})^{i_0} + a_{i_1}(\omega^{r+1})^{i_2} + \cdots + a_{i_{s-1}}(\omega^{r+1})^{i_{s-1}} = 0$$

$$\vdots$$

$$a_{i_0}(\omega^{r+s-1})^{i_0} + a_{i_1}(\omega^{r+s-1})^{i_1} + \cdots + a_{i_{s-1}}(\omega^{r+s-1})^{i_{s-1}} = 0.$$

Therefore, $(a_{i_0}, a_{i_1}, \ldots, a_{i_{s-1}})$ is a solution to the homogeneous system of linear equations

$$(\omega^{i_0})^r x_0 + (\omega^{i_1})^r x_1 + \cdots + (\omega^{i_{s-1}})^r x_{n-1} = 0$$
$$(\omega^{i_0})^{r+1} x_0 + (\omega^{i_1})^{r+1} x_1 + \cdots + (\omega^{i_{s-1}})^{r+1} x_{n-1} = 0$$
$$\vdots$$
$$(\omega^{i_0})^{r+s-1} x_0 + (\omega^{i_1})^{r+s-1} x_1 + \cdots + (\omega^{i_{s-1}})^{r+s-1} x_{n-1} = 0.$$

However, this system has a unique solution, since the determinant of the matrix

$$\begin{pmatrix} (\omega^{i_0})^r & (\omega^{i_1})^r & \cdots & (\omega^{i_{s-1}})^r \\ (\omega^{i_0})^{r+1} & (\omega^{i_1})^{r+1} & \cdots & (\omega^{i_{s-1}})^{r+1} \\ \vdots & \vdots & \ddots & \vdots \\ (\omega^{i_0})^{r+s-1} & (\omega^{i_1})^{r+s-1} & \cdots & (\omega^{i_{s-1}})^{r+s-1} \end{pmatrix}$$

can be shown to be nonzero using Lemma 22.20 on page 366 and the basic properties of determinants (Exercise). Therefore, this solution must be $a_{i_0} = a_{i_1} = \cdots = a_{i_{s-1}} = 0$. ∎

BCH Codes

Some of the most important codes, discovered independently by A. Hocquenghem in 1959 and by R. C. Bose and D. V. Ray-Chaudhuri in 1960, are BCH codes. The European and transatlantic communication systems both use BCH codes. Information words to be encoded are of length 231, and a polynomial of degree 24 is used to generate the code. Since $231 + 24 = 255 = 2^8 - 1$, we are dealing with a $(255, 231)$-block code. This BCH code will detect six errors and has a failure rate of 1 in 16 million. One advantage of BCH codes is that efficient error correction algorithms exist for them.

The idea behind BCH codes is to choose a generator polynomial of smallest degree that has the largest error detection and error correction capabilities. Let $d = 2r + 1$ for some $r \geq 0$. Suppose that ω is a primitive nth root of unity over $\mathbb{Z}_2$, and let $m_i(x)$ be the minimal polynomial over $\mathbb{Z}_2$ of ω^i. If

$$g(x) = \text{lcm}[m_1(x), m_2(x), \ldots, m_{2r}(x)],$$

then the cyclic code $\langle g(t) \rangle$ in R_n is called the *BCH code of length n and distance d*. By Theorem 22.21 on the preceding page, the minimum distance of C is at least d.

Theorem 22.22. Let $C = \langle g(t) \rangle$ be a cyclic code in R_n. The following statements are equivalent.

1. The code C is a BCH code whose minimum distance is at least d.

2. A code polynomial $f(t)$ is in C if and only if $f(\omega^i) = 0$ for $1 \le i < d$.

3. The matrix

$$H = \begin{pmatrix} 1 & \omega & \omega^2 & \cdots & \omega^{n-1} \\ 1 & \omega^2 & \omega^4 & \cdots & \omega^{(n-1)(2)} \\ 1 & \omega^3 & \omega^6 & \cdots & \omega^{(n-1)(3)} \\ \vdots & \vdots & \vdots & \ddots & \vdots \\ 1 & \omega^{2r} & \omega^{4r} & \cdots & \omega^{(n-1)(2r)} \end{pmatrix}$$

is a parity-check matrix for C.

Proof. $(1) \Rightarrow (2)$. If $f(t)$ is in C, then $g(x) \mid f(x)$ in $\mathbb{Z}_2[x]$. Hence, for $i = 1, \ldots, 2r$, $f(\omega^i) = 0$ since $g(\omega^i) = 0$. Conversely, suppose that $f(\omega^i) = 0$ for $1 \le i \le d$. Then $f(x)$ is divisible by each $m_i(x)$, since $m_i(x)$ is the minimal polynomial of ω^i. Therefore, $g(x) \mid f(x)$ by the definition of $g(x)$. Consequently, $f(x)$ is a codeword.

$(2) \Rightarrow (3)$. Let $f(t) = a_0 + a_1 t + \cdots + a_{n-1} v t^{n-1}$ be in R_n. The corresponding n-tuple in $\mathbb{Z}_2^n$ is $\mathbf{x} = (a_0 a_1 \cdots a_{n-1})^t$. By (2),

$$H\mathbf{x} = \begin{pmatrix} a_0 + a_1 \omega + \cdots + a_{n-1} \omega^{n-1} \\ a_0 + a_1 \omega^2 + \cdots + a_{n-1} (\omega^2)^{n-1} \\ \vdots \\ a_0 + a_1 \omega^{2r} + \cdots + a_{n-1} (\omega^{2r})^{n-1} \end{pmatrix} = \begin{pmatrix} f(\omega) \\ f(\omega^2) \\ \vdots \\ f(\omega^{2r}) \end{pmatrix} = 0$$

exactly when $f(t)$ is in C. Thus, H is a parity-check matrix for C.

$(3) \Rightarrow (1)$. By (3), a code polynomial $f(t) = a_0 + a_1 t + \cdots + a_{n-1} t^{n-1}$ is in C exactly when $f(\omega^i) = 0$ for $i = 1, \ldots, 2r$. The smallest such polynomial is $g(t) = \text{lcm}[m_1(t), \ldots, m_{2r}(t)]$. Therefore, $C = \langle g(t) \rangle$. ∎

Example 22.23. It is easy to verify that $x^{15} - 1 \in \mathbb{Z}_2[x]$ has a factorization

$$x^{15} - 1 = (x+1)(x^2 + x + 1)(x^4 + x + 1)(x^4 + x^3 + 1)(x^4 + x^3 + x^2 + x + 1),$$

where each of the factors is an irreducible polynomial. Let ω be a root of $1 + x + x^4$. The Galois field $GF(2^4)$ is

$$\{ a_0 + a_1 \omega + a_2 \omega^2 + a_3 \omega^3 : a_i \in \mathbb{Z}_2 \text{ and } 1 + \omega + \omega^4 = 0 \}.$$

By Example 22.8 on page 359, ω is a primitive 15th root of unity. The minimal polynomial of ω is $m_1(x) = 1 + x + x^4$. It is easy to see that ω^2 and ω^4 are also roots of $m_1(x)$. The minimal polynomial of ω^3 is $m_2(x) = 1 + x + x^2 + x^3 + x^4$. Therefore,

$$g(x) = m_1(x)m_2(x) = 1 + x^4 + x^6 + x^7 + x^8$$

has roots ω, ω^2, ω^3, ω^4. Since both $m_1(x)$ and $m_2(x)$ divide $x^{15} - 1$, the BCH code is a $(15, 7)$-code. If $x^{15} - 1 = g(x)h(x)$, then $h(x) = 1 + x^4 + x^6 + x^7$; therefore, a parity-check matrix for this code is

$$\begin{pmatrix}
0 & 0 & 0 & 0 & 0 & 0 & 0 & 1 & 1 & 0 & 1 & 0 & 0 & 0 & 1 \\
0 & 0 & 0 & 0 & 0 & 0 & 1 & 1 & 0 & 1 & 0 & 0 & 0 & 1 & 0 \\
0 & 0 & 0 & 0 & 0 & 1 & 1 & 0 & 1 & 0 & 0 & 0 & 1 & 0 & 0 \\
0 & 0 & 0 & 0 & 1 & 1 & 0 & 1 & 0 & 0 & 0 & 1 & 0 & 0 & 0 \\
0 & 0 & 0 & 1 & 1 & 0 & 1 & 0 & 0 & 0 & 1 & 0 & 0 & 0 & 0 \\
0 & 0 & 1 & 1 & 0 & 1 & 0 & 0 & 0 & 1 & 0 & 0 & 0 & 0 & 0 \\
0 & 1 & 1 & 0 & 1 & 0 & 0 & 0 & 1 & 0 & 0 & 0 & 0 & 0 & 0 \\
1 & 1 & 0 & 1 & 0 & 0 & 0 & 1 & 0 & 0 & 0 & 0 & 0 & 0 & 0
\end{pmatrix}.$$

$\square$

Sage. Finite fields are important in a variety of applied disciplines, such as cryptography and coding theory (see introductions to these topics in other chapters). Sage has excellent support for finite fields allowing for a wide variety of computations.

22.3 Reading Questions

1. When is a field extension separable?
2. What are the possible orders for subfields of a finite field?
3. What is the structure of the non-zero elements of a finite field?
4. Provide a characterization of finite fields using the concept of a splitting field.
5. Why is a theorem in this chapter titled "The Freshman's Dream?"

22.4 Exercises

1. Calculate each of the following.
   (a) $[GF(3^6) : GF(3^3)]$ (c) $[GF(625) : GF(25)]$
   (b) $[GF(128) : GF(16)]$ (d) $[GF(p^{12}) : GF(p^2)]$
2. Calculate $[GF(p^m) : GF(p^n)]$, where $n \mid m$.

3. What is the lattice of subfields for $GF(p^{30})$?

4. Let α be a zero of $x^3 + x^2 + 1$ over $\mathbb{Z}_2$. Construct a finite field of order 8. Show that $x^3 + x^2 + 1$ splits in $\mathbb{Z}_2(\alpha)$.

5. Construct a finite field of order 27.

6. Prove or disprove: $\mathbb{Q}^*$ is cyclic.

7. Factor each of the following polynomials in $\mathbb{Z}_2[x]$.
   (a) $x^5 - 1$ (c) $x^9 - 1$
   (b) $x^6 + x^5 + x^4 + x^3 + x^2 + x + 1$ (d) $x^4 + x^3 + x^2 + x + 1$

8. Prove or disprove: $\mathbb{Z}_2[x]/\langle x^3 + x + 1\rangle \cong \mathbb{Z}_2[x]/\langle x^3 + x^2 + 1\rangle$.

9. Determine the number of cyclic codes of length n for $n = 6, 7, 8, 10$.

10. Prove that the ideal $\langle t + 1\rangle$ in R_n is the code in $\mathbb{Z}_2^n$ consisting of all words of even parity.

11. Construct all BCH codes of
    (a) length 7.   (b) length 15.

12. Prove or disprove: There exists a finite field that is algebraically closed.

13. Let p be prime. Prove that the field of rational functions $\mathbb{Z}_p(x)$ is an infinite field of characteristic p.

14. Let D be an integral domain of characteristic p. Prove that $(a-b)^{p^n} = a^{p^n} - b^{p^n}$ for all $a, b \in D$.

15. Show that every element in a finite field can be written as the sum of two squares.

16. Let E and F be subfields of a finite field K. If E is isomorphic to F, show that $E = F$.

17. Let $F \subset E \subset K$ be fields. If K is a separable extension of F, show that K is also separable extension of E.

18. Let E be an extension of a finite field F, where F has q elements. Let $\alpha \in E$ be algebraic over F of degree n. Prove that $F(\alpha)$ has q^n elements.

19. Show that every finite extension of a finite field F is simple; that is, if E is a finite extension of a finite field F, prove that there exists an $\alpha \in E$ such that $E = F(\alpha)$.

20. Show that for every n there exists an irreducible polynomial of degree n in $\mathbb{Z}_p[x]$.

21. Prove that the *Frobenius map* $\Phi : GF(p^n) \to GF(p^n)$ given by $\Phi : \alpha \mapsto \alpha^p$ is an automorphism of order n.

22. Show that every element in $GF(p^n)$ can be written in the form a^p for some unique $a \in GF(p^n)$.

23. Let E and F be subfields of $GF(p^n)$. If $|E| = p^r$ and $|F| = p^s$, what is the order of $E \cap F$?

24. **Wilson's Theorem.** Let p be prime. Prove that $(p-1)! \equiv -1 \pmod{p}$.

25. If $g(t)$ is the minimal generator polynomial for a cyclic code C in R_n, prove that the constant term of $g(x)$ is 1.

26. Often it is conceivable that a burst of errors might occur during transmission, as in the case of a power surge. Such a momentary burst of interference might alter several consecutive bits in a codeword. Cyclic codes permit the detection of such error bursts. Let C be an (n, k)-cyclic code. Prove that any error burst up to $n - k$ digits can be detected.

27. Prove that the rings R_n and $\mathbb{Z}_2^n$ are isomorphic as vector spaces.

28. Let C be a code in R_n that is generated by $g(t)$. If $\langle f(t) \rangle$ is another code in R_n, show that $\langle g(t) \rangle \subset \langle f(t) \rangle$ if and only if $f(x)$ divides $g(x)$ in $\mathbb{Z}_2[x]$.

29. Let $C = \langle g(t) \rangle$ be a cyclic code in R_n and suppose that $x^n - 1 = g(x)h(x)$, where $g(x) = g_0 + g_1 x + \cdots + g_{n-k} x^{n-k}$ and $h(x) = h_0 + h_1 x + \cdots + h_k x^k$. Define G to be the $n \times k$ matrix

$$
G = \begin{pmatrix}
g_0 & 0 & \cdots & 0 \\
g_1 & g_0 & \cdots & 0 \\
\vdots & \vdots & \ddots & \vdots \\
g_{n-k} & g_{n-k-1} & \cdots & g_0 \\
0 & g_{n-k} & \cdots & g_1 \\
\vdots & \vdots & \ddots & \vdots \\
0 & 0 & \cdots & g_{n-k}
\end{pmatrix}
$$

and H to be the $(n - k) \times n$ matrix

$$
H = \begin{pmatrix}
0 & \cdots & 0 & 0 & h_k & \cdots & h_0 \\
0 & \cdots & 0 & h_k & \cdots & h_0 & 0 \\
\cdots & \cdots & \cdots & \cdots & \cdots & \cdots & \cdots \\
h_k & \cdots & h_0 & 0 & 0 & \cdots & 0
\end{pmatrix}.
$$

(a) Prove that G is a generator matrix for C.

(b) Prove that H is a parity-check matrix for C.

(c) Show that $HG = 0$.

22.5 Additional Exercises: Error Correction for BCH Codes

BCH codes have very attractive error correction algorithms. Let C be a BCH code in R_n, and suppose that a code polynomial $c(t) = c_0 + c_1t + \cdots + c_{n-1}t^{n-1}$ is transmitted. Let $w(t) = w_0 + w_1t + \cdots w_{n-1}t^{n-1}$ be the polynomial in R_n that is received. If errors have occurred in bits $a_1, \ldots, a_k$, then $w(t) = c(t) + e(t)$, where $e(t) = t^{a_1} + t^{a_2} + \cdots + t^{a_k}$ is the *error polynomial*. The decoder must determine the integers a_i and then recover $c(t)$ from $w(t)$ by flipping the a_ith bit. From $w(t)$ we can compute $w(\omega^i) = s_i$ for $i = 1, \ldots, 2r$, where ω is a primitive nth root of unity over $\mathbb{Z}_2$. We say the *syndrome* of $w(t)$ is $s_1, \ldots, s_{2r}$.

1. Show that $w(t)$ is a code polynomial if and only if $s_i = 0$ for all i.

2. Show that

    $$s_i = w(\omega^i) = e(\omega^i) = \omega^{ia_1} + \omega^{ia_2} + \cdots + \omega^{ia_k}$$

    for $i = 1, \ldots, 2r$. The *error-locator polynomial* is defined to be

    $$s(x) = (x + \omega^{a_1})(x + \omega^{a_2}) \cdots (x + \omega^{a_k}).$$

3. Recall the $(15, 7)$-block BCH code in Example 22.19 on page 365. By Theorem 8.13 on page 120, this code is capable of correcting two errors. Suppose that these errors occur in bits a_1 and a_2. The error-locator polynomial is $s(x) = (x + \omega^{a_1})(x + \omega^{a_2})$. Show that

    $$s(x) = x^2 + s_1 x + \left(s_1^2 + \frac{s_3}{s_1} \right).$$

4. Let $w(t) = 1 + t^2 + t^4 + t^5 + t^7 + t^{12} + t^{13}$. Determine what the originally transmitted code polynomial was.

22.6 References and Suggested Readings

[1] Childs, L. *A Concrete Introduction to Higher Algebra.* 2nd ed. Springer-Verlag, New York, 1995.

[2] Gåding, L. and Tambour, T. *Algebra for Computer Science.* Springer-Verlag, New York, 1988.

[3] Lidl, R. and Pilz, G. *Applied Abstract Algebra.* 2nd ed. Springer, New York, 1998. An excellent presentation of finite fields and their applications.

[4] Mackiw, G. *Applications of Abstract Algebra.* Wiley, New York, 1985.

[5] Roman, S. *Coding and Information Theory.* Springer-Verlag, New York, 1992.

[6] van Lint, J. H. *Introduction to Coding Theory.* Springer, New York, 1999.

Galois Theory

$\mathcal{A}$ classic problem of algebra is to find the solutions of a polynomial equation. The solution to the quadratic equation was known in antiquity. Italian mathematicians found general solutions to the general cubic and quartic equations in the sixteenth century; however, attempts to solve the general fifth-degree, or quintic, polynomial were repulsed for the next three hundred years. Certainly, equations such as $x^5 - 1 = 0$ or $x^6 - x^3 - 6 = 0$ could be solved, but no solution like the quadratic formula was found for the general quintic,

$$ax^5 + bx^4 + cx^3 + dx^2 + ex + f = 0.$$

Finally, at the beginning of the nineteenth century, Ruffini and Abel both found quintics that could not be solved with any formula. It was Galois, however, who provided the full explanation by showing which polynomials could and could not be solved by formulas. He discovered the connection between groups and field extensions. Galois theory demonstrates the strong interdependence of group and field theory, and has had far-reaching implications beyond its original purpose.

In this chapter we will prove the Fundamental Theorem of Galois Theory. This result will be used to establish the insolvability of the quintic and to prove the Fundamental Theorem of Algebra.

23.1 Field Automorphisms

Our first task is to establish a link between group theory and field theory by examining automorphisms of fields.

Proposition 23.1. The set of all automorphisms of a field F is a group under composition of functions.

Proof. If σ and τ are automorphisms of F, then so are $\sigma\tau$ and σ^{-1}. The identity is certainly an automorphism; hence, the set of all automorphisms of a field F is indeed a group. ∎

Proposition 23.2. Let E be a field extension of F. Then the set of all automorphisms of E that fix F elementwise is a group; that is, the set of all automorphisms $\sigma :$ $E \to E$ such that $\sigma(\alpha) = \alpha$ for all $\alpha \in F$ is a group.

Proof. We need only show that the set of automorphisms of E that fix F elementwise is a subgroup of the group of all automorphisms of E. Let σ and τ be two automorphisms of E such that $\sigma(\alpha) = \alpha$ and $\tau(\alpha) = \alpha$ for all $\alpha \in F$. Then $\sigma\tau(\alpha) = \sigma(\alpha) = \alpha$ and $\sigma^{-1}(\alpha) = \alpha$. Since the identity fixes every element of E, the set of automorphisms of E that leave elements of F fixed is a subgroup of the entire group of automorphisms of E.   ∎

Let E be a field extension of F. We will denote the full group of automorphisms of E by $\text{Aut}(E)$. We define the *Galois group* of E over F to be the group of automorphisms of E that fix F elementwise; that is,

$$G(E/F) = \{\sigma \in \text{Aut}(E) : \sigma(\alpha) = \alpha \text{ for all } \alpha \in F\}.$$

If $f(x)$ is a polynomial in $F[x]$ and E is the splitting field of $f(x)$ over F, then we define the Galois group of $f(x)$ to be $G(E/F)$.

Example 23.3. Complex conjugation, defined by $\sigma : a + bi \mapsto a - bi$, is an automorphism of the complex numbers. Since

$$\sigma(a) = \sigma(a + 0i) = a - 0i = a,$$

the automorphism defined by complex conjugation must be in $G(\mathbb{C}/\mathbb{R})$.   □

Example 23.4. Consider the fields $\mathbb{Q} \subset \mathbb{Q}(\sqrt{5}) \subset \mathbb{Q}(\sqrt{3}, \sqrt{5})$. Then for $a, b \in \mathbb{Q}(\sqrt{5})$,

$$\sigma(a + b\sqrt{3}) = a - b\sqrt{3}$$

is an automorphism of $\mathbb{Q}(\sqrt{3}, \sqrt{5})$ leaving $\mathbb{Q}(\sqrt{5})$ fixed. Similarly,

$$\tau(a + b\sqrt{5}) = a - b\sqrt{5}$$

is an automorphism of $\mathbb{Q}(\sqrt{3}, \sqrt{5})$ leaving $\mathbb{Q}(\sqrt{3})$ fixed. The automorphism $\mu = \sigma\tau$ moves both $\sqrt{3}$ and $\sqrt{5}$. It will soon be clear that $\{\text{id}, \sigma, \tau, \mu\}$ is the Galois group of $\mathbb{Q}(\sqrt{3}, \sqrt{5})$ over $\mathbb{Q}$. The following table shows that this group is isomorphic to $\mathbb{Z}_2 \times \mathbb{Z}_2$.

	id	σ	τ	μ
id	id	σ	τ	μ
σ	σ	id	μ	τ
τ	τ	μ	id	σ
μ	μ	τ	σ	id

We may also regard the field $\mathbb{Q}(\sqrt{3}, \sqrt{5})$ as a vector space over $\mathbb{Q}$ that has basis $\{1, \sqrt{3}, \sqrt{5}, \sqrt{15}\}$. It is no coincidence that $|G(\mathbb{Q}(\sqrt{3}, \sqrt{5})/\mathbb{Q})| = [\mathbb{Q}(\sqrt{3}, \sqrt{5}) : \mathbb{Q})] = 4$. □

Proposition 23.5. Let E be a field extension of F and $f(x)$ be a polynomial in $F[x]$. Then any automorphism in $G(E/F)$ defines a permutation of the roots of $f(x)$ that lie in E.

Proof. Let

$$f(x) = a_0 + a_1 x + a_2 x^2 + \cdots + a_n x^n$$

and suppose that $\alpha \in E$ is a zero of $f(x)$. Then for $\sigma \in G(E/F)$,

$$\begin{aligned}
0 &= \sigma(0) \\
&= \sigma(f(\alpha)) \\
&= \sigma(a_0 + a_1\alpha + a_2\alpha^2 + \cdots + a_n\alpha^n) \\
&= a_0 + a_1\sigma(\alpha) + a_2[\sigma(\alpha)]^2 + \cdots + a_n[\sigma(\alpha)]^n;
\end{aligned}$$

therefore, $\sigma(\alpha)$ is also a zero of $f(x)$. ∎

Let E be an algebraic extension of a field F. Two elements $\alpha, \beta \in E$ are *conjugate* over F if they have the same minimal polynomial. For example, in the field $\mathbb{Q}(\sqrt{2})$ the elements $\sqrt{2}$ and $-\sqrt{2}$ are conjugate over $\mathbb{Q}$ since they are both roots of the irreducible polynomial $x^2 - 2$.

A converse of the last proposition exists. The proof follows directly from Lemma 21.32 on page 344.

Proposition 23.6. If α and β are conjugate over F, there exists an isomorphism $\sigma : F(\alpha) \to F(\beta)$ such that σ is the identity when restricted to F.

Theorem 23.7. Let $f(x)$ be a polynomial in $F[x]$ and suppose that E is the splitting field for $f(x)$ over F. If $f(x)$ has no repeated roots, then

$$|G(E/F)| = [E : F].$$

Proof. We will use mathematical induction on $[E : F]$. If $[E : F] = 1$, then $E = F$ and there is nothing to show. If $[E : F] > 1$, let $f(x) = p(x)q(x)$, where $p(x)$ is irreducible of degree d. We may assume that $d > 1$; otherwise, $f(x)$ splits over F and $[E : F] = 1$. Let α be a root of $p(x)$. If $\phi : F(\alpha) \to E$ is any injective homomorphism, then $\phi(\alpha) = \beta$ is a root of $p(x)$, and $\phi : F(\alpha) \to F(\beta)$ is a field automorphism. Since $f(x)$ has no repeated roots, $p(x)$ has exactly d roots $\beta \in E$. By Proposition 23.5, there are exactly d isomorphisms $\phi : F(\alpha) \to F(\beta_i)$ that fix F, one for each root $\beta_1, \ldots, \beta_d$ of $p(x)$ (see Figure 23.8 on the following page).

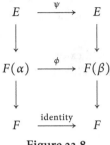

Figure 23.8.

Since E is a splitting field of $f(x)$ over F, it is also a splitting field over $F(\alpha)$. Similarly, E is a splitting field of $f(x)$ over $F(\beta)$. Since $[E : F(\alpha)] = [E : F]/d$, induction shows that each of the d isomorphisms ϕ has exactly $[E : F]/d$ extensions, $\psi : E \to E$, and we have constructed $[E : F]$ isomorphisms that fix F. Finally, suppose that σ is any automorphism fixing F. Then σ restricted to $F(\alpha)$ is ϕ for some $\phi : F(\alpha) \to F(\beta)$.   ∎

Corollary 23.9. Let F be a finite field with a finite extension E such that $[E : F] = k$. Then $G(E/F)$ is cyclic of order k.

Proof. Let p be the characteristic of E and F and assume that the orders of E and F are p^m and p^n, respectively. Then $nk = m$. We can also assume that E is the splitting field of $x^{p^m} - x$ over a subfield of order p. Therefore, E must also be the splitting field of $x^{p^m} - x$ over F. Applying Theorem 23.7 on the previous page, we find that $|G(E/F)| = k$.

To prove that $G(E/F)$ is cyclic, we must find a generator for $G(E/F)$. Let $\sigma : E \to E$ be defined by $\sigma(\alpha) = \alpha^{p^n}$. We claim that σ is the element in $G(E/F)$ that we are seeking. We first need to show that σ is in $\text{Aut}(E)$. If α and β are in E,

$$\sigma(\alpha + \beta) = (\alpha + \beta)^{p^n} = \alpha^{p^n} + \beta^{p^n} = \sigma(\alpha) + \sigma(\beta)$$

by Lemma 22.3 on page 357. Also, it is easy to show that $\sigma(\alpha\beta) = \sigma(\alpha)\sigma(\beta)$. Since σ is a nonzero homomorphism of fields, it must be injective. It must also be onto, since E is a finite field. We know that σ must be in $G(E/F)$, since F is the splitting field of $x^{p^n} - x$ over the base field of order p. This means that σ leaves every element in F fixed. Finally, we must show that the order of σ is k. By Theorem 23.7 on the previous page, we know that

$$\sigma^k(\alpha) = \alpha^{p^{nk}} = \alpha^{p^m} = \alpha$$

is the identity of $G(E/F)$. However, σ^r cannot be the identity for $1 \leq r < k$; otherwise, $x^{p^{nr}} - x$ would have p^m roots, which is impossible. ∎

Example 23.10. We can now confirm that the Galois group of $\mathbb{Q}(\sqrt{3}, \sqrt{5})$ over $\mathbb{Q}$ in Example 23.4 on page 376 is indeed isomorphic to $\mathbb{Z}_2 \times \mathbb{Z}_2$. Certainly the group $H = \{\mathrm{id}, \sigma, \tau, \mu\}$ is a subgroup of $G(\mathbb{Q}(\sqrt{3}, \sqrt{5})/\mathbb{Q})$; however, H must be all of $G(\mathbb{Q}(\sqrt{3}, \sqrt{5})/\mathbb{Q})$, since

$$|H| = [\mathbb{Q}(\sqrt{3}, \sqrt{5}) : \mathbb{Q}] = |G(\mathbb{Q}(\sqrt{3}, \sqrt{5})/\mathbb{Q})| = 4.$$

□

Example 23.11. Let us compute the Galois group of

$$f(x) = x^4 + x^3 + x^2 + x + 1$$

over $\mathbb{Q}$. We know that $f(x)$ is irreducible by Exercise 17.5.20 on page 282. Furthermore, since $(x - 1)f(x) = x^5 - 1$, we can use DeMoivre's Theorem to determine that the roots of $f(x)$ are ω^i, where $i = 1, \ldots, 4$ and

$$\omega = \cos(2\pi/5) + i\sin(2\pi/5).$$

Hence, the splitting field of $f(x)$ must be $\mathbb{Q}(\omega)$. We can define automorphisms σ_i of $\mathbb{Q}(\omega)$ by $\sigma_i(\omega) = \omega^i$ for $i = 1, \ldots, 4$. It is easy to check that these are indeed distinct automorphisms in $G(\mathbb{Q}(\omega)/\mathbb{Q})$. Since

$$[\mathbb{Q}(\omega) : \mathbb{Q}] = |G(\mathbb{Q}(\omega)/\mathbb{Q})| = 4,$$

the σ_i's must be all of $G(\mathbb{Q}(\omega)/\mathbb{Q})$. Therefore, $G(\mathbb{Q}(\omega)/\mathbb{Q}) \cong \mathbb{Z}_4$ since ω is a generator for the Galois group. □

Separable Extensions

Many of the results that we have just proven depend on the fact that a polynomial $f(x)$ in $F[x]$ has no repeated roots in its splitting field. It is evident that we need to know exactly when a polynomial factors into distinct linear factors in its splitting field. Let E be the splitting field of a polynomial $f(x)$ in $F[x]$. Suppose that $f(x)$ factors over E as

$$f(x) = (x - \alpha_1)^{n_1}(x - \alpha_2)^{n_2}\cdots(x - \alpha_r)^{n_r} = \prod_{i=1}^{r}(x - \alpha_i)^{n_i}.$$

We define the ***multiplicity*** of a root α_i of $f(x)$ to be n_i. A root with multiplicity 1 is called a ***simple root***. Recall that a polynomial $f(x) \in F[x]$ of degree n is ***separable*** if it has n distinct roots in its splitting field E. Equivalently, $f(x)$ is separable if it factors into distinct linear factors over $E[x]$. An extension E of F is a *separable extension* of F if every element in E is the root of a separable polynomial in $F[x]$. Also recall that $f(x)$ is separable if and only if $\gcd(f(x), f'(x)) = 1$ (Lemma 22.5 on page 358).

Proposition 23.12. Let $f(x)$ be an irreducible polynomial over F. If the characteristic of F is 0, then $f(x)$ is separable. If the characteristic of F is p and $f(x) \neq g(x^p)$ for some $g(x)$ in $F[x]$, then $f(x)$ is also separable.

Proof. First assume that char $F = 0$. Since $\deg f'(x) < \deg f(x)$ and $f(x)$ is irreducible, the only way $\gcd(f(x), f'(x)) \neq 1$ is if $f'(x)$ is the zero polynomial; however, this is impossible in a field of characteristic zero. If char $F = p$, then $f'(x)$ can be the zero polynomial if every coefficient of $f'(x)$ is a multiple of p. This can happen only if we have a polynomial of the form $f(x) = a_0 + a_1 x^p + a_2 x^{2p} + \cdots + a_n x^{np}$. ∎

Certainly extensions of a field F of the form $F(\alpha)$ are some of the easiest to study and understand. Given a field extension E of F, the obvious question to ask is when it is possible to find an element $\alpha \in E$ such that $E = F(\alpha)$. In this case, α is called a *primitive element*. We already know that primitive elements exist for certain extensions. For example,

$$\mathbb{Q}(\sqrt{3}, \sqrt{5}) = \mathbb{Q}(\sqrt{3} + \sqrt{5})$$

and

$$\mathbb{Q}(\sqrt[3]{5}, \sqrt{5}\, i) = \mathbb{Q}(\sqrt[6]{5}\, i).$$

Corollary 22.12 on page 360 tells us that there exists a primitive element for any finite extension of a finite field. The next theorem tells us that we can often find a primitive element.

Theorem 23.13. Primitive Element Theorem. Let E be a finite separable extension of a field F. Then there exists an $\alpha \in E$ such that $E = F(\alpha)$.

Proof. We already know that there is no problem if F is a finite field. Suppose that E is a finite extension of an infinite field. We will prove the result for $F(\alpha, \beta)$. The general case easily follows when we use mathematical induction. Let $f(x)$ and $g(x)$ be the minimal polynomials of α and β, respectively. Let K be the field in which both $f(x)$ and $g(x)$ split. Suppose that $f(x)$ has zeros $\alpha = \alpha_1, \ldots, \alpha_n$ in K and $g(x)$ has zeros $\beta = \beta_1, \ldots, \beta_m$ in K. All of these zeros have multiplicity 1, since E is separable over F. Since F is infinite, we can find an a in F such that

$$a \neq \frac{\alpha_i - \alpha}{\beta - \beta_j}$$

for all i and j with $j \neq 1$. Therefore, $a(\beta - \beta_j) \neq \alpha_i - \alpha$. Let $\gamma = \alpha + a\beta$. Then

$$\gamma = \alpha + a\beta \neq \alpha_i + a\beta_j;$$

hence, $\gamma - a\beta_j \neq \alpha_i$ for all i, j with $j \neq 1$. Define $h(x) \in F(\gamma)[x]$ by $h(x) = f(\gamma - ax)$. Then $h(\beta) = f(\alpha) = 0$. However, $h(\beta_j) \neq 0$ for $j \neq 1$. Hence, $h(x)$ and $g(x)$ have a single common factor in $F(\gamma)[x]$; that is, the minimal polynomial of β over $F(\gamma)$ must be linear, since β is the only zero common to both $g(x)$ and $h(x)$. So $\beta \in F(\gamma)$ and $\alpha = \gamma - a\beta$ is in $F(\gamma)$. Hence, $F(\alpha, \beta) = F(\gamma)$. ∎

23.2 The Fundamental Theorem

The goal of this section is to prove the Fundamental Theorem of Galois Theory. This theorem explains the connection between the subgroups of $G(E/F)$ and the intermediate fields between E and F.

Proposition 23.14. Let $\{\sigma_i : i \in I\}$ be a collection of automorphisms of a field F. Then

$$F_{\{\sigma_i\}} = \{a \in F : \sigma_i(a) = a \text{ for all } \sigma_i\}$$

is a subfield of F.

Proof. Let $\sigma_i(a) = a$ and $\sigma_i(b) = b$. Then

$$\sigma_i(a \pm b) = \sigma_i(a) \pm \sigma_i(b) = a \pm b$$

and

$$\sigma_i(ab) = \sigma_i(a)\sigma_i(b) = ab.$$

If $a \neq 0$, then $\sigma_i(a^{-1}) = [\sigma_i(a)]^{-1} = a^{-1}$. Finally, $\sigma_i(0) = 0$ and $\sigma_i(1) = 1$ since σ_i is an automorphism. ∎

Corollary 23.15. Let F be a field and let G be a subgroup of $\mathrm{Aut}(F)$. Then

$$F_G = \{\alpha \in F : \sigma(\alpha) = \alpha \text{ for all } \sigma \in G\}$$

is a subfield of F.

The subfield $F_{\{\sigma_i\}}$ of F is called the *fixed field* of $\{\sigma_i\}$. The field fixed by a subgroup G of $\mathrm{Aut}(F)$ will be denoted by F_G.

Example 23.16. Let $\sigma : \mathbb{Q}(\sqrt{3}, \sqrt{5}) \to \mathbb{Q}(\sqrt{3}, \sqrt{5})$ be the automorphism that maps $\sqrt{3}$ to $-\sqrt{3}$. Then $\mathbb{Q}(\sqrt{5})$ is the subfield of $\mathbb{Q}(\sqrt{3}, \sqrt{5})$ left fixed by σ. □

Proposition 23.17. Let E be a splitting field over F of a separable polynomial. Then $E_{G(E/F)} = F$.

Proof. Let $G = G(E/F)$. Clearly, $F \subset E_G \subset E$. Also, E must be a splitting field of E_G and $G(E/F) = G(E/E_G)$. By Theorem 23.7 on page 377,

$$|G| = [E : E_G] = [E : F].$$

Therefore, $[E_G : F] = 1$. Consequently, $E_G = F$.   ∎

A large number of mathematicians first learned Galois theory from Emil Artin's monograph on the subject [1]. The very clever proof of the following lemma is due to Artin.

Lemma 23.18. *Let G be a finite group of automorphisms of E and let $F = E_G$. Then $[E : F] \leq |G|$.*

Proof. Let $|G| = n$. We must show that any set of $n + 1$ elements $\alpha_1, \ldots, \alpha_{n+1}$ in E is linearly dependent over F; that is, we need to find elements $a_i \in F$, not all zero, such that

$$a_1\alpha_1 + a_2\alpha_2 + \cdots + a_{n+1}\alpha_{n+1} = 0.$$

Suppose that $\sigma_1 = \text{id}, \sigma_2, \ldots, \sigma_n$ are the automorphisms in G. The homogeneous system of linear equations

$$\sigma_1(\alpha_1)x_1 + \sigma_1(\alpha_2)x_2 + \cdots + \sigma_1(\alpha_{n+1})x_{n+1} = 0$$
$$\sigma_2(\alpha_1)x_1 + \sigma_2(\alpha_2)x_2 + \cdots + \sigma_2(\alpha_{n+1})x_{n+1} = 0$$
$$\vdots$$
$$\sigma_n(\alpha_1)x_1 + \sigma_n(\alpha_2)x_2 + \cdots + \sigma_n(\alpha_{n+1})x_{n+1} = 0$$

has more unknowns than equations. From linear algebra we know that this system has a nontrivial solution, say $x_i = a_i$ for $i = 1, 2, \ldots, n + 1$. Since σ_1 is the identity, the first equation translates to

$$a_1\alpha_1 + a_2\alpha_2 + \cdots + a_{n+1}\alpha_{n+1} = 0.$$

The problem is that some of the a_i's may be in E but not in F. We must show that this is impossible.

Suppose that at least one of the a_i's is in E but not in F. By rearranging the α_i's we may assume that a_1 is nonzero. Since any nonzero multiple of a solution is also a solution, we can also assume that $a_1 = 1$. Of all possible solutions fitting this description, we choose the one with the smallest number of nonzero terms. Again, by rearranging $\alpha_2, \ldots, \alpha_{n+1}$ if necessary, we can assume that a_2 is in E but not in F. Since F is the subfield of E that is fixed elementwise by G, there exists a σ_i in G such that $\sigma_i(a_2) \neq a_2$. Applying σ_i to each equation in the system, we end up with the same homogeneous system, since G is a group. Therefore, $x_1 = \sigma_i(a_1) = 1$,

$x_2 = \sigma_i(a_2)$, ..., $x_{n+1} = \sigma_i(a_{n+1})$ is also a solution of the original system. We know that a linear combination of two solutions of a homogeneous system is also a solution; consequently,

$$x_1 = 1 - 1 = 0$$
$$x_2 = a_2 - \sigma_i(a_2)$$
$$\vdots$$
$$x_{n+1} = a_{n+1} - \sigma_i(a_{n+1})$$

must be another solution of the system. This is a nontrivial solution because $\sigma_i(a_2) \neq a_2$, and has fewer nonzero entries than our original solution. This is a contradiction, since the number of nonzero solutions to our original solution was assumed to be minimal. We can therefore conclude that $a_1, \ldots, a_{n+1} \in F$.   ∎

Let E be an algebraic extension of F. If every irreducible polynomial in $F[x]$ with a root in E has all of its roots in E, then E is called a **normal extension** of F; that is, every irreducible polynomial in $F[x]$ containing a root in E is the product of linear factors in $E[x]$.

Theorem 23.19. Let E be a field extension of F. Then the following statements are equivalent.

1. E is a finite, normal, separable extension of F.

2. E is a splitting field over F of a separable polynomial.

3. $F = E_G$ for some finite group G of automorphisms of E.

Proof. (1) $\Rightarrow$ (2). Let E be a finite, normal, separable extension of F. By the Primitive Element Theorem, we can find an α in E such that $E = F(\alpha)$. Let $f(x)$ be the minimal polynomial of α over F. The field E must contain all of the roots of $f(x)$ since it is a normal extension F; hence, E is a splitting field for $f(x)$.

(2) $\Rightarrow$ (3). Let E be the splitting field over F of a separable polynomial. By Proposition 23.17 on page 381, $E_{G(E/F)} = F$. Since $|G(E/F)| = [E : F]$, this is a finite group.

(3) $\Rightarrow$ (1). Let $F = E_G$ for some finite group of automorphisms G of E. Since $[E : F] \leq |G|$, E is a finite extension of F. To show that E is a finite, normal extension of F, let $f(x) \in F[x]$ be an irreducible monic polynomial that has a root α in E. We must show that $f(x)$ is the product of distinct linear factors in $E[x]$. By Proposition 23.5 on page 377, automorphisms in G permute the roots of $f(x)$ lying in E. Hence, if we let G act on α, we can obtain distinct roots $\alpha_1 = \alpha, \alpha_2, \ldots, \alpha_n$ in E. Let $g(x) = \prod_{i=1}^{n}(x - \alpha_i)$. Then $g(x)$ is separable over F and $g(\alpha) = 0$.

Any automorphism σ in G permutes the factors of $g(x)$ since it permutes these roots; hence, when σ acts on $g(x)$, it must fix the coefficients of $g(x)$. Therefore, the coefficients of $g(x)$ must be in F. Since $\deg g(x) \le \deg f(x)$ and $f(x)$ is the minimal polynomial of α, $f(x) = g(x)$.   ∎

Corollary 23.20. Let K be a field extension of F such that $F = K_G$ for some finite group of automorphisms G of K. Then $G = G(K/F)$.

Proof. Since $F = K_G$, G is a subgroup of $G(K/F)$. Hence,

$$[K : F] \le |G| \le |G(K/F)| = [K : F].$$

It follows that $G = G(K/F)$, since they must have the same order.   ∎

Before we determine the exact correspondence between field extensions and automorphisms of fields, let us return to a familiar example.

Example 23.21. In Example 23.4 on page 376 we examined the automorphisms of $\mathbb{Q}(\sqrt{3}, \sqrt{5})$ fixing $\mathbb{Q}$. Figure 23.22 compares the lattice of field extensions of $\mathbb{Q}$ with the lattice of subgroups of $G(\mathbb{Q}(\sqrt{3}, \sqrt{5})/\mathbb{Q})$. The Fundamental Theorem of Galois Theory tells us what the relationship is between the two lattices.   □

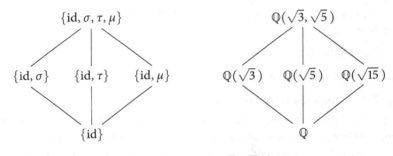

Figure 23.22. $G(\mathbb{Q}(\sqrt{3}, \sqrt{5})/\mathbb{Q})$

We are now ready to state and prove the Fundamental Theorem of Galois Theory.

Theorem 23.23. Fundamental Theorem of Galois Theory. Let F be a finite field or a field of characteristic zero. If E is a finite normal extension of F with Galois group $G(E/F)$, then the following statements are true.

1. The map $K \mapsto G(E/K)$ is a bijection of subfields K of E containing F with the subgroups of $G(E/F)$.

2. If $F \subset K \subset E$, then

$$[E : K] = |G(E/K)| \text{ and } [K : F] = [G(E/F) : G(E/K)].$$

3. $F \subset K \subset L \subset E$ if and only if $\{id\} \subset G(E/L) \subset G(E/K) \subset G(E/F)$.

4. K is a normal extension of F if and only if $G(E/K)$ is a normal subgroup of $G(E/F)$. In this case

$$G(K/F) \cong G(E/F)/G(E/K).$$

Proof. (1) Suppose that $G(E/K) = G(E/L) = G$. Both K and L are fixed fields of G; hence, $K = L$ and the map defined by $K \mapsto G(E/K)$ is one-to-one. To show that the map is onto, let G be a subgroup of $G(E/F)$ and K be the field fixed by G. Then $F \subset K \subset E$; consequently, E is a normal extension of K. Thus, $G(E/K) = G$ and the map $K \mapsto G(E/K)$ is a bijection.

(2) By Theorem Theorem 23.7 on page 377, $|G(E/K)| = [E:K]$; therefore,

$$|G(E/F)| = [G(E/F):G(E/K)] \cdot |G(E/K)| = [E:F] = [E:K][K:F].$$

Thus, $[K:F] = [G(E/F):G(E/K)]$.

Statement (3) is illustrated in Figure 23.24 on the following page. We leave the proof of this property as an exercise.

(4) This part takes a little more work. Let K be a normal extension of F. If σ is in $G(E/F)$ and τ is in $G(E/K)$, we need to show that $\sigma^{-1}\tau\sigma$ is in $G(E/K)$; that is, we need to show that $\sigma^{-1}\tau\sigma(\alpha) = \alpha$ for all $\alpha \in K$. Suppose that $f(x)$ is the minimal polynomial of α over F. Then $\sigma(\alpha)$ is also a root of $f(x)$ lying in K, since K is a normal extension of F. Hence, $\tau(\sigma(\alpha)) = \sigma(\alpha)$ or $\sigma^{-1}\tau\sigma(\alpha) = \alpha$.

Conversely, let $G(E/K)$ be a normal subgroup of $G(E/F)$. We need to show that $F = K_{G(K/F)}$. Let $\tau \in G(E/K)$. For all $\sigma \in G(E/F)$ there exists a $\overline{\tau} \in G(E/K)$ such that $\tau\sigma = \sigma\overline{\tau}$. Consequently, for all $\alpha \in K$

$$\tau(\sigma(\alpha)) = \sigma(\overline{\tau}(\alpha)) = \sigma(\alpha);$$

hence, $\sigma(\alpha)$ must be in the fixed field of $G(E/K)$. Let $\overline{\sigma}$ be the restriction of σ to K. Then $\overline{\sigma}$ is an automorphism of K fixing F, since $\sigma(\alpha) \in K$ for all $\alpha \in K$; hence, $\overline{\sigma} \in G(K/F)$. Next, we will show that the fixed field of $G(K/F)$ is F. Let β be an element in K that is fixed by all automorphisms in $G(K/F)$. In particular, $\overline{\sigma}(\beta) = \beta$ for all $\sigma \in G(E/F)$. Therefore, β belongs to the fixed field F of $G(E/F)$.

Finally, we must show that when K is a normal extension of F,

$$G(K/F) \cong G(E/F)/G(E/K).$$

For $\sigma \in G(E/F)$, let σ_K be the automorphism of K obtained by restricting σ to K. Since K is a normal extension, the argument in the preceding paragraph shows that $\sigma_K \in G(K/F)$. Consequently, we have a map $\phi : G(E/F) \to G(K/F)$ defined by $\sigma \mapsto \sigma_K$. This map is a group homomorphism since

$$\phi(\sigma\tau) = (\sigma\tau)_K = \sigma_K\tau_K = \phi(\sigma)\phi(\tau).$$

The kernel of ϕ is $G(E/K)$. By (2),

$$|G(E/F)|/|G(E/K)| = [K : F] = |G(K/F)|.$$

Hence, the image of ϕ is $G(K/F)$ and ϕ is onto. Applying the First Isomorphism Theorem, we have

$$G(K/F) \cong G(E/F)/G(E/K).$$

■

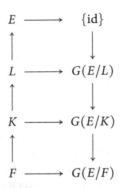

Figure 23.24. Subgroups of $G(E/F)$ and subfields of E

Example 23.25.   In this example we will illustrate the Fundamental Theorem of Galois Theory by determining the lattice of subgroups of the Galois group of $f(x) = x^4 - 2$. We will compare this lattice to the lattice of field extensions of $\mathbb{Q}$ that are contained in the splitting field of $x^4 - 2$. The splitting field of $f(x)$ is $\mathbb{Q}(\sqrt[4]{2}, i)$. To see this, notice that $f(x)$ factors as $(x^2 + \sqrt{2})(x^2 - \sqrt{2})$; hence, the roots of $f(x)$ are $\pm\sqrt[4]{2}$ and $\pm\sqrt[4]{2}\,i$. We first adjoin the root $\sqrt[4]{2}$ to $\mathbb{Q}$ and then adjoin the root i of $x^2 + 1$ to $\mathbb{Q}(\sqrt[4]{2})$. The splitting field of $f(x)$ is then $\mathbb{Q}(\sqrt[4]{2})(i) = \mathbb{Q}(\sqrt[4]{2}, i)$.

Since $[\mathbb{Q}(\sqrt[4]{2}) : \mathbb{Q}] = 4$ and i is not in $\mathbb{Q}(\sqrt[4]{2})$, it must be the case that $[\mathbb{Q}(\sqrt[4]{2}, i) : \mathbb{Q}(\sqrt[4]{2})] = 2$. Hence, $[\mathbb{Q}(\sqrt[4]{2}, i) : \mathbb{Q}] = 8$. The set

$$\{1, \sqrt[4]{2}, (\sqrt[4]{2})^2, (\sqrt[4]{2})^3, i, i\sqrt[4]{2}, i(\sqrt[4]{2})^2, i(\sqrt[4]{2})^3\}$$

is a basis of $\mathbb{Q}(\sqrt[4]{2}, i)$ over $\mathbb{Q}$. The lattice of field extensions of $\mathbb{Q}$ contained in $\mathbb{Q}(\sqrt[4]{2}, i)$ is illustrated in Figure 23.26(a).

The Galois group G of $f(x)$ must be of order 8. Let σ be the automorphism defined by $\sigma(\sqrt[4]{2}) = i\sqrt[4]{2}$ and $\sigma(i) = i$, and τ be the automorphism defined by complex conjugation; that is, $\tau(i) = -i$. Then G has an element of order 4 and an element of order 2. It is easy to verify by direct computation that the elements of G are $\{\text{id}, \sigma, \sigma^2, \sigma^3, \tau, \sigma\tau, \sigma^2\tau, \sigma^3\tau\}$ and that the relations $\tau^2 = \text{id}$, $\sigma^4 = \text{id}$, and $\tau\sigma\tau = \sigma^{-1}$ are satisfied; hence, G must be isomorphic to D_4. The lattice of subgroups of G is illustrated in Figure 23.26(b).   □

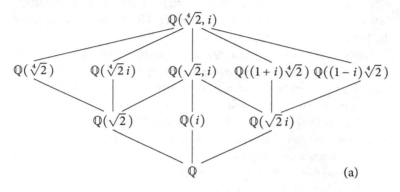

(a)

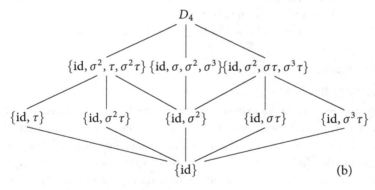

(b)

Figure 23.26. Galois group of $x^4 - 2$

◟ Historical Note ◞

Solutions for the cubic and quartic equations were discovered in the 1500s. Attempts to find solutions for the quintic equations puzzled some of history's best mathematicians. In 1798, P. Ruffini submitted a paper that claimed no such solution could be found; however, the paper was not well received. In 1826, Niels Henrik Abel (1802–1829) finally offered the first correct proof that quintics are not always solvable by radicals.

Abel inspired the work of Évariste Galois. Born in 1811, Galois began to display extraordinary mathematical talent at the age of 14. He applied for entrance to the École Polytechnique several times; however, he had great difficulty meeting the formal entrance requirements, and the examiners failed to recognize his mathematical genius. He was finally accepted at the École Normale in 1829.

Galois worked to develop a theory of solvability for polynomials. In 1829, at the age of 17, Galois presented two papers on the solution of algebraic equations to the Académie des Sciences de Paris. These papers were sent to Cauchy, who subsequently lost them. A third paper was submitted to Fourier, who died before he could read the paper. Another paper was presented, but was not published until 1846.

Galois' democratic sympathies led him into the Revolution of 1830. He was expelled from school and sent to prison for his part in the turmoil. After his release in 1832, he was drawn into a duel possibly over a love affair. Certain that he would be killed, he spent the evening before his death outlining his work and his basic ideas for research in a long letter to his friend Chevalier. He was indeed dead the next day, at the age of 20.

23.3 Applications

Solvability by Radicals

Throughout this section we shall assume that all fields have characteristic zero to ensure that irreducible polynomials do not have multiple roots. The immediate goal of this section is to determine when the roots of a polynomial $f(x)$ can be computed with a finite number of operations on the coefficients of $f(x)$. The allowable operations are addition, subtraction, multiplication, division, and the extraction of nth roots. Certainly the solution to the quadratic equation, $ax^2 + bx + c = 0$, illustrates this process:

$$x = \frac{-b \pm \sqrt{b^2 - 4ac}}{2a}.$$

The only one of these operations that might demand a larger field is the taking of nth roots. We are led to the following definition.

An extension field E of a field F is an ***extension by radicals*** if there exists a chain of subfields

$$F = F_0 \subset F_1 \subset F_2 \subset \cdots \subset F_r = E$$

such for $i = 1, 2, \ldots, r$, we have $F_i = F_{i-1}(\alpha_i)$ and $\alpha_i^{n_i} \in F_{i-1}$ for some positive integer n_i. A polynomial $f(x)$ is ***solvable by radicals*** over F if the splitting field K of $f(x)$ over F is contained in an extension of F by radicals. Our goal is to arrive at criteria that will tell us whether or not a polynomial $f(x)$ is solvable by radicals by examining the Galois group $f(x)$.

The easiest polynomial to solve by radicals is one of the form $x^n - a$. As we discussed in Chapter 4, the roots of $x^n - 1$ are called the ***nth roots of unity***. These roots are a finite subgroup of the splitting field of $x^n - 1$. By Corollary 22.11 on page 360, the nth roots of unity form a cyclic group. Any generator of this group is called a ***primitive nth root of unity***.

Example 23.27. The polynomial $x^n - 1$ is solvable by radicals over $\mathbb{Q}$. The roots of this polynomial are $1, \omega, \omega^2, \ldots, \omega^{n-1}$, where

$$\omega = \cos\left(\frac{2\pi}{n}\right) + i \sin\left(\frac{2\pi}{n}\right).$$

The splitting field of $x^n - 1$ over $\mathbb{Q}$ is $\mathbb{Q}(\omega)$.   □

We shall prove that a polynomial is solvable by radicals if its Galois group is solvable. Recall that a subnormal series of a group G is a finite sequence of subgroups

$$G = H_n \supset H_{n-1} \supset \cdots \supset H_1 \supset H_0 = \{e\},$$

where H_i is normal in H_{i+1}. A group G is solvable if it has a subnormal series $\{H_i\}$ such that all of the factor groups H_{i+1}/H_i are abelian. For example, if we examine the series $\{\text{id}\} \subset A_3 \subset S_3$, we see that S_3 is solvable. On the other hand, S_5 is not solvable, by Theorem 10.11 on page 163.

Lemma 23.28. Let F be a field of characteristic zero and E be the splitting field of $x^n - a$ over F with $a \in F$. Then $G(E/F)$ is a solvable group.

Proof. The roots of $x^n - a$ are $\sqrt[n]{a}, \omega\sqrt[n]{a}, \ldots, \omega^{n-1}\sqrt[n]{a}$, where ω is a primitive nth root of unity. Suppose that F contains all of its nth roots of unity. If ζ is one of the roots of $x^n - a$, then distinct roots of $x^n - a$ are $\zeta, \omega\zeta, \ldots, \omega^{n-1}\zeta$, and $E = F(\zeta)$. Since $G(E/F)$ permutes the roots $x^n - a$, the elements in $G(E/F)$ must be determined by their action on these roots. Let σ and τ be in $G(E/F)$ and suppose that $\sigma(\zeta) = \omega^i \zeta$ and $\tau(\zeta) = \omega^j \zeta$. If F contains the roots of unity, then

$$\sigma\tau(\zeta) = \sigma(\omega^j\zeta) = \omega^j\sigma(\zeta) = \omega^{i+j}\zeta = \omega^i\tau(\zeta) = \tau(\omega^i\zeta) = \tau\sigma(\zeta).$$

Therefore, $\sigma\tau = \tau\sigma$ and $G(E/F)$ is abelian, and $G(E/F)$ must be solvable.

Now suppose that F does not contain a primitive nth root of unity. Let ω be a generator of the cyclic group of the nth roots of unity. Let α be a zero of $x^n - a$. Since α and $\omega\alpha$ are both in the splitting field of $x^n - a$, $\omega = (\omega\alpha)/\alpha$ is also in E. Let $K = F(\omega)$. Then $F \subset K \subset E$. Since K is the splitting field of $x^n - 1$, K is a normal extension of F. Therefore, any automorphism σ in $G(F(\omega)/F)$ is determined by $\sigma(\omega)$. It must be the case that $\sigma(\omega) = \omega^i$ for some integer i since all of the zeros of $x^n - 1$ are powers of ω. If $\tau(\omega) = \omega^j$ is in $G(F(\omega)/F)$, then

$$\sigma\tau(\omega) = \sigma(\omega^j) = [\sigma(\omega)]^j = \omega^{ij} = [\tau(\omega)]^i = \tau(\omega^i) = \tau\sigma(\omega).$$

Therefore, $G(F(\omega)/F)$ is abelian. By the Fundamental Theorem of Galois Theory the series

$$\{\mathrm{id}\} \subset G(E/F(\omega)) \subset G(E/F)$$

is a normal series. By our previous argument, $G(E/F(\omega))$ is abelian. Since

$$G(E/F)/G(E/F(\omega)) \cong G(F(\omega)/F)$$

is also abelian, $G(E/F)$ is solvable.   ∎

Lemma 23.29. Let F be a field of characteristic zero and let

$$F = F_0 \subset F_1 \subset F_2 \subset \cdots \subset F_r = E$$

a radical extension of F. Then there exists a normal radical extension

$$F = K_0 \subset K_1 \subset K_2 \subset \cdots \subset K_r = K$$

such that K that contains E and K_i is a normal extension of K_{i-1}.

Proof. Since E is a radical extension of F, there exists a chain of subfields

$$F = F_0 \subset F_1 \subset F_2 \subset \cdots \subset F_r = E$$

such for $i = 1, 2, \ldots, r$, we have $F_i = F_{i-1}(\alpha_i)$ and $\alpha_i^{n_i} \in F_{i-1}$ for some positive integer n_i. We will build a normal radical extension of F,

$$F = K_0 \subset K_1 \subset K_2 \subset \cdots \subset K_r = K$$

such that $K \supseteq E$. Define K_1 for be the splitting field of $x^{n_1} - \alpha_1^{n_1}$. The roots of this polynomial are $\alpha_1, \alpha_1\omega, \alpha_1\omega^2, \ldots, \alpha_1\omega^{n_1-1}$, where ω is a primitive n_1th root of unity. If F contains all of its n_1 roots of unity, then $K_1 = F(\alpha_1)$. On the other hand, suppose that F does not contain a primitive n_1th root of unity. If β is a root of $x^{n_1} - \alpha_1^{n_1}$, then all of the roots of $x^{n_1} - \alpha_1^{n_1}$ must be $\beta, \omega\beta, \ldots, \omega^{n_1-1}$, where ω is a primitive n_1th root of unity. In this case, $K_1 = F(\omega\beta)$. Thus, K_1 is a normal radical extension of F containing F_1. Continuing in this manner, we obtain

$$F = K_0 \subset K_1 \subset K_2 \subset \cdots \subset K_r = K$$

such that K_i is a normal extension of K_{i-1} and $K_i \supseteq F_i$ for $i = 1, 2, \ldots, r$. ∎

We will now prove the main theorem about solvability by radicals.

Theorem 23.30. Let $f(x)$ be in $F[x]$, where char $F = 0$. If $f(x)$ is solvable by radicals, then the Galois group of $f(x)$ over F is solvable.

Proof. Since $f(x)$ is solvable by radicals there exists an extension E of F by radicals $F = F_0 \subset F_1 \subset \cdots \subset F_n = E$. By Lemma 23.29, we can assume that E is a splitting field $f(x)$ and F_i is normal over F_{i-1}. By the Fundamental Theorem of Galois Theory, $G(E/F_i)$ is a normal subgroup of $G(E/F_{i-1})$. Therefore, we have a subnormal series of subgroups of $G(E/F)$:

$$\{\mathrm{id}\} \subset G(E/F_{n-1}) \subset \cdots \subset G(E/F_1) \subset G(E/F).$$

Again by the Fundamental Theorem of Galois Theory, we know that

$$G(E/F_{i-1})/G(E/F_i) \cong G(F_i/F_{i-1}).$$

By Lemma 23.28 on page 389, $G(F_i/F_{i-1})$ is solvable; hence, $G(E/F)$ is also solvable. ∎

The converse of Theorem 23.30 is also true. For a proof, see any of the references at the end of this chapter.

Insolvability of the Quintic

We are now in a position to find a fifth-degree polynomial that is not solvable by radicals. We merely need to find a polynomial whose Galois group is S_5. We begin by proving a lemma.

Lemma 23.31. If p is prime, then any subgroup of S_p that contains a transposition and a cycle of length p must be all of S_p.

Proof. Let G be a subgroup of S_p that contains a transposition σ and τ a cycle of length p. We may assume that $\sigma = (1\,2)$. The order of τ is p and τ^n must be a cycle of length p for $1 \leq n < p$. Therefore, we may assume that $\mu = \tau^n = (1, 2, i_3, \ldots, i_p)$ for some n, where $1 \leq n < p$ (see Exercise 5.4.13 on page 88). Noting that $(12)(1, 2, i_3, \ldots, i_p) = (2, i_3, \ldots, i_p)$ and $(2, i_3, \ldots, i_p)^k (12)(2, i_3, \ldots, i_p)^{-k} = (1\,i_k)$, we can obtain all the transpositions of the form $(1n)$ for $1 \leq n < p$. However, these transpositions generate all transpositions in S_p, since $(1\,j)(1\,i)(1\,j) = (i\,j)$. The transpositions generate S_p. ∎

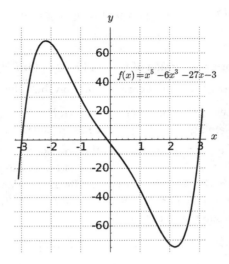

Figure 23.32. The graph of $f(x) = x^5 - 6x^3 - 27x - 3$

Example 23.33. We will show that $f(x) = x^5 - 6x^3 - 27x - 3 \in \mathbb{Q}[x]$ is not solvable. We claim that the Galois group of $f(x)$ over $\mathbb{Q}$ is S_5. By Eisenstein's Criterion, $f(x)$ is irreducible and, therefore, must be separable. The derivative of $f(x)$ is $f'(x) = 5x^4 - 18x^2 - 27$; hence, setting $f'(x) = 0$ and solving, we find that the only real roots of $f'(x)$ are

$$x = \pm \sqrt{\frac{6\sqrt{6} + 9}{5}}.$$

Therefore, $f(x)$ can have at most one maximum and one minimum. It is easy to show that $f(x)$ changes sign between -3 and -2, between -2 and 0, and once again between 0 and 4 (Figure 23.32). Therefore, $f(x)$ has exactly three distinct real roots. The remaining two roots of $f(x)$ must be complex conjugates. Let K be the splitting field of $f(x)$. Since $f(x)$ has five distinct roots in K and every automorphism of K fixing $\mathbb{Q}$ is determined by the way it permutes the roots of $f(x)$, we know that $G(K/\mathbb{Q})$ is a subgroup of S_5. Since f is irreducible, there is an element in $\sigma \in G(K/\mathbb{Q})$ such that $\sigma(a) = b$ for two roots a and b of $f(x)$. The automorphism of $\mathbb{C}$ that takes $a + bi \mapsto a - bi$ leaves the real roots fixed and interchanges the complex roots; consequently, $G(K/\mathbb{Q})$ contains a transposition. If α is one of the real roots of $f(x)$, then $[\mathbb{Q}(\alpha) : \mathbb{Q}] = 5$ by Exercise 21.5.28 on page 355. Since $\mathbb{Q}(\alpha)$ is a subfield of K, it must be the case that $[K : \mathbb{Q}]$ is divisible by 5. Since $[K : \mathbb{Q}] = |G(K/\mathbb{Q})|$ and $G(K/\mathbb{Q}) \subset S_5$, we know that $G(K/\mathbb{Q})$ contains a cycle of length 5. By Lemma 23.31 on the previous page, S_5 is generated by a transposition and an element of order 5; therefore, $G(K/\mathbb{Q})$ must be all of S_5. By Theorem 10.11 on page 163, S_5 is not solvable. Consequently, $f(x)$ cannot be solved by radicals.   □

The Fundamental Theorem of Algebra

It seems fitting that the last theorem that we will state and prove is the Fundamental Theorem of Algebra. This theorem was first proven by Gauss in his doctoral thesis. Prior to Gauss's proof, mathematicians suspected that there might exist polynomials over the real and complex numbers having no solutions. The Fundamental Theorem of Algebra states that every polynomial over the complex numbers factors into distinct linear factors.

Theorem 23.34. Fundamental Theorem of Algebra. The field of complex numbers is algebraically closed; that is, every polynomial in $\mathbb{C}[x]$ has a root in $\mathbb{C}$.

Proof. Suppose that E is a proper finite field extension of the complex numbers. Since any finite extension of a field of characteristic zero is a simple extension, there exists an $\alpha \in E$ such that $E = \mathbb{C}(\alpha)$ with α the root of an irreducible polynomial $f(x)$ in $\mathbb{C}[x]$. The splitting field L of $f(x)$ is a finite normal separable extension of $\mathbb{C}$ that contains E. We must show that it is impossible for L to be a proper extension of $\mathbb{C}$.

Suppose that L is a proper extension of $\mathbb{C}$. Since L is the splitting field of $f(x) = (x^2 + 1)$ over $\mathbb{R}$, L is a finite normal separable extension of $\mathbb{R}$. Let K be the fixed field of a Sylow 2-subgroup G of $G(L/\mathbb{R})$. Then $L \supset K \supset \mathbb{R}$ and $|G(L/K)| = [L : K]$. Since $[L : \mathbb{R}] = [L : K][K : \mathbb{R}]$, we know that $[K : \mathbb{R}]$ must be odd. Consequently, $K = \mathbb{R}(\beta)$ with β having a minimal polynomial $f(x)$ of odd degree. Therefore, $K = \mathbb{R}$.

We now know that $G(L/\mathbb{R})$ must be a 2-group. It follows that $G(L/\mathbb{C})$ is a 2-group. We have assumed that $L \neq \mathbb{C}$; therefore, $|G(L/\mathbb{C})| \geq 2$. By the first Sylow Theorem and the Fundamental Theorem of Galois Theory, there exists a subgroup G of $G(L/\mathbb{C})$ of index 2 and a field E fixed elementwise by G. Then $[E : \mathbb{C}] = 2$ and there exists an element $\gamma \in E$ with minimal polynomial $x^2 + bx + c$ in $\mathbb{C}[x]$. This polynomial has roots $(-b \pm \sqrt{b^2 - 4c})/2$ that are in $\mathbb{C}$, since $b^2 - 4c$ is in $\mathbb{C}$. This is impossible; hence, $L = \mathbb{C}$.   ■

Although our proof was strictly algebraic, we were forced to rely on results from calculus. It is necessary to assume the completeness axiom from analysis to show that every polynomial of odd degree has a real root and that every positive real number has a square root. It seems that there is no possible way to avoid this difficulty and formulate a purely algebraic argument. It is somewhat amazing that there are several elegant proofs of the Fundamental Theorem of Algebra that use complex analysis. It is also interesting to note that we can obtain a proof of such an important theorem from two very different fields of mathematics.

Sage. Fields, field extensions, roots of polynomials, and group theory — Sage has it all, and so it is possible to carefully study very complicated examples from Galois Theory with Sage.

23.4 Reading Questions

1. What is the Galois group of a field extension?

2. When are two elements of a field extension conjugate? (In other words, what is the definition?)

3. Summarize the nature and importance of the Fundamental Theorem of Galois Theory. Capture the essence of the result without getting bogged down in too many details.

4. Why are "solvable" groups so named? Paraphrasing the relevant theorem would be a good answer.

5. Argue the following statement, both pro and con. Which side wins the debate?

> Everything we have done in this entire course has been in preparation for this chapter.

23.5 Exercises

1. Compute each of the following Galois groups. Which of these field extensions are normal field extensions? If the extension is not normal, find a normal extension of $\mathbb{Q}$ in which the extension field is contained.

   (a) $G(\mathbb{Q}(\sqrt{30})/\mathbb{Q})$

   (b) $G(\mathbb{Q}(\sqrt[4]{5})/\mathbb{Q})$

   (c) $G(\mathbb{Q}(\sqrt{2}, \sqrt{3}, \sqrt{5})/\mathbb{Q})$

   (d) $G(\mathbb{Q}(\sqrt{2}, \sqrt[3]{2}, i)/\mathbb{Q})$

   (e) $G(\mathbb{Q}(\sqrt{6}, i)/\mathbb{Q})$

2. Determine the separability of each of the following polynomials.

   (a) $x^3 + 2x^2 - x - 2$ over $\mathbb{Q}$

   (b) $x^4 + 2x^2 + 1$ over $\mathbb{Q}$

   (c) $x^4 + x^2 + 1$ over $\mathbb{Z}_3$

   (d) $x^3 + x^2 + 1$ over $\mathbb{Z}_2$

3. Give the order and describe a generator of the Galois group of GF(729) over GF(9).

4. Determine the Galois groups of each of the following polynomials in $\mathbb{Q}[x]$; hence, determine the solvability by radicals of each of the polynomials.

   (a) $x^5 - 12x^2 + 2$

   (b) $x^5 - 4x^4 + 2x + 2$

   (c) $x^3 - 5$

   (d) $x^4 - x^2 - 6$

   (e) $x^5 + 1$

   (f) $(x^2 - 2)(x^2 + 2)$

   (g) $x^8 - 1$

   (h) $x^8 + 1$

   (i) $x^4 - 3x^2 - 10$

5. Find a primitive element in the splitting field of each of the following polynomials in $\mathbb{Q}[x]$.

   (a) $x^4 - 1$

   (b) $x^4 - 8x^2 + 15$

   (c) $x^4 - 2x^2 - 15$

   (d) $x^3 - 2$

6. Prove that the Galois group of an irreducible quadratic polynomial is isomorphic to $\mathbb{Z}_2$.

7. Prove that the Galois group of an irreducible cubic polynomial is isomorphic to S_3 or $\mathbb{Z}_3$.

8. Let $F \subset K \subset E$ be fields. If E is a normal extension of F, show that E must also be a normal extension of K.

9. Let G be the Galois group of a polynomial of degree n. Prove that $|G|$ divides $n!$.

10. Let $F \subset E$. If $f(x)$ is solvable over F, show that $f(x)$ is also solvable over E.

11. Construct a polynomial $f(x)$ in $\mathbb{Q}[x]$ of degree 7 that is not solvable by radicals.

12. Let p be prime. Prove that there exists a polynomial $f(x) \in \mathbb{Q}[x]$ of degree p with Galois group isomorphic to S_p. Conclude that for each prime p with $p \geq 5$ there exists a polynomial of degree p that is not solvable by radicals.

13. Let p be a prime and $\mathbb{Z}_p(t)$ be the field of rational functions over $\mathbb{Z}_p$. Prove that $f(x) = x^p - t$ is an irreducible polynomial in $\mathbb{Z}_p(t)[x]$. Show that $f(x)$ is not separable.

14. Let E be an extension field of F. Suppose that K and L are two intermediate fields. If there exists an element $\sigma \in G(E/F)$ such that $\sigma(K) = L$, then K and L are said to be *conjugate fields.* Prove that K and L are conjugate if and only if $G(E/K)$ and $G(E/L)$ are conjugate subgroups of $G(E/F)$.

15. Let $\sigma \in \mathrm{Aut}(\mathbb{R})$. If a is a positive real number, show that $\sigma(a) > 0$.

16. Let K be the splitting field of $x^3 + x^2 + 1 \in \mathbb{Z}_2[x]$. Prove or disprove that K is an extension by radicals.

17. Let F be a field such that $\text{char}(F) \neq 2$. Prove that the splitting field of $f(x) = ax^2 + bx + c$ is $F(\sqrt{\alpha})$, where $\alpha = b^2 - 4ac$.

18. Prove or disprove: Two different subgroups of a Galois group will have different fixed fields.

19. Let K be the splitting field of a polynomial over F. If E is a field extension of F contained in K and $[E : F] = 2$, then E is the splitting field of some polynomial in $F[x]$.

20. We know that the cyclotomic polynomial

$$\Phi_p(x) = \frac{x^p - 1}{x - 1} = x^{p-1} + x^{p-2} + \cdots + x + 1$$

is irreducible over $\mathbb{Q}$ for every prime p. Let ω be a zero of $\Phi_p(x)$, and consider the field $\mathbb{Q}(\omega)$.

   (a) Show that $\omega, \omega^2, \ldots, \omega^{p-1}$ are distinct zeros of $\Phi_p(x)$, and conclude that they are all the zeros of $\Phi_p(x)$.

   (b) Show that $G(\mathbb{Q}(\omega)/\mathbb{Q})$ is abelian of order $p - 1$.

   (c) Show that the fixed field of $G(\mathbb{Q}(\omega)/\mathbb{Q})$ is $\mathbb{Q}$.

21. Let F be a finite field or a field of characteristic zero. Let E be a finite normal extension of F with Galois group $G(E/F)$. Prove that $F \subset K \subset L \subset E$ if and only if $\{\text{id}\} \subset G(E/L) \subset G(E/K) \subset G(E/F)$.

22. Let F be a field of characteristic zero and let $f(x) \in F[x]$ be a separable polynomial of degree n. If E is the splitting field of $f(x)$, let $\alpha_1, \ldots, \alpha_n$ be the roots of $f(x)$ in E. Let $\Delta = \prod_{i<j}(\alpha_i - \alpha_j)$. We define the **discriminant** of $f(x)$ to be Δ^2.

  (a) If $f(x) = x^2 + bx + c$, show that $\Delta^2 = b^2 - 4c$.

  (b) If $f(x) = x^3 + px + q$, show that $\Delta^2 = -4p^3 - 27q^2$.

  (c) Prove that Δ^2 is in F.

  (d) If $\sigma \in G(E/F)$ is a transposition of two roots of $f(x)$, show that $\sigma(\Delta) = -\Delta$.

  (e) If $\sigma \in G(E/F)$ is an even permutation of the roots of $f(x)$, show that $\sigma(\Delta) = \Delta$.

  (f) Prove that $G(E/F)$ is isomorphic to a subgroup of A_n if and only if $\Delta \in F$.

  (g) Determine the Galois groups of $x^3 + 2x - 4$ and $x^3 + x - 3$.

23.6 References and Suggested Readings

[1] Artin, E. *Theory: Lectures Delivered at the University of Notre Dame (Notre Dame Mathematical Lectures, Number 2)*. Dover, Mineola, NY, 1997.

[2] Edwards, H. M. *Galois Theory*. Springer-Verlag, New York, 1984.

[3] Fraleigh, J. B. *A First Course in Abstract Algebra*. 7th ed. Pearson, Upper Saddle River, NJ, 2003.

[4] Gaal, L. *Classical Galois Theory with Examples*. American Mathematical Society, Providence, 1979.

[5] Garling, D. J. H. *A Course in Galois Theory*. Cambridge University Press, Cambridge, 1986.

[6] Kaplansky, I. *Fields and Rings*. 2nd ed. University of Chicago Press, Chicago, 1972.

[7] Rothman, T. "The Short Life of Évariste Galois," *Scientific American*, April 1982, 136–49.

Hints and Answers to Selected Exercises

1 ◆ Preliminaries

1.4.1. Hint. (a) $A \cap B = \{2\}$; (b) $B \cap C = \{5\}$.

1.4.2. Hint. (a) $A \times B = \{(a,1),(a,2),(a,3),(b,1),(b,2),(b,3),(c,1),(c,2),(c,3)\}$; (d) $A \times D = \varnothing$.

1.4.6. Hint. Observe that $x \in A \cup B$ if and only if $x \in A$ or $x \in B$. Equivalently, $x \in B$ or $x \in A$, which is the same as $x \in B \cup A$. Therefore, $A \cup B = B \cup A$.

1.4.10. Hint. $(A \cap B) \cup (A \setminus B) \cup (B \setminus A) = (A \cap B) \cup (A \cap B') \cup (B \cap A') = [A \cap (B \cup B')] \cup (B \cap A') = A \cup (B \cap A') = (A \cup B) \cap (A \cup A') = A \cup B$.

1.4.14. Hint. $A \setminus (B \cup C) = A \cap (B \cup C)' = (A \cap A) \cap (B' \cap C') = (A \cap B') \cap (A \cap C') = (A \setminus B) \cap (A \setminus C)$.

1.4.17. Hint. (a) Not a map since $f(2/3)$ is undefined; (b) this is a map; (c) not a map, since $f(1/2) = 3/4$ but $f(2/4) = 3/8$; (d) this is a map.

1.4.18. Hint. (a) f is one-to-one but not onto. $f(\mathbb{R}) = \{x \in \mathbb{R} : x > 0\}$. (c) f is neither one-to-one nor onto. $f(\mathbb{R}) = \{x : -1 \leq x \leq 1\}$.

1.4.20. Hint. (a) $f(n) = n + 1$.

1.4.22. Hint. (a) Let $x, y \in A$. Then $g(f(x)) = (g \circ f)(x) = (g \circ f)(y) = g(f(y))$. Thus, $f(x) = f(y)$ and $x = y$, so $g \circ f$ is one-to-one. (b) Let $c \in C$, then $c = (g \circ f)(x) = g(f(x))$ for some $x \in A$. Since $f(x) \in B$, g is onto.

1.4.23. Hint. $f^{-1}(x) = (x + 1)/(x - 1)$.

1.4.24. Hint. (a) Let $y \in f(A_1 \cup A_2)$. Then there exists an $x \in A_1 \cup A_2$ such that $f(x) = y$. Hence, $y \in f(A_1)$ or $f(A_2)$. Therefore, $y \in f(A_1) \cup f(A_2)$. Consequently, $f(A_1 \cup A_2) \subset f(A_1) \cup f(A_2)$. Conversely, if $y \in f(A_1) \cup f(A_2)$, then $y \in f(A_1)$ or $f(A_2)$. Hence, there exists an x in A_1 or A_2 such that $f(x) = y$. Thus, there exists an $x \in A_1 \cup A_2$ such that $f(x) = y$. Therefore, $f(A_1) \cup f(A_2) \subset f(A_1 \cup A_2)$, and $f(A_1 \cup A_2) = f(A_1) \cup f(A_2)$.

1.4.25. Hint. (a) The relation fails to be symmetric. (b) The relation is not reflexive, since 0 is not equivalent to itself. (c) The relation is not transitive.

1.4.28. Hint. Let $X = \mathbb{N} \cup \{\sqrt{2}\}$ and define $x \sim y$ if $x + y \in \mathbb{N}$.

2 ◆ The Integers

2.4.1. Hint. The base case, $S(1) : [1(1+1)(2(1)+1)]/6 = 1 = 1^2$ is true. Assume that $S(k) : 1^2 + 2^2 + \cdots + k^2 = [k(k+1)(2k+1)]/6$ is true. Then

$$1^2 + 2^2 + \cdots + k^2 + (k+1)^2 = [k(k+1)(2k+1)]/6 + (k+1)^2$$
$$= [(k+1)((k+1)+1)(2(k+1)+1)]/6,$$

and so $S(k+1)$ is true. Thus, $S(n)$ is true for all positive integers n.

2.4.3. Hint. The base case, $S(4) : 4! = 24 > 16 = 2^4$ is true. Assume $S(k) : k! > 2^k$ is true. Then $(k+1)! = k!(k+1) > 2^k \cdot 2 = 2^{k+1}$, so $S(k+1)$ is true. Thus, $S(n)$ is true for all positive integers n.

2.4.8. Hint. Follow the proof in Example 2.4 on page 23.

2.4.11. Hint. The base case, $S(0) : (1+x)^0 - 1 = 0 \geq 0 = 0 \cdot x$ is true. Assume $S(k) :$ $(1+x)^k - 1 \geq kx$ is true. Then

$$(1+x)^{k+1} - 1 = (1+x)(1+x)^k - 1$$
$$= (1+x)^k + x(1+x)^k - 1$$
$$\geq kx + x(1+x)^k$$
$$\geq kx + x$$
$$= (k+1)x,$$

so $S(k+1)$ is true. Therefore, $S(n)$ is true for all positive integers n.

2.4.17. Fibonacci Numbers. Hint. For (a) and (b) use mathematical induction. (c) Show that $f_1 = 1$, $f_2 = 1$, and $f_{n+2} = f_{n+1} + f_n$. (e) Use part (b) and Exercise 2.4.16 on page 32.

2.4.19. Hint. Use the Fundamental Theorem of Arithmetic.

2.4.23. Hint. Use the Principle of Well-Ordering and the division algorithm.

2.4.27. Hint. Since $\gcd(a,b) = 1$, there exist integers r and s such that $ar + bs = 1$. Thus, $acr + bcs = c$.

2.4.29. Hint. Every prime must be of the form 2, 3, $6n + 1$, or $6n + 5$. Suppose there are only finitely many primes of the form $6k + 5$.

3 ◆ Groups

3.5.1. Hint. (a) $3 + 7\mathbb{Z} = \{\ldots, -4, 3, 10, \ldots\}$; (c) $18 + 26\mathbb{Z}$; (e) $5 + 6\mathbb{Z}$.

3.5.2. Hint. (a) Not a group; (c) a group.

3.5.6. Hint.

·	1	5	7	11
1	1	5	7	11
5	5	1	11	7
7	7	11	1	5
11	11	7	5	1

3.5.8. Hint. Pick two matrices. Almost any pair will work.

3.5.15. Hint. There is a nonabelian group containing six elements.

3.5.16. Hint. Look at the symmetry group of an equilateral triangle or a square.

3.5.17. Hint. The are five different groups of order 8.

3.5.18. Hint. Let

$$\sigma = \begin{pmatrix} 1 & 2 & \cdots & n \\ a_1 & a_2 & \cdots & a_n \end{pmatrix}$$

be in S_n. All of the a_is must be distinct. There are n ways to choose a_1, $n-1$ ways to choose $a_2, \ldots$, 2 ways to choose a_{n-1}, and only one way to choose a_n. Therefore, we can form σ in $n(n-1)\cdots 2 \cdot 1 = n!$ ways.

3.5.25. Hint.

$$
\begin{aligned}
(aba^{-1})^n &= (aba^{-1})(aba^{-1})\cdots(aba^{-1}) \\
&= ab(aa^{-1})b(aa^{-1})b\cdots b(aa^{-1})ba^{-1} \\
&= ab^n a^{-1}.
\end{aligned}
$$

3.5.31. Hint. Since $abab = (ab)^2 = e = a^2 b^2 = aabb$, we know that $ba = ab$.

3.5.35. Hint. $H_1 = \{\text{id}\}$, $H_2 = \{\text{id}, \rho_1, \rho_2\}$, $H_3 = \{\text{id}, \mu_1\}$, $H_4 = \{\text{id}, \mu_2\}$, $H_5 = \{\text{id}, \mu_3\}$, S_3.

3.5.41. Hint. The identity of G is $1 = 1 + 0\sqrt{2}$. Since $(a + b\sqrt{2})(c + d\sqrt{2}) = (ac + 2bd) + (ad + bc)\sqrt{2}$, G is closed under multiplication. Finally, $(a + b\sqrt{2})^{-1} = a/(a^2 - 2b^2) - b\sqrt{2}/(a^2 - 2b^2)$.

3.5.46. Hint. Look at S_3.

3.5.49. Hint. $ba = a^4 b = a^3 ab = ab$

4 ◆ Cyclic Groups

4.5.1. Hint. (a) False; (c) false; (e) true.

4.5.2. Hint. (a) 12; (c) infinite; (e) 10.

4.5.3. Hint. (a) $7\mathbb{Z} = \{\ldots, -7, 0, 7, 14, \ldots\}$; (b) $\{0, 3, 6, 9, 12, 15, 18, 21\}$; (c) $\{0\}$, $\{0, 6\}$, $\{0, 4, 8\}$, $\{0, 3, 6, 9\}$, $\{0, 2, 4, 6, 8, 10\}$; (g) $\{1, 3, 7, 9\}$; (j) $\{1, -1, i, -i\}$.

4.5.4. Hint. (a)

$$\begin{pmatrix} 1 & 0 \\ 0 & 1 \end{pmatrix}, \begin{pmatrix} -1 & 0 \\ 0 & -1 \end{pmatrix}, \begin{pmatrix} 0 & -1 \\ 1 & 0 \end{pmatrix}, \begin{pmatrix} 0 & 1 \\ -1 & 0 \end{pmatrix}.$$

(c)

$$\begin{pmatrix} 1 & 0 \\ 0 & 1 \end{pmatrix}, \begin{pmatrix} 1 & -1 \\ 1 & 0 \end{pmatrix}, \begin{pmatrix} -1 & 1 \\ -1 & 0 \end{pmatrix}, \begin{pmatrix} 0 & 1 \\ -1 & 1 \end{pmatrix}, \begin{pmatrix} 0 & -1 \\ 1 & -1 \end{pmatrix}, \begin{pmatrix} -1 & 0 \\ 0 & -1 \end{pmatrix}.$$

4.5.10. Hint. (a) 0; (b) 1, −1.

4.5.11. Hint. 1, 2, 3, 4, 6, 8, 12, 24.

4.5.15. Hint. (a) $-3 + 3i$; (c) $43 - 18i$; (e) i

4.5.16. Hint. (a) $\sqrt{3} + i$; (c) -3.

4.5.17. Hint. (a) $\sqrt{2} \operatorname{cis}(7\pi/4)$; (c) $2\sqrt{2} \operatorname{cis}(\pi/4)$; (e) $3 \operatorname{cis}(3\pi/2)$.

4.5.18. Hint. (a) $(1 - i)/2$; (c) $16(i - \sqrt{3})$; (e) $-1/4$.

4.5.22. Hint. (a) 292; (c) 1523.

4.5.27. Hint. $|\langle g \rangle \cap \langle h \rangle| = 1$.

4.5.31. Hint. The identity element in any group has finite order. Let $g, h \in G$ have orders m and n, respectively. Since $(g^{-1})^m = e$ and $(gh)^{mn} = e$, the elements of finite order in G form a subgroup of G.

4.5.37. Hint. If g is an element distinct from the identity in G, g must generate G; otherwise, $\langle g \rangle$ is a nontrivial proper subgroup of G.

5 ◆ Permutation Groups

5.4.1. Hint. (a) $(1\,2\,4\,5\,3)$; (c) $(1\,3)(2\,5)$.

5.4.2. Hint. (a) $(1\,3\,5)(2\,4)$; (c) $(1\,4)(2\,3)$; (e) $(1\,3\,2\,4)$; (g) $(1\,3\,4)(2\,5)$; (n) $(1\,7\,3\,5\,2)$.

5.4.3. Hint. (a) $(1\,6)(1\,5)(1\,3)(1\,4)$; (c) $(1\,6)(1\,4)(1\,2)$.

5.4.4. Hint. $(a_1, a_2, \ldots, a_n)^{-1} = (a_1, a_n, a_{n-1}, \ldots, a_2)$

5.4.5. Hint. (a) $\{(1\,3), (1\,3)(2\,4), (1\,3\,2), (1\,3\,4), (1\,3\,2\,4), (1\,3\,4\,2)\}$ is not a subgroup.

5.4.8. Hint. $(1\,2\,3\,4\,5)(6\,7\,8)$.

5.4.11. Hint. Permutations of the form

$$(1), (a_1, a_2)(a_3, a_4), (a_1, a_2, a_3), (a_1, a_2, a_3, a_4, a_5)$$

are possible for A_5.

5.4.17. Hint. Calculate $(1\,2\,3)(1\,2)$ and $(1\,2)(1\,2\,3)$.

5.4.25. Hint. Consider the cases $(a, b)(b, c)$ and $(a, b)(c, d)$.

5.4.29. Hint. Show that the center of D_n consists of the identity if n is odd and consists of the identity and a $180°$ rotation if n is even.

5.4.30. Hint. For (a), show that $\sigma\tau\sigma^{-1}(\sigma(a_i)) = \sigma(a_{i+1})$.

6 ◆ Cosets and Lagrange's Theorem

6.5.1. Hint. The order of g and the order h must both divide the order of G.

6.5.2. Hint. The possible orders must divide 60.

6.5.3. Hint. This is true for every proper nontrivial subgroup.

6.5.4. Hint. False.

6.5.5. Hint. (a) $\langle 8 \rangle, 1 + \langle 8 \rangle, 2 + \langle 8 \rangle, 3 + \langle 8 \rangle, 4 + \langle 8 \rangle, 5 + \langle 8 \rangle, 6 + \langle 8 \rangle$, and $7 + \langle 8 \rangle$; (c) $3\mathbb{Z}$, $1 + 3\mathbb{Z}$, and $2 + 3\mathbb{Z}$.

6.5.7. Hint. $4^{\phi(15)} \equiv 4^8 \equiv 1 \pmod{15}$.

6.5.12. Hint. Let $g_1 \in gH$. Show that $g_1 \in Hg$ and thus $gH \subset Hg$.

6.5.19. Hint. Show that $g(H \cap K) = gH \cap gK$.

6.5.22. Hint. If $\gcd(m, n) = 1$, then $\phi(mn) = \phi(m)\phi(n)$ (Exercise 2.4.26 on page 33).

7 ◆ Introduction to Cryptography

7.4.1. Hint. LAORYHAPDWK

7.4.3. Hint. Hint: V = E, E = X (also used for spaces and punctuation), K = R.

7.4.4. Hint. $26! - 1$

7.4.7. Hint. (a) 2791; (c) 11213525032442.

7.4.9. Hint. (a) 31 (c) 14.

7.4.10. Hint. (a) $n = 11 \cdot 41$; (c) $n = 8779 \cdot 4327$.

8 ◆ Algebraic Coding Theory

8.6.2. Hint. This cannot be a group code since $(0000) \notin C$.

8.6.3. Hint. (a) 2; (c) 2.

8.6.4. Hint. (a) 3; (c) 4.

8.6.6. Hint. (a) $d_{min} = 2$; (c) $d_{min} = 1$.

8.6.7. Hint.

   (a) $(00000), (00101), (10011), (10110)$

$$G = \begin{pmatrix} 0 & 1 \\ 0 & 0 \\ 1 & 0 \\ 0 & 1 \\ 1 & 1 \end{pmatrix}$$

   (b) $(000000), (010111), (101101), (111010)$

$$G = \begin{pmatrix} 1 & 0 \\ 0 & 1 \\ 1 & 0 \\ 1 & 1 \\ 0 & 1 \\ 1 & 1 \end{pmatrix}$$

8.6.9. Hint. Multiple errors occur in one of the received words.

8.6.11. Hint. (a) A canonical parity-check matrix with standard generator matrix

$$G = \begin{pmatrix} 1 \\ 1 \\ 0 \\ 0 \\ 1 \end{pmatrix}.$$

   (c) A canonical parity-check matrix with standard generator matrix

$$G = \begin{pmatrix} 1 & 0 \\ 0 & 1 \\ 1 & 1 \\ 1 & 0 \end{pmatrix}.$$

8.6.12. Hint. (a) All possible syndromes occur.

8.6.15. Hint. (a) C, $(10000) + C$, $(01000) + C$, $(00100) + C$, $(00010) + C$, $(11000) + C$, $(01100) + C$, $(01010) + C$. A decoding table does not exist for C since this is only a single error-detecting code.

8.6.19. Hint. Let $\mathbf{x} \in C$ have odd weight and define a map from the set of odd codewords to the set of even codewords by $\mathbf{y} \mapsto \mathbf{x} + \mathbf{y}$. Show that this map is a bijection.

8.6.23. Hint. For 20 information positions, at least 6 check bits are needed to ensure an error-correcting code.

9 ◆ Isomorphisms

9.4.1. Hint. Every infinite cyclic group is isomorphic to $\mathbb{Z}$ by Theorem 9.7 on page 144.

9.4.2. Hint. Define $\phi : \mathbb{C}^* \to GL_2(\mathbb{R})$ by

$$\phi(a + bi) = \begin{pmatrix} a & b \\ -b & a \end{pmatrix}.$$

9.4.3. Hint. False.

9.4.6. Hint. Define a map from $\mathbb{Z}_n$ into the nth roots of unity by $k \mapsto \mathrm{cis}(2k\pi/n)$.

9.4.8. Hint. Assume that $\mathbb{Q}$ is cyclic and try to find a generator.

9.4.11. Hint. There are two nonabelian and three abelian groups that are not isomorphic.

9.4.16. Hint. (a) 12; (c) 5.

9.4.19. Hint. Draw the picture.

9.4.20. Hint. True.

9.4.25. Hint. True.

9.4.27. Hint. Let a be a generator for G. If $\phi : G \to H$ is an isomorphism, show that $\phi(a)$ is a generator for H.

9.4.38. Hint. Any automorphism of $\mathbb{Z}_6$ must send 1 to another generator of $\mathbb{Z}_6$.

9.4.45. Hint. To show that ϕ is one-to-one, let $g_1 = h_1 k_1$ and $g_2 = h_2 k_2$ and consider $\phi(g_1) = \phi(g_2)$.

10 ◆ Normal Subgroups and Factor Groups

10.4.1. Hint. (a)

	A_4	$(12)A_4$
A_4	A_4	$(12)A_4$
$(12)A_4$	$(12)A_4$	A_4

(c) D_4 is not normal in S_4.

10.4.8. Hint. If $a \in G$ is a generator for G, then aH is a generator for G/H.

10.4.11. Hint. For any $g \in G$, show that the map $i_g : G \to G$ defined by $i_g : x \mapsto gxg^{-1}$ is an isomorphism of G with itself. Then consider $i_g(H)$.

10.4.12. Hint. Suppose that $\langle g \rangle$ is normal in G and let y be an arbitrary element of G. If $x \in C(g)$, we must show that yxy^{-1} is also in $C(g)$. Show that $(yxy^{-1})g = g(yxy^{-1})$.

10.4.14. Hint. (a) Let $g \in G$ and $h \in G'$. If $h = aba^{-1}b^{-1}$, then

$$ghg^{-1} = gaba^{-1}b^{-1}g^{-1}$$
$$= (gag^{-1})(gbg^{-1})(ga^{-1}g^{-1})(gb^{-1}g^{-1})$$
$$= (gag^{-1})(gbg^{-1})(gag^{-1})^{-1}(gbg^{-1})^{-1}.$$

We also need to show that if $h = h_1 \cdots h_n$ with $h_i = a_i b_i a_i^{-1} b_i^{-1}$, then ghg^{-1} is a product of elements of the same type. However, $ghg^{-1} = gh_1 \cdots h_n g^{-1} = (gh_1 g^{-1})(gh_2 g^{-1}) \cdots (gh_n g^{-1})$.

11 ♦ Homomorphisms

11.4.2. Hint. (a) is a homomorphism with kernel $\{1\}$; (c) is not a homomorphism.

11.4.4. Hint. Since $\phi(m+n) = 7(m+n) = 7m+7n = \phi(m)+\phi(n)$, ϕ is a homomorphism.

11.4.5. Hint. For any homomorphism $\phi : \mathbb{Z}_{24} \to \mathbb{Z}_{18}$, the kernel of ϕ must be a subgroup of $\mathbb{Z}_{24}$ and the image of ϕ must be a subgroup of $\mathbb{Z}_{18}$. Now use the fact that a generator must map to a generator.

11.4.9. Hint. Let $a, b \in G$. Then $\phi(a)\phi(b) = \phi(ab) = \phi(ba) = \phi(b)\phi(a)$.

11.4.17. Hint. Find a counterexample.

12 ♦ Matrix Groups and Symmetry

12.4.1. Hint.

$$\frac{1}{2}\left[\|\mathbf{x}+\mathbf{y}\|^2 + \|\mathbf{x}\|^2 - \|\mathbf{y}\|^2\right] = \frac{1}{2}\left[\langle x+y, x+y\rangle - \|\mathbf{x}\|^2 - \|\mathbf{y}\|^2\right]$$

$$= \frac{1}{2}\left[\|\mathbf{x}\|^2 + 2\langle x, y\rangle + \|\mathbf{y}\|^2 - \|\mathbf{x}\|^2 - \|\mathbf{y}\|^2\right]$$

$$= \langle \mathbf{x}, \mathbf{y}\rangle.$$

12.4.3. Hint. (a) is in $SO(2)$; (c) is not in $O(3)$.

12.4.5. Hint. (a) $\langle \mathbf{x}, \mathbf{y}\rangle = \langle \mathbf{y}, \mathbf{x}\rangle$.

12.4.7. Hint. Use the unimodular matrix

$$\begin{pmatrix} 5 & 2 \\ 2 & 1 \end{pmatrix}.$$

12.4.10. Hint. Show that the kernel of the map $\det : O(n) \to \mathbb{R}^*$ is $SO(n)$.

12.4.13. Hint. True.

12.4.17. Hint. $p6m$

13 ♦ The Structure of Groups

13.4.1. Hint. There are three possible groups.

13.4.4. Hint. (a) $\{0\} \subset \langle 6\rangle \subset \langle 3\rangle \subset \mathbb{Z}_{12}$; (e) $\{(1)\} \times \{0\} \subset \{(1),(123),(132)\} \times \{0\} \subset S_3 \times \{0\} \subset S_3 \times \langle 2\rangle \subset S_3 \times \mathbb{Z}_4$.

13.4.7. Hint. Use the Fundamental Theorem of Finitely Generated Abelian Groups.

13.4.12. Hint. If N and G/N are solvable, then they have solvable series

$$N = N_n \supset N_{n-1} \supset \cdots \supset N_1 \supset N_0 = \{e\}$$

$$G/N = G_n/N \supset G_{n-1}/N \supset \cdots G_1/N \supset G_0/N = \{N\}.$$

13.4.16. Hint. Use the fact that D_n has a cyclic subgroup of index 2.

13.4.21. Hint. G/G' is abelian.

14 ♦ Group Actions

14.5.1. Hint. Example 14.1 on page 211: 0, $\mathbb{R}^2 \setminus \{0\}$. Example 14.2 on page 211: $X = \{1, 2, 3, 4\}$.

14.5.2. Hint. (a) $X_{(1)} = \{1, 2, 3\}$, $X_{(12)} = \{3\}$, $X_{(13)} = \{2\}$, $X_{(23)} = \{1\}$, $X_{(123)} = X_{(132)} = \varnothing$. $G_1 = \{(1),(23)\}$, $G_2 = \{(1),(13)\}$, $G_3 = \{(1),(12)\}$.

14.5.3. Hint. (a) $\mathcal{O}_1 = \mathcal{O}_2 = \mathcal{O}_3 = \{1, 2, 3\}$.

14.5.6. Hint. The conjugacy classes for S_4 are

$$\mathcal{O}_{(1)} = \{(1)\},$$
$$\mathcal{O}_{(12)} = \{(12), (13), (14), (23), (24), (34)\},$$
$$\mathcal{O}_{(12)(34)} = \{(12)(34), (13)(24), (14)(23)\},$$
$$\mathcal{O}_{(123)} = \{(123), (132), (124), (142), (134), (143), (234), (243)\},$$
$$\mathcal{O}_{(1234)} = \{(1234), (1243), (1324), (1342), (1423), (1432)\}.$$

The class equation is $1 + 3 + 6 + 6 + 8 = 24$.

14.5.8. Hint. $(3^4 + 3^1 + 3^2 + 3^1 + 3^2 + 3^2 + 3^3 + 3^3)/8 = 21$.

14.5.11. Hint. The group of rigid motions of the cube can be described by the allowable permutations of the six faces and is isomorphic to S_4. There are the identity cycle, 6 permutations with the structure $(abcd)$ that correspond to the quarter turns, 3 permutations with the structure $(ab)(cd)$ that correspond to the half turns, 6 permutations with the structure $(ab)(cd)(ef)$ that correspond to rotating the cube about the centers of opposite edges, and 8 permutations with the structure $(abc)(def)$ that correspond to rotating the cube about opposite vertices.

14.5.15. Hint. $(1 \cdot 2^6 + 3 \cdot 2^4 + 4 \cdot 2^3 + 2 \cdot 2^2 + 2 \cdot 2^1)/12 = 13$.

14.5.17. Hint. $(1 \cdot 2^8 + 3 \cdot 2^6 + 2 \cdot 2^4)/6 = 80$.

14.5.22. Hint. Use the fact that $x \in gC(a)g^{-1}$ if and only if $g^{-1}xg \in C(a)$.

15 ◆ The Sylow Theorems

15.4.1. Hint. If $|G| = 18 = 2 \cdot 3^2$, then the order of a Sylow 2-subgroup is 2, and the order of a Sylow 3-subgroup is 9.

15.4.2. Hint. The four Sylow 3-subgroups of S_4 are $P_1 = \{(1), (123), (132)\}$, $P_2 = \{(1), (124), (142)\}$, $P_3 = \{(1), (134), (143)\}$, $P_4 = \{(1), (234), (243)\}$.

15.4.5. Hint. Since $|G| = 96 = 2^5 \cdot 3$, G has either one or three Sylow 2-subgroups by the Third Sylow Theorem. If there is only one subgroup, we are done. If there are three Sylow 2-subgroups, let H and K be two of them. Therefore, $|H \cap K| \geq 16$; otherwise, HK would have $(32 \cdot 32)/8 = 128$ elements, which is impossible. Thus, $H \cap K$ is normal in both H and K since it has index 2 in both groups.

15.4.8. Hint. Show that G has a normal Sylow p-subgroup of order p^2 and a normal Sylow q-subgroup of order q^2.

15.4.10. Hint. False.

15.4.17. Hint. If G is abelian, then G is cyclic, since $|G| = 3 \cdot 5 \cdot 17$. Now look at Example 15.14 on page 233.

15.4.23. Hint. Define a mapping between the right cosets of $N(H)$ in G and the conjugates of H in G by $N(H)g \mapsto g^{-1}Hg$. Prove that this map is a bijection.

15.4.26. Hint. Let $aG', bG' \in G/G'$. Then $(aG')(bG') = abG' = ab(b^{-1}a^{-1}ba)G' = (abb^{-1}a^{-1})baG' = baG'$.

16 ◆ Rings

16.7.1. Hint. (a) $7\mathbb{Z}$ is a ring but not a field; (c) $\mathbb{Q}(\sqrt{2})$ is a field; (f) R is not a ring.

16.7.3. Hint. (a) $\{1,3,7,9\}$; (c) $\{1,2,3,4,5,6\}$; (e)

$$\left\{ \begin{pmatrix} 1 & 0 \\ 0 & 1 \end{pmatrix}, \begin{pmatrix} 1 & 1 \\ 0 & 1 \end{pmatrix}, \begin{pmatrix} 1 & 0 \\ 1 & 1 \end{pmatrix}, \begin{pmatrix} 0 & 1 \\ 1 & 0 \end{pmatrix}, \begin{pmatrix} 1 & 1 \\ 1 & 0 \end{pmatrix}, \begin{pmatrix} 0 & 1 \\ 1 & 1 \end{pmatrix}, \right\}.$$

16.7.4. Hint. (a) $\{0\}$, $\{0,9\}$, $\{0,6,12\}$, $\{0,3,6,9,12,15\}$, $\{0,2,4,6,8,10,12,14,16\}$; (c) there are no nontrivial ideals.

16.7.7. Hint. Assume there is an isomorphism $\phi : \mathbb{C} \to \mathbb{R}$ with $\phi(i) = a$.

16.7.8. Hint. False. Assume there is an isomorphism $\phi : \mathbb{Q}(\sqrt{2}) \to \mathbb{Q}(\sqrt{3})$ such that $\phi(\sqrt{2}) = a$.

16.7.13. Hint. (a) $x \equiv 17 \pmod{55}$; (c) $x \equiv 214 \pmod{2772}$.

16.7.16. Hint. If $I \neq \{0\}$, show that $1 \in I$.

16.7.18. Hint. (a) $\phi(a)\phi(b) = \phi(ab) = \phi(ba) = \phi(b)\phi(a)$.

16.7.26. Hint. Let $a \in R$ with $a \neq 0$. Then the principal ideal generated by a is R. Thus, there exists a $b \in R$ such that $ab = 1$.

16.7.28. Hint. Compute $(a + b)^2$ and $(-ab)^2$.

16.7.33. Hint. Let $a/b, c/d \in \mathbb{Z}_{(p)}$. Then $a/b + c/d = (ad + bc)/bd$ and $(a/b) \cdot (c/d) = (ac)/(bd)$ are both in $\mathbb{Z}_{(p)}$, since $\gcd(bd, p) = 1$.

16.7.37. Hint. Suppose that $x^2 = x$ and $x \neq 0$. Since R is an integral domain, $x = 1$. To find a nontrivial idempotent, look in $\mathbb{M}_2(\mathbb{R})$.

17 ◆ Polynomials

17.5.2. Hint. (a) $9x^2 + 2x + 5$; (b) $8x^4 + 7x^3 + 2x^2 + 7x$.

17.5.3. Hint. (a) $5x^3 + 6x^2 - 3x + 4 = (5x^2 + 2x + 1)(x - 2) + 6$; (c) $4x^5 - x^3 + x^2 + 4 = (4x^2 + 4)(x^3 + 3) + 4x^2 + 2$.

17.5.5. Hint. (a) No zeros in $\mathbb{Z}_{12}$; (c) 3, 4.

17.5.7. Hint. Look at $(2x + 1)$.

17.5.8. Hint. (a) Reducible; (c) irreducible.

17.5.10. Hint. One factorization is $x^2 + x + 8 = (x + 2)(x + 9)$.

17.5.13. Hint. The integers $\mathbb{Z}$ do not form a field.

17.5.14. Hint. False.

17.5.16. Hint. Let $\phi : R \to S$ be an isomorphism. Define $\overline{\phi} : R[x] \to S[x]$ by $\overline{\phi}(a_0 + a_1x + \cdots + a_nx^n) = \phi(a_0) + \phi(a_1)x + \cdots + \phi(a_n)x^n$.

17.5.20. Cyclotomic Polynomials. Hint. The polynomial

$$\Phi_n(x) = \frac{x^n - 1}{x - 1} = x^{n-1} + x^{n-2} + \cdots + x + 1$$

is called the *cyclotomic polynomial*. Show that $\Phi_p(x)$ is irreducible over $\mathbb{Q}$ for any prime p.

17.5.26. Hint. Find a nontrivial proper ideal in $F[x]$.

18 ◆ Integral Domains

18.4.1. Hint. Note that $z^{-1} = 1/(a + b\sqrt{3}\,i) = (a - b\sqrt{3}\,i)/(a^2 + 3b^2)$ is in $\mathbb{Z}[\sqrt{3}\,i]$ if and only if $a^2 + 3b^2 = 1$. The only integer solutions to the equation are $a = \pm 1, b = 0$.

18.4.2. Hint. (a) $5 = -i(1 + 2i)(2 + i)$; (c) $6 + 8i = -i(1 + i)^2(2 + i)^2$.

18.4.4. Hint. True.

18.4.9. Hint. Let $z = a + bi$ and $w = c + di \neq 0$ be in $\mathbb{Z}[i]$. Prove that $z/w \in \mathbb{Q}(i)$.

18.4.15. Hint. Let $a = ub$ with u a unit. Then $v(b) \leq v(ub) \leq v(a)$. Similarly, $v(a) \leq v(b)$.

18.4.16. Hint. Show that 21 can be factored in two different ways.

19 ◆ Lattices and Boolean Algebras

19.5.2. Hint.

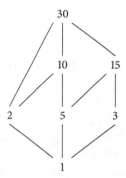

19.5.4. Hint. What are the atoms of B?

19.5.5. Hint. False.

19.5.6. Hint. (a) $(a \vee b \vee a') \wedge a$

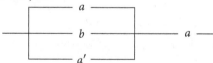

(c) $a \vee (a \wedge b)$

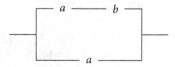

19.5.8. Hint. Not equivalent.

19.5.10. Hint. (a) $a' \wedge [(a \wedge b') \vee b] = a \wedge (a \vee b)$.

19.5.14. Hint. Let I, J be ideals in R. We need to show that $I + J = \{r + s : r \in I$ and $s \in J\}$ is the smallest ideal in R containing both I and J. If $r_1, r_2 \in I$ and $s_1, s_2 \in J$, then

$(r_1 + s_1) + (r_2 + s_2) = (r_1 + r_2) + (s_1 + s_2)$ is in $I + J$. For $a \in R$, $a(r_1 + s_1) = ar_1 + as_1 \in I + J$; hence, $I + J$ is an ideal in R.

19.5.18. Hint. (a) No.

19.5.20. Hint. ($\Rightarrow$). $a = b \Rightarrow (a \wedge b') \vee (a' \wedge b) = (a \wedge a') \vee (a' \wedge a) = O \vee O = O$.
($\Leftarrow$). $(a \wedge b') \vee (a' \wedge b) = O \Rightarrow a \vee b = (a \vee a) \vee b = a \vee (a \vee b) = a \vee [I \wedge (a \vee b)] = a \vee [(a \vee a') \wedge (a \vee b)] = [a \vee (a \wedge b')] \vee [a \vee (a' \wedge b)] = a \vee [(a \wedge b') \vee (a' \wedge b)] = a \vee O = a$.
A symmetric argument shows that $a \vee b = b$.

20 ◆ Vector Spaces

20.5.3. Hint. $\mathbb{Q}(\sqrt{2}, \sqrt{3})$ has basis $\{1, \sqrt{2}, \sqrt{3}, \sqrt{6}\}$ over $\mathbb{Q}$.

20.5.5. Hint. The set $\{1, x, x^2, \ldots, x^{n-1}\}$ is a basis for P_n.

20.5.7. Hint. (a) Subspace of dimension 2 with basis $\{(1, 0, -3), (0, 1, 2)\}$; (d) not a subspace

20.5.10. Hint. Since $0 = \alpha 0 = \alpha(-v + v) = \alpha(-v) + \alpha v$, it follows that $-\alpha v = \alpha(-v)$.

20.5.12. Hint. Let $v_0 = 0, v_1, \ldots, v_n \in V$ and $\alpha_0 \neq 0, \alpha_1, \ldots, \alpha_n \in F$. Then $\alpha_0 v_0 + \cdots + \alpha_n v_n = 0$.

20.5.15. Linear Transformations. Hint. (a) Let $u, v \in \ker(T)$ and $\alpha \in F$. Then

$$T(u + v) = T(u) + T(v) = 0$$
$$T(\alpha v) = \alpha T(v) = \alpha 0 = 0.$$

Hence, $u + v, \alpha v \in \ker(T)$, and $\ker(T)$ is a subspace of V.

(c) The statement that $T(u) = T(v)$ is equivalent to $T(u - v) = T(u) - T(v) = 0$, which is true if and only if $u - v = 0$ or $u = v$.

20.5.17. Direct Sums. Hint. (a) Let $u, u' \in U$ and $v, v' \in V$. Then

$$(u + v) + (u' + v') = (u + u') + (v + v') \in U + V$$
$$\alpha(u + v) = \alpha u + \alpha v \in U + V.$$

21 ◆ Fields

21.5.1. Hint. (a) $x^4 - (2/3)x^2 - 62/9$; (c) $x^4 - 2x^2 + 25$.

21.5.2. Hint. (a) $\{1, \sqrt{2}, \sqrt{3}, \sqrt{6}\}$; (c) $\{1, i, \sqrt{2}, \sqrt{2}i\}$; (e) $\{1, 2^{1/6}, 2^{1/3}, 2^{1/2}, 2^{2/3}, 2^{5/6}\}$.

21.5.3. Hint. (a) $\mathbb{Q}(\sqrt{3}, \sqrt{7})$.

21.5.5. Hint. Use the fact that the elements of $\mathbb{Z}_2[x]/\langle x^3 + x + 1 \rangle$ are $0, 1, \alpha, 1 + \alpha, \alpha^2, 1 + \alpha^2, \alpha + \alpha^2, 1 + \alpha + \alpha^2$ and the fact that $\alpha^3 + \alpha + 1 = 0$.

21.5.8. Hint. False.

21.5.14. Hint. Suppose that E is algebraic over F and K is algebraic over E. Let $\alpha \in K$. It suffices to show that α is algebraic over some finite extension of F. Since α is algebraic over E, it must be the zero of some polynomial $p(x) = \beta_0 + \beta_1 x + \cdots + \beta_n x^n$ in $E[x]$. Hence α is algebraic over $F(\beta_0, \ldots, \beta_n)$.

21.5.22. Hint. Since $\{1, \sqrt{3}, \sqrt{7}, \sqrt{21}\}$ is a basis for $\mathbb{Q}(\sqrt{3}, \sqrt{7})$ over $\mathbb{Q}$, $\mathbb{Q}(\sqrt{3}, \sqrt{7}) \supset \mathbb{Q}(\sqrt{3} + \sqrt{7})$. Since $[\mathbb{Q}(\sqrt{3}, \sqrt{7}) : \mathbb{Q}] = 4$, $[\mathbb{Q}(\sqrt{3} + \sqrt{7}) : \mathbb{Q}] = 2$ or 4. Since the degree of the minimal polynomial of $\sqrt{3} + \sqrt{7}$ is 4, $\mathbb{Q}(\sqrt{3}, \sqrt{7}) = \mathbb{Q}(\sqrt{3} + \sqrt{7})$.

21.5.27. Hint. Let $\beta \in F(\alpha)$ not in F. Then $\beta = p(\alpha)/q(\alpha)$, where p and q are polynomials in α with $q(\alpha) \neq 0$ and coefficients in F. If β is algebraic over F, then there exists a polynomial $f(x) \in F[x]$ such that $f(\beta) = 0$. Let $f(x) = a_0 + a_1x + \cdots + a_nx^n$. Then

$$0 = f(\beta) = f\left(\frac{p(\alpha)}{q(\alpha)}\right) = a_0 + a_1\left(\frac{p(\alpha)}{q(\alpha)}\right) + \cdots + a_n\left(\frac{p(\alpha)}{q(\alpha)}\right)^n.$$

Now multiply both sides by $q(\alpha)^n$ to show that there is a polynomial in $F[x]$ that has α as a zero.

21.5.28. Hint. See the comments following Theorem 21.13 on page 337.

22 ◆ Finite Fields

22.4.1. Hint. Make sure that you have a field extension.

22.4.4. Hint. There are eight elements in $\mathbb{Z}_2(\alpha)$. Exhibit two more zeros of $x^3 + x^2 + 1$ other than α in these eight elements.

22.4.5. Hint. Find an irreducible polynomial $p(x)$ in $\mathbb{Z}_3[x]$ of degree 3 and show that $\mathbb{Z}_3[x]/\langle p(x)\rangle$ has 27 elements.

22.4.7. Hint. (a) $x^5 - 1 = (x+1)(x^4+x^3+x^2+x+1)$; (c) $x^9 - 1 = (x+1)(x^2+x+1)(x^6+x^3+1)$.

22.4.8. Hint. True.

22.4.11. Hint. (a) Use the fact that $x^7 - 1 = (x+1)(x^3+x+1)(x^3+x^2+1)$.

22.4.12. Hint. False.

22.4.17. Hint. If $p(x) \in F[x]$, then $p(x) \in E[x]$.

22.4.18. Hint. Since α is algebraic over F of degree n, we can write any element $\beta \in F(\alpha)$ uniquely as $\beta = a_0 + a_1\alpha + \cdots + a_{n-1}\alpha^{n-1}$ with $a_i \in F$. There are q^n possible n-tuples $(a_0, a_1, \ldots, a_{n-1})$.

22.4.24. Wilson's Theorem. Hint. Factor $x^{p-1} - 1$ over $\mathbb{Z}_p$.

23 ◆ Galois Theory

23.5.1. Hint. (a) $\mathbb{Z}_2$; (c) $\mathbb{Z}_2 \times \mathbb{Z}_2 \times \mathbb{Z}_2$.

23.5.2. Hint. (a) Separable over $\mathbb{Q}$ since $x^3 + 2x^2 - x - 2 = (x-1)(x+1)(x+2)$; (c) not separable over $\mathbb{Z}_3$ since $x^4 + x^2 + 1 = (x+1)^2(x+2)^2$.

23.5.3. Hint. If

$$[\mathrm{GF}(729) : \mathrm{GF}(9)] = [\mathrm{GF}(729) : \mathrm{GF}(3)]/[\mathrm{GF}(9) : \mathrm{GF}(3)] = 6/2 = 3,$$

then $G(\mathrm{GF}(729)/\mathrm{GF}(9)) \cong \mathbb{Z}_3$. A generator for $G(\mathrm{GF}(729)/\mathrm{GF}(9))$ is σ, where $\sigma_{3^6}(\alpha) = \alpha^{3^6} = \alpha^{729}$ for $\alpha \in \mathrm{GF}(729)$.

23.5.4. Hint. (a) S_5; (c) S_3; (g) see Example 23.11 on page 379.

23.5.5. Hint. (a) $\mathbb{Q}(i)$

23.5.7. Hint. Let E be the splitting field of a cubic polynomial in $F[x]$. Show that $[E : F]$ is less than or equal to 6 and is divisible by 3. Since $G(E/F)$ is a subgroup of S_3 whose order is divisible by 3, conclude that this group must be isomorphic to $\mathbb{Z}_3$ or S_3.

23.5.9. Hint. G is a subgroup of S_n.

23.5.16. Hint. True.

23.5.20. Hint.

(a) Clearly $\omega, \omega^2, \ldots, \omega^{p-1}$ are distinct since $\omega \neq 1$ or 0. To show that ω^i is a zero of Φ_p, calculate $\Phi_p(\omega^i)$.

(b) The conjugates of ω are $\omega, \omega^2, \ldots, \omega^{p-1}$. Define a map $\phi_i : \mathbb{Q}(\omega) \to \mathbb{Q}(\omega^i)$ by

$$\phi_i(a_0 + a_1\omega + \cdots + a_{p-2}\omega^{p-2}) = a_0 + a_1\omega^i + \cdots + c_{p-2}(\omega^i)^{p-2},$$

where $a_i \in \mathbb{Q}$. Prove that ϕ_i is an isomorphism of fields. Show that ϕ_2 generates $G(\mathbb{Q}(\omega)/\mathbb{Q})$.

(c) Show that $\{\omega, \omega^2, \ldots, \omega^{p-1}\}$ is a basis for $\mathbb{Q}(\omega)$ over $\mathbb{Q}$, and consider which linear combinations of $\omega, \omega^2, \ldots, \omega^{p-1}$ are left fixed by all elements of $G(\mathbb{Q}(\omega)/\mathbb{Q})$.

Notation

The following table defines the notation used in this book. Page numbers or references refer to the first appearance of each symbol.

Symbol	Description	Page
$a \in A$	a is in the set A	4
$\mathbb{N}$	the natural numbers	5
$\mathbb{Z}$	the integers	5
$\mathbb{Q}$	the rational numbers	5
$\mathbb{R}$	the real numbers	5
$\mathbb{C}$	the complex numbers	5
$A \subset B$	A is a subset of B	5
$\varnothing$	the empty set	5
$A \cup B$	the union of sets A and B	5
$A \cap B$	the intersection of sets A and B	5
A'	complement of the set A	6
$A \setminus B$	difference between sets A and B	6
$A \times B$	Cartesian product of sets A and B	8
A^n	$A \times \cdots \times A$ (n times)	8
id	identity mapping	12
f^{-1}	inverse of the function f	12
$a \equiv b \pmod{n}$	a is congruent to b modulo n	16
$n!$	n factorial	23
$\binom{n}{k}$	binomial coefficient $n!/(k!(n-k)!)$	23
$a \mid b$	a divides b	26
$\gcd(a, b)$	greatest common divisor of a and b	26
$\mathcal{P}(X)$	power set of X	32
$\mathrm{lcm}(m, n)$	the least common multiple of m and n	33
$\mathbb{Z}_n$	the integers modulo n	36
$U(n)$	group of units in $\mathbb{Z}_n$	43
$\mathbb{M}_n(\mathbb{R})$	the $n \times n$ matrices with entries in $\mathbb{R}$	44
$\det A$	the determinant of A	44
$GL_n(\mathbb{R})$	the general linear group	44
Q_8	the group of quaternions	44

(Continued on next page)

Symbol	Description	Page
$\mathbb{C}^*$	the multiplicative group of complex numbers	44
$\lvert G \rvert$	the order of a group	45
$\mathbb{R}^*$	the multiplicative group of real numbers	47
$\mathbb{Q}^*$	the multiplicative group of rational numbers	48
$SL_n(\mathbb{R})$	the special linear group	48
$Z(G)$	the center of a group	54
$\langle a \rangle$	cyclic group generated by a	58
$\lvert a \rvert$	the order of an element a	59
$\operatorname{cis} \theta$	$\cos \theta + i \sin \theta$	64
$\mathbb{T}$	the circle group	65
S_n	the symmetric group on n letters	74
$(a_1, a_2, \ldots, a_k)$	cycle of length k	76
A_n	the alternating group on n letters	81
D_n	the dihedral group	82
$[G : H]$	index of a subgroup H in a group G	93
$\mathcal{L}_H$	the set of left cosets of a subgroup H in a group G	93
$\mathcal{R}_H$	the set of right cosets of a subgroup H in a group G	93
$a \nmid b$	a does not divide b	96
$d(\mathbf{x}, \mathbf{y})$	Hamming distance between $\mathbf{x}$ and $\mathbf{y}$	118
$d_{\min}$	the minimum distance of a code	118
$w(\mathbf{x})$	the weight of $\mathbf{x}$	118
$\mathbb{M}_{m \times n}(\mathbf{Z}_2)$	the set of $m \times n$ matrices with entries in $\mathbb{Z}_2$	123
$\operatorname{Null}(H)$	null space of a matrix H	123
δ_{ij}	Kronecker delta	128
$G \cong H$	G is isomorphic to a group H	142
$\operatorname{Aut}(G)$	automorphism group of a group G	154
i_g	$i_g(x) = gxg^{-1}$	154
$\operatorname{Inn}(G)$	inner automorphism group of a group G	154
ρ_g	right regular representation	155
G/N	factor group of G mod N	158
G'	commutator subgroup of G	166
$\ker \phi$	kernel of ϕ	169
(a_{ij})	matrix	178
$O(n)$	orthogonal group	181
$\lVert \mathbf{x} \rVert$	length of a vector $\mathbf{x}$	181
$SO(n)$	special orthogonal group	185
$E(n)$	Euclidean group	185
$\mathcal{O}_x$	orbit of x	213
X_g	fixed point set of g	213
G_x	isotropy subgroup of x	213

(Continued on next page)

Symbol	Description	Page
$N(H)$	normalizer of s subgroup H	230
$\mathbb{H}$	the ring of quaternions	242
$\mathbb{Z}[i]$	the Gaussian integers	245
char R	characteristic of a ring R	246
$\mathbb{Z}_{(p)}$	ring of integers localized at p	262
$\deg f(x)$	degree of a polynomial	266
$R[x]$	ring of polynomials over a ring R	266
$R[x_1, x_2, \ldots, x_n]$	ring of polynomials in n indeterminants	269
ϕ_α	evaluation homomorphism at α	269
$\mathbb{Q}(x)$	field of rational functions over $\mathbb{Q}$	290
$v(a)$	Euclidean valuation of a	294
$F(x)$	field of rational functions in x	300
$F(x_1, \ldots, x_n)$	field of rational functions in $x_1, \ldots, x_n$	300
$a \le b$	a is less than b	304
$a \vee b$	join of a and b	306
$a \wedge b$	meet of a and b	306
I	largest element in a lattice	308
O	smallest element in a lattice	308
a'	complement of a in a lattice	308
$\dim V$	dimension of a vector space V	327
$U \oplus V$	direct sum of vector spaces U and V	330
$\mathrm{Hom}(V, W)$	set of all linear transformations from U into V	330
V^*	dual of a vector space V	330
$F(\alpha_1, \ldots, \alpha_n)$	smallest field containing F and $\alpha_1, \ldots, \alpha_n$	335
$[E : F]$	dimension of a field extension of E over F	338
$\mathrm{GF}(p^n)$	Galois field of order p^n	359
F^*	multiplicative group of a field F	360
$G(E/F)$	Galois group of E over F	376
$F_{\{\sigma_i\}}$	field fixed by the automorphism σ_i	381
F_G	field fixed by the automorphism group G	381
Δ^2	discriminant of a polynomial	397

Index

Colophon

This book was authored and produced with PreTeXt.

CPSIA information can be obtained
at www.ICGtesting.com
Printed in the USA
LVHW080853110822
725644LV00004B/67

9 781944 325145